HUMAN ANATOMY AND PHYSIOLOGY

This book provides a simple but inclusive account of the structure of the human body, the way in which it works, and the related biochemical and biophysical processes. It is well within the grasp of those with a knowledge of elementary biology and chemistry, and is profusely illustrated to help the general reader who has no opportunity for dissection. At the same time, it should prove useful as an introductory manual to medical and nursing students, and to others in allied professions.

HUMAN ANATOMY AND PHYSIOLOGY

David Le Vay, M.S., F.R.C.S.

TEACH YOURSELF BOOKS
Hodder and Stoughton

First printed 1974
Sixth impression 1980

ISBN 0 340 15250 8

Printed and bound in Great Britain
for Hodder and Stoughton Paperbacks,
a division of Hodder and Stoughton Ltd,
Mill Road, Dunton Green, Sevenoaks, Kent
(Editorial Office: 47 Bedford Square, London WC1 3DP) by
Richard Clay (The Chaucer Press) Ltd,
Bungay, Suffolk

Contents

List of Colour Plates

Preface

This new book brings together in a single volume the material previously published in the volumes *Anatomy* and *Physiology* in the Teach Yourself series, but thoroughly revised and rewritten and brought completely up to date.

I repeat my former acknowledgments for illustrations and material to Messrs. Longmans, Green & Co. as regards *Gray's Anatomy*; to Messrs. J. & A. Churchill in respect of their *Principles of Human Physiology*; and to Professor Young and the Oxford University Press regarding *Life of Vertebrates*. But in the course of this revision I have leaned heavily on that invaluable work of reference, *A Companion to Medical Studies, Volume 1*, ed. R. Passmore and J. S. Robson, published by Blackwell Scientific Publications, and the figures reproduced therefrom are individually acknowledged in the text.

1 Definitions, applications, history

Anatomy

The literal meaning of the word 'anatomy' is the cutting up of the body to examine its parts and their relations one to another. Knowledge gained by dissection in this way is essentially *regional* or *topographical*; that is, one gains a familiarity with each part, such as the arm or leg, similar to one's acquaintanceship with the particular districts of a city, and this, in fact, is how the medical student first learns his way about the human body. But every part is found to contain the same *kinds* of organs—blood vessels, nerves, bones, etc.—so that, superimposed on regional anatomy, there is a *systemic* aspect, in which the body is considered as being made up of several co-ordinated systems, vascular, nervous, skeletal and so on, each of which has a definite integrity of its own. Gross naked-eye observations of this kind are known as *macroscopic* anatomy, which is contrasted with microscopic anatomy, or *histology*, the study of the fine structure of individual cells and tissues.

All this implies a static view of adult anatomy as an unchanging entity, a very necessary practical knowledge for such people as the surgeon. But we have also to consider developmental anatomy; this consists, first, of *embryology*, the growth of the individual within the womb from the single cell formed by the fusion of ovum and spermatozoön, and, secondly, the *postnatal* development through infancy and childhood to maturity. And towards the end of life certain changes of *senescence* appear, though these are usually regarded as pathological rather than anatomical; similarly, *morbid anatomy*, the study of diseased organs and tissues, really comes within the separate science of pathology.

The development of the individual is known as *ontogeny*, as contrasted with *phylogeny*, the evolutional development of the race from more primitive forms. To some extent the individual recapitulates phylogeny in a condensed and distorted form during his own development. *Comparative anatomy* is the study of the human body as a finished product in relation to the structure of animals; we sometimes use the word *morphology* for the study of differences and resemblances in the structure, use and evolutionary transformation of corresponding organs in man and animals. For instance, the human foot and the horse's hoof are basically similar structures that have evolved on differing lines to have differing final uses.

Methods

The oldest and most basic method of acquiring anatomical knowledge is by *dissection*, a process extended in recent centuries by *microscopy*. But most of the bodies used for this purpose are those of shrunken, elderly people, distorted by embalming, and we need to check this information against that obtainable from the living body.

Surface anatomy deals with the relation of the superficial landmarks of the body to the deeper structures and is of special value to the artist. Inspection of the living body reveals the bulge of muscles, their activity when they contract, the pulsation of arteries and the course of veins, the position of bony prominences and superficial tendons.

Palpation, manipulation and *percussion* reveal respectively the consistence and resistance of the deep structures, the movements of joints and the boundaries of hollow, air-containing or solid organs, while listening with a stethoscope—*auscultation*—locates by sound organs such as the heart, lungs and bowel. *Endoscopy* is the introduction of an instrument for visualizing the internal aspects of structures such as the ear or stomach with the eye or the camera via more or less complicated optical systems, and *surgical exploration* is a useful source of information about the living body.

In *radiographic anatomy* X-rays are used to show up the skeletal system and also to demonstrate hollow organs—intestines, canals, blood vessels— when these are filled with substances opaque to the rays; solid organs can be studied thus if outlined by air blown in by injection. *Cineradiography* demonstrates the actual movement and function in life of the joints, heart, lungs and viscera. Radiographic anatomy is particularly valuable because in life the positions of organs such as the stomach undergo great variation with changes of posture and emotional disturbance, and need not correspond at all to the standard views based on dissection of formalin-hardened bodies.

Variation

A general principle in anatomy is that of the essential similarity of human beings; but against this background there is a continuous variation from the standard pattern as far as inessential details are concerned. Just as we vary outwardly in height and colouring of eyes and hair, so internally there may be minor aberrations in the arrangement of nerves and blood vessels, the exact pattern of bile ducts or bronchi, and the particular relation structures bear to each other. But just as the general descriptive catalogue— 'item, two lips, indifferent red; item, two eyes, with lids to them; item, one neck, one chin and so forth'— is always correct, so we may be sure that the femoral artery or the biceps muscle is where we expect to find it. In fact, the average anatomical arrangement is less of a fiction than the average man. There are, of course, occasional grosser developmental errors which are not within the normal range of variation, but which represent a failure or distortion of normal developmental processes—the absence of part of a limb, or the transposition of heart or viscera in mirror-image fashion.

Physiology

It is obvious that the study of the structure of body systems and tissues can only be artificially separated from the investigation of their actual behaviour during life and the related biochemical and biophysical processes, i.e. physiology. If function is not always borne in mind, the mere study of form is sterile. Physiology is this science of bodily function, and it is the complement to anatomy. Anatomy is essentially a descriptive discipline, physiology essentially experimental. We have to describe both the structures and what they appear to do, and, for this, experiment is essential.

Observation and experiment may be performed on both man and animals. Indeed, without animal experiment, modern physiology—and, for that matter, modern medicine—could never have evolved. Fortunately, the basic behaviour of living matter is the same throughout the animal world, whether in microscopic unicellular creatures or in the myriads of cells of the human body, and the organs and systems of other mammals are essentially similar to those of man.

The human body carries on its functions in a constantly changing *external environment*. Man lives at the Poles or in deserts, on mountains or in valleys, on Earth or in outer space; he has to struggle against physical forces and living enemies, to exist on very varied diets and to withstand the effects of drugs. It is only possible for him to do so if his *internal environment*—e.g. the composition of the blood and body fluids, the aeration of the blood—is maintained at a constant level regardless of external changes. This steady regulation of the inner state, whatever outer influences may be at work, is an underlying factor of immense importance in physiology and is known as *homeostasis* (see pages 9, 25). It can only be achieved because the physical and chemical processes of life are largely reversible. The body maintains its internal equilibrium as does a gyroscope, whatever the inclination of the supporting base, until it eventually runs

down and topples over in the ultimate, irreversible changes
of death.

History

A brief historical review may be of value before we proceed
to a detailed account of human anatomy and physiology.
We may assume some prehistoric anatomical instinct, in
that carvings and cave drawings are uncannily accurate in
contour and indicate an acquaintance with underlying
structures. And the Egyptians probably had a considerable
body of knowledge derived from embalming, various surgical procedures and ritual divination using the viscera of
animals. But it is to the Greeks that we owe the real development of the idea of anatomical structure as a matter for
deliberate investigation, coloured though their views were
by speculative philosophical ideas on humours and vital
spirits, and by a misconception of the working of the
vascular system. The superb surface anatomy of Greek
sculpture is well known, and various operations, such as
trephining for head injuries, were common practice. **Aristotle** and **Hippocrates** are the two great names of this
period. Aristotle, who founded comparative anatomy, probably never did any human dissection, but he made many
accurate observations on adult and embryo animals.
Hippocrates worked from a more practical, surgical viewpoint in connection with fractures and dislocations.

Physiology, too, began with the ancient Greeks, who were
often surprisingly accurate in their observations, though they
were also misled by ideas, such as Aristotle's theory in the
fourth century B.C. that the heart was the organ of reason.
The variations of the pulse were carefully remarked for their
prognostic significance by the physicians of the time, and
they were also aware of the differences between arteries
and veins; yet their speculations about disease were little
removed from the magical.

After the death of Aristotle, the Hellenistic School at

Alexandria continued the Greek tradition, establishing an anatomical discipline that included the public dissection of human bodies; in fact, there is some dispute as to whether human vivisection was practised at that time! **Eristratus** recognized the role of muscles in locomotion and the association of human intelligence with the intricate convolutions of man's brain. But he produced a confusing theory, based on the realities of respiration, of the permeation of the blood vessels and nerves by an essential spirit, or *pneuma*, diffused to the organs by the action of the heart.

During the Roman period there was a medical school in the capital, and observations on the structure of bones and organs began to be systematically recorded. In the second century A.D., **Galen** stands out as the greatest physician and anatomist of antiquity; his writings dominated medical knowledge for a thousand years, surviving even the Dark Ages which extinguished scientific activity for centuries, and were restored to Europe in translations from the Arab scholars. Galen described exhaustively the naked-eye structure of the body, though his subjects were apes rather than men; but, once again, speculation replaced experiment to mislead him in assessing function.

The Middle Ages witnessed the rise of the great Italian universities and their medical schools—as at Bologna and Padua—at which regular dissection of the human body began to produce a firmer basis for physiological thought, though Galen's writ still ran unquestioned. The great figure of **Leonardo da Vinci** stands out at this time. A brilliant dilettante artist and anatomist, he established sound principles of medical illustration, furthered by the invention of printing, and made numerous observations on the actions of muscles and the nature of vision. But he failed to publicize his findings and it was left to others to remake his discoveries. In the sixteenth century **Andreas Vesalius**, a young Belgian working at Padua, rejuvenated the medical sciences. Rejecting the traditional reliance on Galen, he insisted on the importance of personal observation and experiment, and in his masterpiece *On the Fabric of the Human Body* gave the first complete account of human anatomy. He recognized that muscles were stimulated via their nerves, that all muscles had antagonists which acted reciprocally and that arterial pulsation was derived from the heartbeat. He also watched the movements of the lungs in dogs by removing portions of the ribs. Before Vesalius, the nature of the circulation had been obscured by Galen's insistence that blood passed from one side of the heart to the other through invisible apertures in the intervening septum; but Vesalius asserted that the septum was impermeable. This resulted in **Servetus'** demonstration that the transfer of blood is effected through the pulmonary vessels and their ramifications in the lungs, and led to the discovery of the true nature of the circulation by the immortal **William Harvey**.

Harvey was an Englishman, and dissection is not known to have begun in England until the very end of the fifteenth century, though some demonstrations on the corpses of criminals were authorized in Henry VIII's reign. Harvey's birth in 1578 heralded a flourishing period of anatomical and physiological research, ushered in by his discovery of the circulation of the blood. He succeeded in grasping its essentials, although, as yet, he was unable to see the capillaries, the fine channels connecting arteries and veins, which were necessary to his theory. His evidence, though indirect, was conclusive. What clinched the matter was the output of the heart, which at every beat pumped about 2 ounces of blood into the aorta, the main artery of the body. The heart beats some seventy-two times a minute, or 4320 times an hour. So over 241 kg of blood left the heart every hour, clearly an impossibility unless there *was* a circulation. Harvey's methods, combined with the growth of chemical and physical knowledge in the seventeenth century and the evolution of the microscope, led to rapid advances in physiology. His postulated capillaries were seen and verified, the red corpuscles of the blood were discovered and blood transfusion in animals was carried out in

1660 (as Pepys notes in his Diary). The nature of respiration was clarified by **Boyle**'s demonstration that a part of the air—oxygen—was as essential to life as to combustion. Inspired air mixed with the blood in the lungs to give it its red arterial colour; deoxygenation in the rest of the body rendered it dark and venous. A country clergyman, **Stephen Hales**, stands out in the eighteenth century as one of a long line of distinguished amateur scientists in England. His best-known achievement was the direct measurement of the blood pressure in horses, using glass tubes several feet high. Hales' early training was in mathematics and physics, and so he was able to lay the foundations of our knowledge of the hydrodynamics of the body.

During the middle and later parts of the eighteenth century, the famous surgical brothers, **William** and **John Hunter**, gave an immense impetus to dissection, and to pure and comparative anatomy and their applications to surgery; John's specimens became the nucleus of the great Hunterian collection of the Royal College of Surgeons. By the close of the century, anatomy was recognized as a basic subject to be taught to new students at the medical schools. They also made numerous physiological observations on digestion, the coagulation of the blood and the growth of bone. About this time, too, **Galvani** investigated electrical phenomena within the body and showed that muscles could be made to contract as the result of electrical stimulation. The discovery of *vitamins* was implicit in the Admiralty order of 1795 for the issue of lime juice to check the ravages of scurvy in His Majesty's ships.

A romantic passage in our own anatomical history is connected with the difficulty in obtaining subjects for dissection in the early nineteenth century. Only the bodies of hanged criminals were available—in fact, their dissection was compulsory by law; but, this source proving inadequate, the snatching of freshly-buried corpses by the 'resurrection men' became prevalent, a lucrative profession for the ruffians concerned. Increasing public anger was aroused, especially when murders were committed specifically to obtain bodies for sale, culminating in the infamous Burke and Hare episode in Edinburgh. The shocking disclosures at the trial in 1828 were ultimately responsible for the Anatomy Act of 1831, which regularized the conduct of anatomy schools and authorized the use of unclaimed bodies for dissection. It has been well said that 'the early anatomists in their desire for knowledge had to overcome many obstacles, to suppress their own prejudices, to overcome the natural disgust raised by the smell of a corrupting body, to brave human laws, to withstand the ban of the Christian church, to rob the grave in the dead of night and pursue their studies in secrecy and peril'.*

For physiology, the nineteenth century marked the end of the period of slow development by individual effort. The recognition of the cell as the essential unit of biological structure and function; the discovery of diffusion and osmosis, crystalloids and colloids; the bridging of the gap between inorganic and organic matter by the artificial synthesis of urea in 1823: all these were of great significance. Furthermore, the conception of metabolism was established —the breaking down of foodstuffs by oxidation within the body to yield carbon dioxide and other wastes, together with the liberation of heat, a process associated with as exact an interchange of energy as any reaction carried out in a crucible and equally subject to the laws of thermodynamics.

A picturesque episode is that of **Beaumont**, in America, studying gastric function in Alexis St. Martin, a trapper shot in the abdomen who had acquired a permanent opening, or fistula, between the stomach and the surface of the abdomen. This allowed Dr. Beaumont to observe directly for the first time the reactions of the stomach to food and emotion, as indicated by changes in secretion and motility.

It is to a Frenchman, **Claude Bernard**, that we owe the idea of the constancy of our internal environment, upon

* R. H. Hunter, *A Short History of Anatomy*, London 1931.

which man's relatively independent life depends. He saw the body as a mechanism always reacting so as to neutralize the effects of any change in the external environment and maintaining an internal equilibrium, in which cells and tissues remained unchanged in essential properties—a regulation achieved by the diffusion of water, oxygen, food and heat through the circulatory system, the whole co-ordinated and modified by the nervous system, and also modified by the internal secretions or hormones, such as those of the thyroid gland.

Finally, there has been the development of our knowledge of the nervous system. **Sir Charles Sherrington**, at Oxford, demonstrated the self-regulating nature of bodily posture and the activities appropriate to each level of the brain. The great Russian physiologist, **Pavlov**, demonstrated clear links between mind and body through his studies of conditioned reflexes—bodily responses produced by the brain reacting not to the original stimuli but to others it has learned to associate with them. Indeed, Pavlov and American psychologists of the behaviourist school went so far as to suggest that human behaviour might be considered as a mass of such reflexes. More practically, the psychologists have entered the field of physiology by demonstrating the immense influence of the emotions and unconscious life in general on the most basic physiological activities, such as heartbeat and digestion.

2 Cells and tissues

Cells

The unit of living tissue is the microscopic *cell*; even the fertilized ovum, the largest of human cells, is only just visible to the naked eye. The cell is the basis of life and has its own capacity for life. The body is composed of different *tissues*, each of which is an aggregation of the same kind of cell, together with a varying amount of intercellular substance. In addition, there is much material produced by cellular

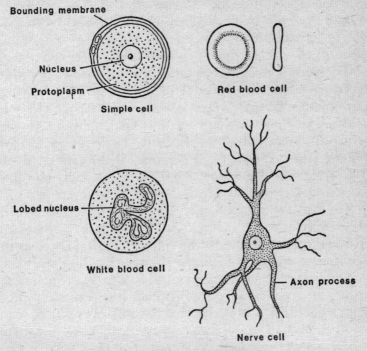

Fig. 1. Various types of cells.

activity but no longer living, e.g. the mineral content of bone, the hair and nails. All the cells of the body have sprung from the division of the female germ cell (ovum) after fertilization by the male germ cell (spermatozoön).

In the adult there are specialized cells that cannot divide and whose loss is irreplaceable, e.g. nerve and muscle cells; cells that divide slowly in health but can be stimulated to rapid growth in case of need—connective tissue in repair after injury, glands with added functional demand; and cells that reproduce rapidly and often to replace specialized cells with a short life-span, e.g. the precursors of red blood cells, the epithelium of the skin and the lining cells of the gut. Many cells have bizarre shapes. Some nerve cells, though themselves minute, send their fibres almost the length of the body. Muscles are made up of multinucleate cylinders often as long as the muscle itself. The red blood cell, or erythrocyte, is modified into a flattened biconcave disc to concentrate the blood pigment (haemoglobin) at the periphery where maximum gaseous interchange can occur.

Cell study is usually by light microscopy, which depends on fixing, embedding and staining the tissue. The resolving power is insufficient for finer details, and although electron microscopy shows more detail it is not applicable to the living cell. Also, cells or tissues can be made radioactive for photographic study, or fluorescent with various dyes. Finally, X-ray diffraction can be used to study the diffraction patterns of crystalline biological materials *in situ*.

The essential parts of any cell are the *bounding membrane*, the protoplasmic substance, or *cytoplasm*, and the *nucleus*. The bounding membrane is permeable to simple ions; it is in active motion and sometimes extrudes pseudopodia. The cell *surface* carries identification marks characteristic of the tissue, the species and the individual; it also contains proteins which act so as to reject foreign cell surfaces, e.g. material grafted from other persons or animals—i.e. there is an *immunological* function. The *cytoplasm* is all the material within the cell except the nucleus. It is a mass of colloidal

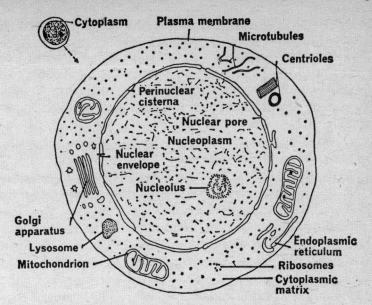

Fig. 2. A section of a cell containing the structures that are essential for its survival. (From *A Companion to Medical Studies, Vol. 1.*)

proteins, carbohydrates and solutions of smaller molecules, and contains most of the ribonucleic acid (RNA) of the cell. It also contains a number of discrete arrangements, or *organelles*, of various kinds:

(1) The *Golgi apparatus*, near the nucleus, function unknown.

(2) The *mitochondria*, thread-like bodies—thousands in a single cell—in constant motion, which contain enzymes for the oxidation of carbohydrates. They are the furnaces where cell fuel is burnt and energy stored, the main sites for energy conversion, and are prominent in very active tissues such as the heart muscle and the kidney tubules.

(3) The *lysosomes* contain enzymes and are concerned with digesting absorbed material and scavenging generally.

(4) The *ribosomes*, dense particles of RNA-protein complexes; points of production of protein molecules.

(5) The *centrioles*, a pair of barrel-shaped bodies at right

angles to each other near the nucleus, which have an important function in cell division.

The *nucleus* is sharply defined and can be separated or centrifuged out of the cell. It contains the *chromosomes* (see below) with RNA and metre-long molecules of DNA (deoxyribonucleic acid). There are also small *nucleoli*, aggregates of RNA. Certain very large cells are multinucleate, e.g. bone osteoclasts and skeletal muscle fibres. The nuclei of the polymorph white blood cells are lobulated.

Cells stick together, i.e. they have a capacity for *adhesion*, and this facilitates their firm, organized arrangement in tissues. It is a characteristic of cancer cells that they lose this adhesion and become free to invade the body. Some, but not all, cells have the power of *phagocytosis*, i.e. they can surround and ingest foreign matter, bacteria and dead cells; this property is notable in the polymorphs of the blood and the cells of connective tissue. The chemical constituents of cells include proteins—some in colloidal solution in the cytoplasm, some particulate in the organelles—droplets of fats or lipids, and carbohydrates as soluble simple compounds or granules of complex polysaccharides. There are also *pigments*: the red haemoglobin of blood, its pink derivative in muscle, and its green and brown degradation products in bile and faeces; the photosensitive visual purple in the rods and cones of the retina; and the brown ageing pigment of fat and iron, lipofuscin.

The cells of different tissues may be greatly modified for their specialized functions, e.g. the red cell of the blood has lost its nucleus, the nerve cell has an enormously elongated axon process to carry stimuli and the striated muscle cell possesses the capacity to contract in length. But despite these modifications, in each case the essential components remain recognizable.

Cell division

Every cell in the body (except for the germ cells) carries within its nucleus the forty-six chromosomes in twenty-

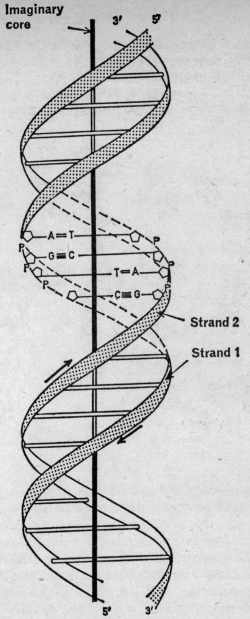

Fig. 3. DNA thread of two strands round an imaginary core in an anti-parallel manner. A, T, G and C represent complementary base parts. (From *A Companion to Medical Studies, Vol. 1.*)

three pairs characteristic of the human species; there are twenty-two pairs not involved in sex determination—the *autosomes*—and one pair of *sex chromosomes*. These are not clearly visible as such until the preliminary phase of cell division. Each DNA molecule contained in the chromosomes

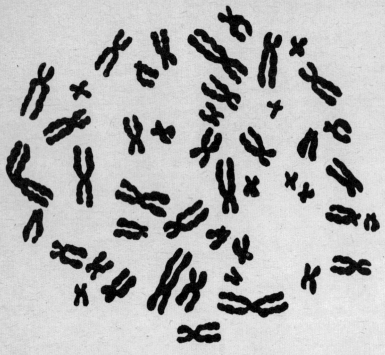

Fig. 4. Chromosomes of a normal human male cell from a culture of peripheral blood. (Adapted from *A Companion to Medical Studies, Vol. 1*.)

consists of a long pair of spirally-wound strands—the double helix. This consists of a backbone of deoxyribose (sugar) and phosphate molecules with purine or pyrimidine bases attached—adenine, thymine, guanine and cytosine. The bases on one strand face those of the other in precise and regularly repeated fashion, and the order of these pairs of bases contains the information or genetic code controlling protein synthesis, ensuring that the constituent amino-acids are inserted in the correct places in the growing polypeptide chains. The organization is complex, and is effected via 'messenger' RNA and 'transfer' RNA molecules. As we have seen, the DNA of the cell is not confined to the nucleus. Some is in the cytoplasm, concerned with protein synthesis, but that of the nucleus is contained in the separate blocks or chromosomes, coiled structures not to be confused with the spiral coils of the DNA molecules themselves. Normal cell division in man is an elaborate process of *mitosis*, designed to ensure an exact duplication of the original cell and an exact sharing of the genetic material represented by the chromosomes and their DNA, for it is this material that contains the information required for the synthesis of new cells in the body.

In the resting phase, or *interphase*, the chromosomes are not visible as separate structures in the nucleus. As mitosis approaches, they become visible as threads within the nucleus (*prophase*). The centrioles divide and move to opposite poles of the cell, where they act as foci from which radiates a controlling spindle of protein fibres. The nuclear membrane disappears. The chromosomes are now seen to be double and align themselves in the plane of the spindle between the centrioles. The two halves of each chromosome separate and are pulled by contracting spindle fibres to opposite ends of the cell to form the new chromosomes of the two daughter nuclei. The cytoplasm of the cell cleaves in two and new nuclear membranes form around each separated bundle of chromosomes, reconstituting two daughter cells and two daughter nuclei genetically identical to the parent nucleus.

Tissues

There are only four essential tissues: epithelium, connective, muscular and nervous. These may be modified further in various ways, but together they compose the structure of the

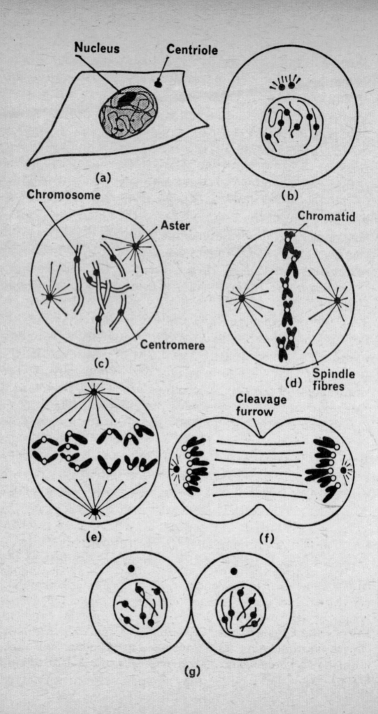

(a)

Nucleus Centriole

(b)

Chromosome

Aster

Centromere

(c)

Chromatid

Spindle
fibres

(d)

(e)

Cleavage
furrow

(f)

(g)

body. They differ in the nature of their component cells and in the particular kind of secreted intercellular ground substance or *matrix*.

Connective tissue

In its various forms this makes up the main supporting framework of the body and the packing round the internal organs. It is characterized by the large amount of intercellular substance, and the exact nature of a connective tissue depends on the particular kind of matrix and on the cells and fibres it contains. It includes such widely different tissues as bone, cartilage, fibrous tissue, elastic tissue, blood and lymph.

In simple *areolar* tissue—the loose, web-like packing in clefts and spaces, seen very well between the skin and deeper structures as the *subcutaneous layer*—there is a semi-fluid matrix containing mucopolysaccharides in which run strands of protein fibres—white (unyielding) and yellow (elastic)—together with rather sparse cells, an arrangement allowing free gliding of part on part. This tissue contains a good deal of the *extracellular fluid* of the body. In *adipose* or *fatty* tissue the cells are modified and distended to carry fat globules; this is found everywhere, except within the cranial cavity, but is most conspicuous under the skin and in the membranes within the abdominal cavity. A predominance of white fibres is found in the firm, white,

---

Fig. 5. Mitosis. (a) Interphase (forty-six chromosomes). (b) Early prophase. Centriole divides, chromosomes appear as threads. (c) Late prophase. Centrioles separate. Aster and spindles form, chromosomes seen to be double, nuclear membrane disappears. (d) Metaphase. Chromosomes line up in equatorial plane. (e) Early anaphase. Centromeres divide, chromatids begin to separate. (f) Late anaphase. Separation of chromatids continues, cleavage furrow forms. (g) Telophase. Two daughter cells each with forty-six chromosomes. (From *A Companion to Medical Studies, Vol. 1.*)

Adipose connective tissue

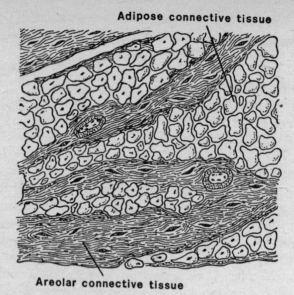

Areolar connective tissue

Fig. 6. A simple connective tissue.

fibrous tissue of tendons and ligaments and sheets of membrane (*fasciae*). Elastic fibres predominate where stretch and
recoil are important, as in the walls of the larger arteries.
Adipose tissue is a store of triglycerides and thus a source of
energy. It insulates against cold and mechanical damage,
buffers the internal organs, or *viscera*, and is more developed
in the female sex. It may disappear almost entirely in the
last stages of starvation. It is divided into fatty lobules by
fibrous septa and is closely packed with cells, each of which
has its nucleus at one side and is distended by a lipid
globule.

In *blood* and *lymph* the matrix is completely fluid, while
the cells have become modified either to carry dissolved
respiratory gases or to deal with invading micro-organisms.

In *cartilage* and *bone* the matrix has become hard and
solid by impregnation with mineral salts.

In general, the cells of connective tissue fall into three
types: *fibroblasts*, which manufacture the fibres; *mast cells*,

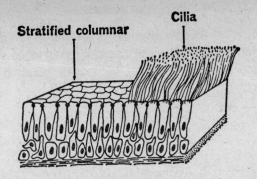

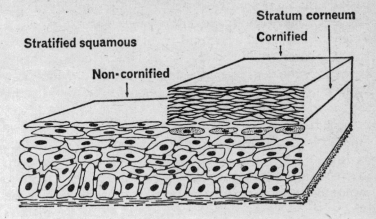

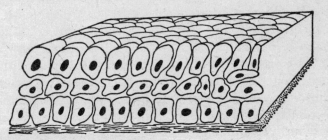

Fig. 7. The stratified epithelia. (From *A Companion to Medical Studies, Vol. 1.*)

which are concerned with carbohydrate metabolism; and *phagocytes*, which scavenge and ingest foreign material and the debris of dead cells, and may have the power of movement.

Epithelium

Epithelia are sheets of cells covering all free body surfaces, not only the external aspect of the skin but also the internal surfaces which line the body cavities. In addition, they provide the linings of the vessels, of the respiratory, digestive and urinary tracts, and of the glands opening into these systems. It is essential to realize that both internal and external surfaces of the body possess this intact epithelial envelope, which is normally unbreached and stands as a barrier between the outside world (including, paradoxically, the contents of the bowel and hollow organs) and the deeper intermediate structures of the body. These inner and outer linings blend with each other at the orifices of the various cavities—at the lips, anus, nose and urethral apertures, where the damp linings or *mucous membranes* of the digestive, respiratory and urinary tracts become continuous with the skin.

Epithelium is typically very cellular and simple in arrangement, with little ground substance. The simplest form is one cell thick, with a basement membrane. It may be squamous, arranged in mosaic fashion, as at the skin surface, or *stratified* in row upon row of cells (the cells may be flattened, cubical or columnar).

Glands. These consist of simple or complex arrangements of epithelial cells forming substances useful to the body in secretory cells. The simplest form is the *goblet cell* of the intestinal mucous membrane. But they may be more complex: tubular—either simple, branched or coiled—or compound, with a branching duct system ending in blind pouches, or *acini*. When acini are present, these are either mucous or serous. The cells of mucous acini appear empty;

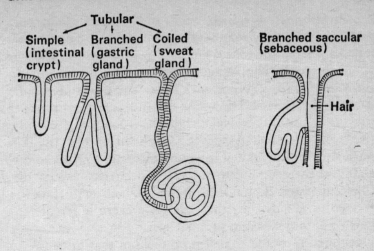

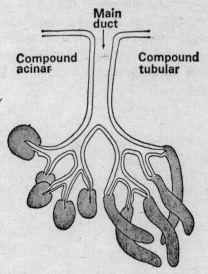

Fig. 8. Types of glands. (Adapted from *A Companion to Medical Studies*, Vol. *1*.)

they have flattened basal nuclei and their clear cytoplasm contains polysaccharides. The serous acinar cells are granular, have a large central nucleus and manufacture protein enzymes. As well as mucus and enzymes, glands may produce salt and solute (sweat glands), acid (gastric mucosa), alkali (duodenal mucosa) or grease (subcutaneous glands of skin). They may secrete continuously or only in response to some stimulus, such as the entry of food into the stomach. They may be relatively independent or under varying nervous, chemical or hormonal control. If, like the sweat and digestive glands, they are connected to the skin surface or the hollow of an internal organ and there discharge their secretion, they are called *exocrine*. If they discharge their secretion directly into the bloodstream, they are endocrine glands and their *hormones* exert a remote influence of target organs at a distance.

It remains to add that it is epithelium that gives rise to the hair, nails and teeth; that *endothelium* is the name given to the simple smooth linings of blood vessels; that the glistening layers that line the great body cavities, pleural and peritoneal, are sometimes called *mesothelia*; and that a *carcinoma* is a form of cancer in which epithelial cells have lost their mutual adhesion and have invaded the subjacent tissues.

3 Basic biophysics and biochemistry

Homeostasis

The human body is an open system, exposed to a constantly changing external environment. These changes are buffered and neutralized by the *internal* environment—the blood, lymph and tissue fluids that bathe the cells and protect them from change. This stability is the object of all the vital mechanisms. Let us take some examples.

Exercise produces heat and the body temperature tends to rise; but perspiration occurs and there is heat loss due to the latent heat of evaporation of water, and this compensates for the raised temperature—i.e. sweating is a physiological cooling mechanism that begins to operate whenever the body temperature rises, whether from exertion or fever.

Again, the blood is normally slightly alkaline. Although during exercise the muscles produce carbon dioxide, there is little or no rise in acidity of the blood (a) because of its chemical buffering properties, and (b) because the excess CO_2 is exhaled as it is formed; hence the rapid breathing during effort and the subsequent panting as the oxygen debt is paid off.

A third instance of homeostasis is that the blood level of sugar (glucose) remains constant during fasting, even though sugar is being burnt by the tissues and none is being ingested; this is because of the formation of sugar from fat and protein in the body stores, mainly in the liver.

Finally, most of the water drunk enters the bloodstream; yet drinking even very large quantities of fluid does not normally lower the osmotic pressure of the blood by dilution. The reason is that any tendency in this direction is sensed by special *osmoreceptors* at the base of the brain which

affect the output of a pituitary hormone that controls the
output of water by the kidney. Normally, excretion of water
in the urine keeps pace with its absorption from the
intestine.

These examples illustrate how the internal environment
is kept constant despite external stresses. Were these
elaborate mechanisms to fail, the health and life of the
organism would be in danger, e.g. in very high fever due to
infections, or when the blood sugar falls to a very low level
if there is a tumour of the pancreas secreting insulin (see
page 340).

Body water

Water is the universal fluid medium of the body. It provides the means of transport within the body of food,
wastes and respiratory gases, and its properties allow these
substances to diffuse into and out of the cells. Water constitutes up to 95% of the cell protoplasm and 60% of body
weight, and no chemical substance can influence the behaviour of living cells until it is first brought into watery
solution.

The *total body water* can be measured by a dilution tech-
nique, i.e. by injecting identifiable substances into the
bloodstream and measuring the concentration of the substance in the plasma. If Q be the quantity given, and C be
the concentration in the blood plasma, then V, the total
body water, equals Q/C, a quantity of some 40 litres in a
normal healthy adult man, or some two-thirds of the body
weight. The total body water includes the water in free
fluids, like the blood plasma and the fluids of the tissue
spaces, as well as the fluid within the cells. If we repeat the
experiment using substances, such as insulin, that cannot
cross cell membranes, we can measure the extracellular
water, and this is some 15 litres, or one-fifth of the body
weight. Three litres of this are contained in the plasma,

12 litres in the interstitial fluid. And, by difference, the cell water amounts to 25 litres:

Total body water   Intracellular 25 l
     40 l   Extracellular 15 l

Interstitial fluid   Plasma
   12 l   3 l

The total body content of extracellular fluid (ECF) varies a little from time to time, and because of this the body weight measure on rising in the morning (after urination) fluctuates by amounts of up to 1 kilogramme. The ECF is reduced by lack of water intake, or by increased output due to sweating, vomiting or diarrhoea. It becomes increased in waterlogging of the tissues (oedema), as may occur in starvation, heart failure or kidney diseases.

There are certain important differences in the chemical composition of the ECF and the intracellular fluid (ICF). The *cations* of the ECF are mainly sodium, a little potassium, calcium and magnesium; those of the ICF are mainly potassium, some sodium and calcium, and rather more magnesium than in the ECF. The *anions* of the ECF are mainly chloride, with some bicarbonate; those of the ICF include some chloride, bicarbonate and sulphate, but'they are mainly organic acids, phosphate and proteins. Thus the principal chemical difference is that sodium is the chief cation outside the cells and potassium inside; and this difference is essential and constant. The ECF, or internal environment, is seen to be principally a solution of NaCl, which has been taken to confirm the supposed marine origin of mammals. Nevertheless, the salt content of sea-water is three times that of the ECF, a load the kidneys are incapable of dealing with, hence the inability to drink sea-water in shipwreck. The bicarbonate of the ECF is controlled by respiration, which expels CO_2 and so regulates the acidity of the fluid.

Body water is derived from the water content of solid

food, the water drunk and the *metabolic water*, i.e. that formed in the tissues by oxidation of the hydrogen in the food. Water is lost in the urine, from the bowel with the faeces, and by evaporation from the skin and lungs. The first two of these may vary greatly, but the last is obligatory; the expired air is always saturated with water vapour. The *water balance* in an adult sedentary male in a temperate climate may be set out as follows:

Intake: food 800 g; drink 1300 g; metabolic 250 g = 2350 g.
Output: urine 1500 g; faeces 25 g; evaporation 825 g = 2350 g.

The evaporated water plays an important part in cooling the body, i.e. in disposing of waste heat, and accounts for a quarter of the total heat loss. Because the water loss from skin and lungs is obligatory, and because there is little or no true body reserve of water, a negative balance is easily produced if the intake falls, if there is excess loss from the bowel in diarrhoea and vomiting, if there is excess sweating (this may amount of 10 kg daily in a hot climate), if there is excess loss from expiration in cold, dry (mountain) conditions or if kidney disease eliminates water excessively. For these reasons death will occur in less than a week in an individual totally deprived of water to drink. On the other hand, it is not possible to create a positive balance by drinking large amounts, for the water lost in the urine keeps in step with intake.

From the *biophysical* angle, the essence of bodily function is concerned with the behaviour of molecules in aqueous solution, to which the ordinary laws of physics apply. Thus, by diffusion, the salts in the blood or the sugar content of the cerebrospinal fluid are uniformly concentrated and the amount of these substances in solution is the same wherever we take our sample.

Certain membranes allow water and dissolved substances to pass through them freely; if such a membrane separates two solutions of different strengths, their concentration is ultimately equalized by the transfer of water and solute in reverse directions. This process is called *dialysis*. Membranes vary in the size of the molecules they will allow to pass, and most of the cell membranes of the body are only semi-permeable; they will allow a free passage to water but are permeable only to some solutes. If such a semi-permeable membrane separates two solutions, the one stronger than the other, of a substance that cannot traverse the membrane, then water is drawn through from the weaker to the stronger side of the partition. This attraction of water is known as *osmosis*. The *osmotic pressure* of the solution is the expression of the energy gradient set up by the tendency of water always to dialyze so as to dilute the stronger solution. Thus a pig's bladder filled with strong sugar solution and immersed in water, or in a weaker sugar solution, attracts water in and swells up.

The behaviour of living body cells, whether aggregated in the tissues or free floating in the blood, is intimately affected by the relative osmotic pressures of the surrounding medium and of the cell substance itself, for the intervening cell membrane is semi-permeable and behaves like the pig's bladder in the experiment. Thus a red blood cell, immersed in strong salt solution, shrivels up as water is sucked out of it, but it will swell up and burst in a weaker solution or in water.

There is obviously an intermediate osmotic pressure for the surrounding medium which is identical to that of the protoplasm of the cell; a solution of such an osmotic pressure leaves the cell unaffected and is known as an *isotonic* solution. The stability of behaviour of the cells is such that all their osmotic pressures are identical and isotonic with a 0·9% solution of sodium chloride, and this is spoken of as *normal saline*. Normal saline is thus identical osmotically to the blood and lymph, and is used experimentally to bathe cells and tissues without harming them. It can also be given as an intravenous infusion whenever it is desired to restore a depleted blood volume, as in surgical shock.

Whatever the substances concerned, solutions containing the same number of molecules per unit volume have the same osmotic pressure. And, therefore, solutions of different substances of the same *percentage* strength have osmotic pressures inversely proportional to the sizes of their molecules, or molecular weights—e.g. a 1% solution of protein, which has enormously large and complex molecules, contains far fewer particles per cubic centimetre than a 1% solution of sugar, and its osmotic pressure is correspondingly less. Further, many substances called electrolytes *ionize* in solution, i.e. their molecules split or dissociate into two or more electrically charged *ions*. The osmotic pressure of such solutions is greater than would be expected from the size of the molecules, for it is proportional to the total number of *particles* present.

Solutions have a further property of great physiological importance which results from the tendency of some substances to ionize. This property is their *chemical reaction*, whether acid, alkaline or neutral. Acidity is due to the presence of hydrogen ions (H) and alkalinity to that of hydroxyl ions (OH); the reaction of water is neutral because, when it dissociates, it produces these ions in equal amounts. In any solution the product of hydrogen and hydroxyl ions is constant, the two being inversely proportional to each other. For this reason the acidity or alkalinity of a solution may be estimated by measuring its hydrogen-ion concentration alone. This is always a small figure (a litre of water contains only 10 g of such ions) and it is therefore customary to use the power of 10 as a positive number—the hydrogen-ion exponent or pH.

Thus the pH of water is 7, and this is the centre point of chemical neutrality. The pH of a solution falls as its acidity rises and increases as it becomes alkaline, a single step from, say, pH 7 to 8, indicating a tenfold increase of alkalinity. The reaction of the blood is normally just on the alkaline side of neutral, at pH 7·4. Cell functions are profoundly modified by quite small changes in chemical reaction; quite a small increase in acidity or alkalinity may suffice to suppress their activities—e.g. to bring an organ such as the heart to a stop. Nevertheless, small amounts of acid and alkali are constantly being formed in cellular life, and the effects of these are minimized by certain *buffer salts*, such as sodium bicarbonate, in the body fluids and cell protoplasm which take them up automatically to prevent any change in the general reaction. The main buffer in the body fluids is sodium phosphate and that of the cells sodium bicarbonate, each of which can so vary between its own acid and alkaline forms as to compensate for any change in the reaction of the medium. Proteins themselves can function either as weak acids or as bases and so stabilize the pH of the cells they compose.

Crystalloids and colloids

All substances fall into two classes: *crystalloids*, simple crystallizable compounds like sugar and salt which diffuse easily in water and traverse animal membranes; and *colloids*, complex materials like gelatin and egg-white which crystallize with difficulty or not at all, diffuse slowly and cannot cross membranes. Obviously, the two can be separated by the process of dialysis already referred to, and the essential difference between the classes lies in the size of their molecules; the molecular weight of salt is 58·5, while that of proteins is of the order of 100 000.

Colloids may exist either as *sols* in apparent solution or as *gels*, a solid jelly state with water. In fact, colloidal solutions are really only suspensions of particles, a disperse phase in the continuous phase of some fluid medium. These so-called phases are reversible, as when milk, a suspension of oil in water, becomes butter, a suspension of water in oil.

One feature of colloidal 'solutions' is that their surface layer has a much greater concentration of the colloid substance than the mass of the solution, and this is the basis of the formation of the bounding membrane of living cells, the membrane whose semi-permeability is responsible for the osmotic transfers already discussed. This membrane is also responsible for maintaining certain differences between the contained protoplasm and the surrounding medium.

Thus there is constantly more potassium than sodium in red blood cells, though the reverse is the case in the plasma itself, the fluid component of the blood. And, again, most cells have a slightly acid reaction of pH 6·8, although immersed in neutral or slightly alkaline surrounding media. These differences are maintained by the selective action of the cell membrane. But this selective action is not purely physical, for the outer layer contains fatty materials and can therefore be penetrated by chemical agents soluble in fats, such as ether and alcohol; such substances act as anaesthetics by influencing the brain cells in this way and will also kill individual cells on direct application. Colloidal solutions are unstable, and the suspended particles may be precipitated by heat, by changes in reaction or by the addition of salts; the *coagulation* of protein sols produced by heat is familiar to us in the case of the albumin of egg-white. A further important property of colloids in relation to their behaviour in living matter is their enormous power of *adsorption* of other substances due to the immense surface area presented by the dispersed particles. Adsorption is the capacity to pick up and retain a substance without entering into chemical combination with it.

We may now review the basic physical factors that control the transference of substances across cell membranes. These are:

(1) A difference of *hydrostatic pressure* between the two sides, i.e. the difference in pressure between two unequal columns of water connected at their lower ends. Thus, when excess water is eliminated from the blood via the kidneys, the osmotic pressure of the proteins dissolved in the blood plasma resists its outward passage from the capillaries into the kidney tubules. This osmotic pressure is the equivalent of 30 mm of mercury. However, this is normally overcome

by the hydrostatic effect of the blood pressure—say, 130 mm of mercury—which drives the water across. But if the blood pressure falls below 30 mm, because of disease or shock, urine ceases to be formed.

(2) The ordinary laws of *diffusion* and *osmosis*, depending on the relative strengths of the solutions on either side.

(3) Differences of *electric potential*.

(4) The varying *permeabilities* of the membranes themselves. Only rarely are they absolutely impermeable, and none is completely semi-permeable, i.e. none allows the passage of water only. Some allow crystalloids, certain colloids and water to cross, but the usual semi-permeable variety permits water and crystalloids only. In certain cases, substances will only pass in one direction, due to changes they produce in the membrane. A final point is that, in the dialysis of a compound of one diffusible and one non-diffusible ion, a difference of chemical reaction develops between the solutions on the two sides of the membrane; this phenomenon provides the basis for the secretion of the highly acid and alkaline digestive juices of the stomach and pancreas respectively.

Energy transformations within the body

The essential chemical elements of protoplasm are carbon, hydrogen, nitrogen, sulphur and phosphorus. Simple compounds of these, such as water and carbon dioxide, require an elaborate synthesis for their conversion into organic or living matter; this is effected by green plants under the influence of sunlight. Neither man nor animals can achieve this synthesis for themselves; they are dependent on the consumption of vegetable matter, either directly or indirectly, after it has been utilized by other animals lower in the food chain which serve as food in their turn. Ultimately, we are all vegetarians, and all flesh is grass. Thus, the energy transformations of living matter begin with light absorbed by plants and end with the heat waste of both plants and animals. In its passage this energy is subject to the laws of thermodynamics and can be utilized to do *work*. (Note that the living organism, as far as its body-building activities are concerned, has temporarily reversed the universal tendency for energy to run down and dissipate in the environment as heat.)

In the final processes of digestion and absorption, the constituents of the food are temporarily broken down into simpler organic substances, which are reassembled in the tissues in a manner that is never quite the same for different animal species. The finished products—whether subcutaneous fat or heart muscle—differ slightly in chemical composition as between man and animals, and, indeed, to some extent as between one man and another. Thus man depends for body-building material and for sources of energy on the complex organic compounds—fats, carbohydrates and proteins—that have been built up in living plants. The only chemical element he takes up as such in the free state is the oxygen in the inspired air (although oxygen is also used in combination). Oxygen frees the potential energy of the foodstuffs by burning or oxidizing them in the cells of the body to form carbon dioxide and water, at the same time liberating energy. Thus this last step is the complete reversal of what the green plant originally achieved. Since the wastes and breakdown products of animal tissues are used again by plants, there is a continuous cycle. Both plants and animals depend on this cycle. And the energy for our body heat and movements has been ultimately derived from the sun.

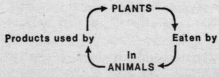

Energy (sunlight) + water + carbon dioxide = organic matter

in

PLANTS

Products used by ⟶ Eaten by

in

ANIMALS

Organic matter + oxgen = carbon dioxide + water + heat and energy

Of course, the energy concentration of the green leaf is very low, hence the large amounts that have to be eaten by herbivores. Nevertheless, this energy can be concentrated by herbivores into fat and meat for use in human diet.

This cycle is only concerned with carbon, hydrogen and oxygen, but similar cycles apply to the other essential elements of the food, such as nitrogen. This is mainly synthesized into organic matter by minute plants, the soil bacteria on which the plant kingdom as a whole is dependent; then the plants pass it on to man and animals; and these return it in the urea and ammonia of their excreta to be used in the soil again. A number of substances—metallic ions—also essential to the body are taken in the form of simple mineral salts: sodium, potassium, calcium, magnesium and iron. Essential basic radicles are phosphates and chlorides.

Most natural foods contain but little *sodium*, and few races are ignorant of the importance of salt in the diet. Civilized man is very dependent on it, and, besides what is taken at table, it is added to bread, butter and canned meat. The salt-free diet requisite for some heart and kidney disorders is felt to be a severe and unpalatable restriction. Although a healthy individual does not retain excess salt, the salt content of the body is maintained well above that of the environment by the kidney. Sweat contains salt, and severe salt loss may occur in hot or humid conditions; the miner avoids cramps by putting salt in his beer and salt tablets are taken in the tropics.

Potassium is abundant in the cells of plant and animal food, and dietary deficiency is very rare. Severe potassium loss may, however, occur in vomiting, diarrhoea and wasting diseases, and it causes muscle weakness and thirst.

Calcium is essential for the formation of bone during growth and for the maintenance of total bone mass in the adult, since there is a constant turnover of the kilogramme of calcium contained in the skeleton. It so happens that most of the calcium in the diet is lost in the faeces in the form of insoluble salts; the problem is one of absorption. Milk is a rich source and cheese, therefore, even more so. The calcium of bread is more readily absorbed if the latter is white, for in wholemeal bread the phytic acid in the husk of the grain forms insoluble calcium phytate which is excreted with the faeces. Thus it is reasonable to enrich high-extraction bread with calcium lactate. Calcium is necessary for the normal electrochemical excitability of tissues; in severe calcium deficiency there is twitching and spasm of the muscles, known as *tetany*.

Magnesium is the most important intracellular cation after potassium and plays an essential part in certain enzyme reactions.

Iron is a special instance of a mineral essential to life, for it is an essential part of the molecule of the red pigment of the blood, haemoglobin, which is responsible for the transference of oxygen from the lungs to the tissues. The body contains 4–5 g of iron, half of which is contained in haemoglobin. It is in constant demand for the synthesis of the pigment in new red cells. A good deal is lost at each menstruation, and it is needed in pregnancy for the foetus and later for the mother's milk. So it is not surprising that the iron requirements of a woman of child-bearing age are twice those of a man and that iron-deficiency anaemia is very common in women as opposed to men. Like calcium, most of the dietary iron is lost in the faeces; the amount absorbed is small and only marginally adequate for women. Rich sources are liver, offal, meat, fruit and vegetables, and iron may be absorbed from cooking pots and in beverages. It is an odd fact that the body is quite unable to excrete iron, and an excessive intake may lead to the condition known as *siderosis*, in which iron is deposited in the tissues.

Lastly, there is a group of what are known as *trace elements*, which, although present in the tissues in minute quantities, are in some cases essential dietary constituents. These include iodine, fluorine, copper, cobalt, zinc and manganese. *Iodine* is an essential constituent of the secretion of the thyroid gland, which will enlarge to form a swelling in the neck, or *goitre*, if the supply of iodine is inadequate. Goitres occur in peoples living in isolated continental or mountainous areas far from the sea, for seafood is rich in iodine, and mainly in adolescent girls. As goitre is associated with subnormal thyroid activity, it is prevented in susceptible areas by adding potassium iodide to table salt. *Fluorine* is necessary for proper hardening of the enamel of the teeth; an excess causes a brown mottling of the teeth and increased density of the bones. Where the fluoride content of drinking water is low, dental caries is commoner and can be countered by adding fluoride to the water supply, though the wisdom of such compulsory mass-medication is to be doubted. *Copper* has been thought to have some importance in the formation of haemoglobin; athough deficiency states have never been recorded in man, deprivation in animals causes anaemia. *Cobalt* forms part of the molecule of the B_{12} vitamin and is therefore essential; and *zinc* is an essential component of some enzyme systems and a factor in the healing of wounds. *Manganese* is also important to certain enzyme systems.

Nutrients

Complex though the varieties of the constituents of living matter and foodstuffs may appear, they can be grouped into three great classes of compounds. These are the fats, carbohydrates and proteins.

Fats

The *fats* occur in the adipose tissue found beneath the skin and elsewhere, and as discrete globules in the cells everywhere, particularly in the liver. They are essentially combinations of glycerol (or glycerine) with three fatty acids—oleic, palmitic and stearic. It is on the relative proportions of those fatty acids it contains that the solidity and melting point of a particular fat depend, e.g. the fluidity of human fat compared with the solidity of mutton fat at body temperatures. Oleic is the most fluid of the three fatty acids, palmitic and stearic being relatively hard. In general, fats are insoluble in water, dissolve easily in ether and alcohol, can be emulsified in water as fine dispersed droplets, and are capable during digestion of being broken down into their constituent fatty acids, prior to rearrangement by the tissues.

Fats provide the second largest source of energy to carbohydrates, and in very concentrated form, so they are valuable in the diet for manual workers and in cold climates. Unfortunately, they are generally expensive. They are vehicles for the fat-soluble vitamins A, D, E and K. A distinction is made between certain essential fatty acids known as polyunsaturated, which have to be ingested as they cannot be synthesized, and the animal triglyceride fats solid at body temperatures. It is now believed that the unsaturated fatty acids are potent protective agents against degenerative arterial disease; they are abundant in the fluid vegetable and marine oils. Fats are oxidized mostly to carbon dioxide and water, though this process is interfered with if the diet is deficient in carbohydrates; then the breakdown of fats may be incomplete and intermediate metabolic products appear in the urine—the condition known as *ketosis* because of the presence of ketones.

The grease of the skin, the oil of the hair and the wax of the ear are rather more complex fats known as *sterols*; and even more complicated combinations of fats with phosphorus are found in nervous tissue and in the bounding membranes of cells generally, where they are responsible for some of the surface phenomena alluded to in the previous chapter.

Carbohydrates

Although these may yield half the total calories in civilized communities, some peoples are healthy with virtually none in the diet, so they are obviously not absolutely essential, if only because the body can form glucose from fats and protein. They are the immediate products in plants of the synthesis of water and carbon dioxide in photosynthesis, and are important in animals as a very ready and available source of energy. They are compounds of carbon, hydrogen and oxygen, and their essential saccharide unit, CH_2O, is used as a building brick in the body to form compounds of increasing complexity.

The *monosaccharides* include glucose, the sugar of fruits, also formed in the body by the digestion of cane sugar and starch, and fructose.

The *disaccharides* include cane sugar, the maltose of fermenting grain and barley, and the lactose of milk.

The *polysaccharides* are complex substances of high molecular weight. They include *starch*, found in all vegetable foods, potatoes, root crops and cereals as grains within the plant cells and ultimately converted by digestion into glucose. *Glycogen*, or animal starch, is very important. Resembling starch, it is stored as granules in all the tissues, but particularly in the liver and muscles. And it is the liver glycogen which is the great energy reserve of the body, mobilized when required by conversion into glucose which re-enters the bloodstream. There is also the rough *cellulose* of plants, digestible by herbivorous animals but not to any extent by man.

Proteins

The *proteins* are the most important constituents of protoplasm and the body cannot long endure deprivation of them in the food. They occur in quantity in lean meat, cheese and the vegetable pulses. They are extremely complex unions of carbon, hydrogen, oxygen and nitrogen, and usually also of sulphur; they are colloid substances, soluble in water and coagulating irreversibly on heating. The essential unit of the protein molecule is the *amino-acid*, and though there is an enormous variety of different animal and vegetable proteins, the only chemical difference lies in the particular amino-acids contained in the protein molecule, their properties and the way their peptide linkages are arranged. This is of some importance to man as a consumer, as the more essential amino-acids are only found in the 'first-class proteins', such as those of meat.

The protein of each animal and vegetable species is quite different from all others; on this specificity, and also on the reaction a foreign protein excites in the living tissues of man, depend the phenomena of *immunization* (e.g. against diphtheria) and *allergy* (e.g. nettle rash). Protein is, in fact, the basic chemical unit of protoplasm in all living cells, and different tissues possess their own characteristic proteins: the red haemoglobin of blood, the slimy mucoproteins of mucous secretions, the casein of milk, egg albumen, muscle protein and the complex nucleoproteins linked with phosphorus contained in cell nuclei. We have already seen that proteins as a whole can function both as weak acids and as weak alkalis, and thus maintain a buffering neutrality to changes in the reaction of tissue fluids.

The proteins are not completely broken down in the tissues, and a considerable part of their nitrogen content is lost in the urine as urea, uric acid and other constituents. Only about 70% of the energy theoretically available in protein is actually utilized by the tissues, in contrast to the position with fats and carbohydrates where breakdown is total and the energy yield is exactly what would have been obtained by physical combustion in a calorimeter in the laboratory. Nitrogen balance is between daily intake and daily loss in the urine (only small amounts are lost in the sweat, from the skin, hair and nails). The balance is negative in starvation and wasting diseases, and also for a time after major injury or operation (this can now be corrected by the direct intravenous infusion of amino-acid complexes). The balance is positive—i.e. there is nitrogen retention—in growing children and convalescent patients on an adequate diet.

Within limits, fats and carbohydrates are interchangeable in the diet as sources of energy, but the diet *must* contain protein to supply nitrogen and certain essential amino-acids that the body cannot itself synthesize. In some primitive, mainly grain-eating, peoples the diet is mainly carbohydrate; others—such as the Eskimo and the Masai—live primarily on protein and fat. In western countries the energy intake is provided as to about 10% by protein, 40% by fat and 50% by carbohydrate.

Enzymes

It is a striking fact that the complex changes of chemical breakdown and resynthesis are performed in the body much more rapidly and easily than in the laboratory; indeed, artificial synthesis is not yet possible in some cases. Thus starch is completely converted into maltose by the saliva within the space of a minute, though it would take the chemist several hours' boiling to achieve this.

This facilitation of chemical changes in the body is due to what are known as *enzymes*. We are familiar in inorganic chemistry with substances called catalysts; these need only be present in minute amounts to accelerate enormously certain reactions in which they are not themselves consumed. Enzymes are the catalysts of the organic reactions of the body. They are complex, colloidal materials, sensitive to changes in the environment and preferring a definite degree of acidity or alkalinity for their optimum performance—e.g. the pepsin of the acid gastric juice, the trypsin of alkaline pancreatic secretion. They facilitate breakdown, synthesis and oxidation processes, but especially a process known as *hydrolysis*, the addition of water to the molecule of the substance to be acted on, a common preliminary to its disintegration.

Enzymes are specific, each acting only on a particular material, known as its *substrate*, e.g. the amylase of saliva acts only on the starch of the food; and all the digestive juices contain different enzymes designed to act on the various constituents of the food—lipases for fats, proteases for proteins, carbohydrases for the sugars and starches. These do not act at all on other substances, or even on other members of the same group of food substance. Most enzyme reactions are reversible, the enzyme being capable of pushing the chain of events either way. Enzyme reactions fall into two main groups: the processes of digestion occurring in the cavity of the bowel from the outpouring of digestive juices; and the more intimate reactions, such as oxidation, occurring within the individual cells of the tissues.

The majority of metabolic reactions, within or outside the cells, occur only in the presence of an appropriate enzyme, and the rate of reaction is proportional not merely to the concentration of substrate and of enzyme, the local pH and temperature but also to the presence of certain *co-enzymes* and of *activators*, which can be metallic ions, such as zinc or magnesium, or organic molecules. There are other substances that act as inhibitors or poisons. Because they are proteins, enzymes can be *denatured* by heat, excessive acidity or alkalinity and certain metal ions. Many have been obtained in the pure or crystalline state. They have been classified according to their effects on molecules in the body as follows:

(1) oxidation/reduction: oxidoreductases;

(2) transfer of chemical groupings from one molecule to another: transferases;

(3) hydrolysis of compounds into simpler molecules: hydroxylases;

(4) splitting: lyases;

(5) joining: ligases;

(6) rearrangement of molecules to form an isomer: isomerases.

In many instances, an enzyme combines more than one of these activities.

Metabolism

Physiologically, the life of man can be regarded as a continual production of energy by oxidation (burning) of the food he consumes, and the expenditure of this energy in (*a*) maintaining a constant body temperature above that of his surroundings, and (*b*) movement.

We have already pointed out that the balance sheet for this energy intake and output is an exact one; food burnt within the body liberates the same energy as it does when oxidized experimentally in the laboratory. However, although the physical laws of thermodynamics are as applicable to the body as to the stoking of a furnace with coal, the human arrangements are complicated by the fact that the food is not burnt *as such* but only after it has been digested and assimilated into the living tissues. It is as if the coal had to become a part of the furnace itself before being used up, for only a few exceptional substances like alcohol can be burnt directly without first becoming part of the living protoplasm.

Thus there is a recurrent cycle of activities, known overall as *metabolism*. Part of this is the process of building up or repair, of assimilation of food into the tissues—*anabolism*; part is the breaking down of these tissues with the liberation of energy and the excretion of wastes—*katabolism*. Both are going on at the same time, all the time, though their relative proportions obviously vary under different circumstances. Anabolism preponderates during the active growth of children; katabolism during starvation, when the body draws on its reserves, and during senescence.

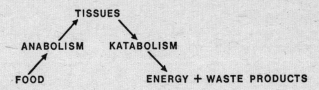

The waste products are excreted by the kidneys, bowel, lungs and skin. They consist mainly of water, carbon dioxide and certain nitrogenous breakdown products of protein, such as urea, which are mostly expelled in the urine, though also to a lesser extent in the faeces.

During fasting, if the body is at complete rest, there is an essential minimum of energy output required to maintain its warmth and the movements of respiration and of the heart. This minimum can be measured and is then known as the *basic metabolic rate*. Although of the same order of magnitude for everyone, it varies from person to person by virtue of their different sizes. This is because most of the energy is expended in making up for heat loss from the surface of the body and is therefore directly proportional to the surface area. Basal metabolism is rather higher in men than in women and higher still in proportion to body weight in young children. Each individual has his normal metabolic rate, predictable from height and weight, and this is much increased in fever, excitement and anxiety, and also by the overactivity of the thyroid gland, which has an intimate control over metabolism. In thyroid deficiency, as in cretinism, the metabolism is considerably below normal.

From the basal rate we can estimate the calorie value of the food required to maintain the essential activities of the body without allowing actual starvation; this is in the region of 2000 calories per day. Below this level the body begins to fall back on its own reserves; the fat and glycogen stores are first consumed, and then the less essential tissues, such as the muscles and glands, digest themselves to feed the brain and heart—with eventual death when the latter organ fails.

4   The skeleton

Cartilage

Cartilage, with bone, is a major component of the skeleton, and the greater part of the skeleton is actually preformed in cartilage during intra-uterine life before being converted to bone. It is found where rigidity and resilience are needed in combination, i.e. at the joint surfaces, the front ends of the ribs, and in the supporting framework of trachea and bronchi, nose and ears. The main types of cartilage are:

(1) the smooth, clear *hyaline* cartilage of joint surfaces, marked by a very low coefficient of friction which produces a very slippery surface;

(2) tough *fibrocartilage* containing white fibrous tissue, found in the intervertebral discs and in the plates or menisci that project into certain joints (e.g. the semilunar cartilages of the knee);

(3) *elastic* cartilage in flapping structures like the ear and epiglottis.

Cartilage is a gristly tissue, largely without blood supply, and because of this cannot repair itself as such after injury, being replaced by fibrous scar tissue; hence the disabling effects on the smooth working of joints of an injury to their articulating surfaces. Basically, it is a connective tissue whose matrix has become solidified as a firm gel bound together by fibrous proteins (collagen and elastin). Its organic content includes the complex molecule of chondromucoprotein, linked with the polysaccharide chondroitin sulphate. The chondroblast or typical cartilage cell is plump and rounded, and is distributed throughout the cartilage substance in small groups. It is the cell responsible for the actual formation of cartilage matrix, and is nourished by diffusion from the fluid of the neighbouring joint and, to some extent, from the underlying bone. Disuse and immobilization tend to lead towards the degeneration and destruction of articular cartilage, whereas activity has a pumping effect which aids diffusion of nutrients.

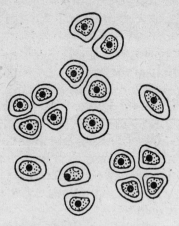

Fig. 9.   Cartilage cells.

Cartilage has the property of *calcification*, i.e. the deposition within it of calcium salts rendering it tough, opaque and less resilient. Calcification is a normal process during growth, occurring as a preliminary to ossification of the cartilage precursors of the bones. But it is also an ageing process that may affect the rib cartilages and the articular cartilage of joints. In late middle life or old age there may be fibrillary degeneration of joint cartilage, a wear-and-tear process that gives rise to osteoarthritis.

Bone

Bone is also a specialized connective tissue of mainly mechanical function, though it also has an important part to play in connection with mineral metabolism and houses the marrow in which new blood cells are formed.

It is the hardest body tissue with the exception of the teeth. It provides the framework and support of the body, protects the internal organs, and gives attachments to the tendons and muscles, as well as constituting the levers these muscles move. The structure of bone allows great strength combined with great economy of material, its internal struts or *trabeculae* being disposed to meet maximum load. There is a considerable safety margin, e.g. the upper end of the femur in an active adult can support a vertical load of a tonne and has to bear three times the body weight at every step. Bone is subject to compression, tension, twisting and bending strains, and is able to withstand these not only because it is strong but also by virtue of a certain *elasticity*. In old age and in a number of diseases this strength is impaired and fractures readily occur.

Bone is a blend of (*a*) an organic fibrocellular matrix or *osteoid*, and (*b*) a *mineral* matrix, consisting of inorganic mineral salts—calcium phosphate and carbonate, magnesium and fluoride—in crystalline form. The organic component can be removed by burning, leaving a brittle mineral skeleton; the inorganic matter can be dissolved by acids, leaving decalcified bone that is perfectly flexible. The osteoid consists mainly of collagen fibres, plus some inter-fibrillary ground substance containing mucopolysaccharides and protein. The mineral matrix has a rigid crystalline structure, known as *hydroxyapatite*, whose composition can be represented as $Ca_{10}(PO_4)_6(OH)_2$, with variable amounts of of other ions—magnesium, sodium, carbonate, citrate and fluoride. The collagen matrix acts as a site for the crystalliza-tion of bone minerals from the dissolved calcium and phos-phate of the tissue fluids under the influence of an enzyme, phosphatase, that concentrates phosphate ions locally and initiates crystal growth in the presence of calcium ions. In health the whole of the bone substance is mineralized, and unmineralized osteoid is only found at sites where bone is being formed rapidly, e.g. at fracture sites. In certain disease states such as rickets mineralization is defective, and the bone is soft and yields easily under stress. Sometimes, as in endemic fluorosis, mineralization is excessive, and the bone is dense and hard but possibly brittle. The salts of bone are in constant dynamic interchange with the ions of the body fluids, and this ionic interchange is the basis of the constant process of remoulding and replacement of bone substance that continues throughout life. The calcium of blood is in equilibrium with that of bone, nearly 1 g being exchanged every day, a process influenced by the secretions of the thyroid and parathyroid glands, and by vitamin D and the local pH. Bone contains 99% of the total calcium of the body, 88% of the phosphate, 70% of the citrate, 80% of the carbonate and 50% of the magnesium. Bone is constantly subject to the influences of simultaneous formation and re-sorption. Both are very active during growth, with bone formation in the ascendant. During adult and early middle life the total bone mass remains fairly constant in health, but in later life resorption leads and the actual amount of bone decreases so that it appears more translucent in an X-ray—osteoporosis. This process begins earlier in women, and accounts for the relative frequency of hip and other fractures from slight injury in the aged.

Microscopically, the substance of the dense outer cortical bone of the shafts of long bones is arranged in layers, or lamellae, containing small clefts, or lacunae, occupied in life by the bone cells, or *osteoblasts*, whose processes ramify around. The lamellae are arranged concentrically around a central Haversian canal containing small blood vessels and nerves, and the bone as a whole is composed of a great many such Haversian systems, or *osteones*. This is modified near the surface of the shaft, where the lamellae run parallel and there are no canals. The osteoblasts are specialized connective tissue cells helping to form the organic matrix, and they contain alkaline phosphatase, which assists their bone-forming activity. In addition, wherever bone is being absorbed or remodelled, there is another type of cell, the *osteoclast*, a giant multinucleate cell that actively reabsorbs and erodes bone substance. The two types of cell work in collaboration so as to shape the internal architecture of bone in response to mechanical stresses.

The *periosteum* is a tough supporting and ensheathing membrane surrounding bone, except where it is covered by

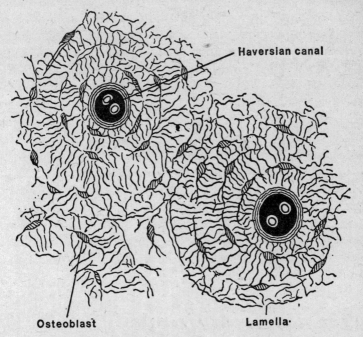

Fig. 10. Microscopic structure of bone.

articular cartilage. Its outer fibrous layer is mainly supporting; it carries part of the bone's blood supply and gives attachment to muscles and ligaments. From it there run fibres penetrating into the bone. In the deeper layer of the periosteum there exist osteoblasts which are responsible for the increase in girth of the bone during growth. During growth, the periosteum is rather like a loose skin that can easily be peeled off the underlying bone; such peeling off may, indeed, occur if there is injury or infection. The attachment is much firmer in adult life.

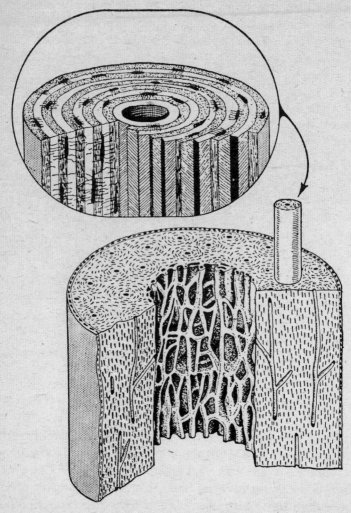

Fig. 11. Diagram of cortical bone showing the structure of an osteone. These rod-shaped units of bone structure are made up of series of lamellae. In each layer the collagen fibres are orientated in a direction different from those in the adjacent layers. (From 'Electrical Effects in Bone' by C. Andrew L. Bassett. Copyright © 1965 by Scientific American Inc. All rights reserved.)

Blood supply of bone

A typical long bone has four main sets of vessels (Fig. 12).
The *epiphyseal* vessels supply the epiphyses, or centres of
ossification at the growing ends of the bone. Then there is a

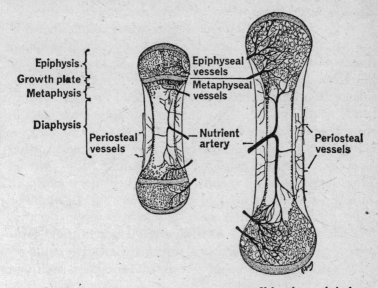

Fig. 12. Diagram showing the arrangement of blood vessels in bone
before and after the growth period. Note that the growth plate
isolates the epiphyseal from the metaphyseal circulation during the
growing period, but that they anastomose freely after growth has
ceased. The periosteal vessels supply the osteogenic layer on the
bone surface during the growth period, but penetrate the cortex
and supplement the nutrient artery in adult life. (From *A Companion to Medical Studies, Vol. 1.*)

separate set of *metaphyseal* vessels, which are very important
in connection with the growth process. The main shaft is
supplied by one or more large *nutrient arteries*, which
penetrate the cortex and enter the marrow cavity, within
which they ramify to supply the marrow and the shaft itself.
Finally, the *periosteal* vessels contribute to the nutrition of
the outer layers of the shaft. If the nutrient artery is damaged,

or if the periosteum is stripped off by suppurative infection, much of the shaft may die.

Types of bone

In gross appearance, there are five main types of bone:

(1) *Long bones* are those of the limbs, e.g. the humerus in the arm, the femur in the thigh. They have a shaft which is roughly cylindrical but sometimes polygonal or triangular in section and never quite straight, and two expanded ends, sometimes rounded off as a definite head or widened into condylar masses. The bone ends are articular, take part in the adjacent joints and are covered with smooth articular cartilage to facilitate movement; the two ends concerned in forming a joint are enclosed in a common joint capsule. Thus the periosteum covering the shaft becomes continuous with the capsule at the end of the bone (Fig. 13(b)). The bone surface is dotted with numerous tiny apertures (*foramina*) for its blood vessels, with a larger nutrient foramen for the main artery near the midshaft. The ends receive a double blood supply, both from the main shaft and from neighbouring vessels.

Cross-section of a long bone shows the outer *cortex* of bony substance and the central *medullary cavity*. The cortex has an outer portion of dense compact bone surrounding a looser meshwork of spongy or cancellous bone; at the ends, where the medullary cavity stops after traversing the shaft, there is a solid mass of spongy bone within a thin compact shell (Fig. 13(a)).

Fatty *yellow marrow* occupies the medullary cavity in the adult and the interstices of the spongy bone at the ends are filled with *red marrow*, responsible for the formation of red and white blood corpuscles. At birth, and in the early years of life, red marrow occupies the whole shaft but retreats to the ends with growth. This process can be reversed in later life if anaemia or haemorrhage make extra demands on blood formation, so that in chronic anaemia the shafts of

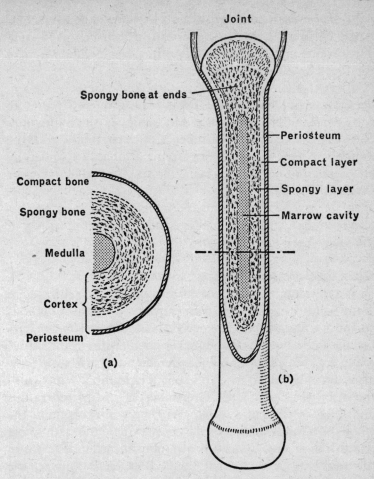

Fig. 13. (a) Cross-section of a long bone. (b) Longitudinal section of a long bone.

all the bones are permanently occupied with red marrow working overtime.

In adult life active red marrow is found mainly in the flat bones, the skull, ribs, sternum (breast bone) and pelvis; it is also found in the vertebrae of the spinal column.

(2) *Short long bones* are simply long bones in miniature, and are found in the hands and feet.

(3) *Short bones* are squat, cuboidal or irregular, composed entirely of spongy bone with a thin compact shell. They occur in the wrist (*carpus*), the corresponding part of the foot (*tarsus*) and in the vertebrae.

(4) *Flat bones* are those of the skull, vault, ribs and scapula (shoulder blade). They consist of two plates of compact bone sandwiching a thin spongy layer; in the skull these layers are called the inner and outer tables, with the diploë between. Certain bones of the skull are expanded by air-containing cavities replacing this spongy layer—the air sinuses.

(5) *Sesamoid bones* are tiny, rounded masses found in certain tendons at points of friction; the largest is the patella at the knee (knee-cap).

Development and growth of bone

In the developing embryo of a few weeks, the central core of the primitive gelatinous connective tissue of the limb buds becomes transformed into an axial rod of cartilage traversing the limb. This rod is absorbed at the sites of the future joints, e.g. the elbow and knee, demarcating arm from forearm and thigh from leg, and split longitudinally in two in the distal segment as the forerunner of the paired radius and ulna, or tibia and fibula. With certain exceptions, the whole skeleton is well formed in cartilage by the sixth week, and at the seventh week a centre of ossification develops in the midshaft of each long bone. Bone cells appear, the matrix is impregnated with calcium salts, and ossification spreads up and down the shaft until, at birth, the long bones are entirely ossified except for their cartilaginous ends.

Epiphyses. There now develops an arrangement, peculiar to mammals, allowing continuous growth in length during the years preceding maturity. In the first years of life separate secondary centres of ossification appear in the bone ends, which ossify completely except for a thin cartilage plate separating them from the main shaft. The rounded end is called the *epiphysis* and the cartilaginous plate; the epiphyseal plate. This plate consists of longitudinal columns of cartilage cells reproducing themselves continuously on the shaft side of the plate and as continuously turned into new bone, which is pushed away into the main shaft, enabling

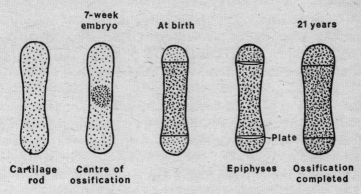

Fig. 14. Stages in the ossification of a long bone.

it to grow into length. Meanwhile, the epiphyseal plate retains its integrity, until at maturity the epiphysis fuses with the shaft by ossifying directly across the intervening cartilage—at the age of eighteen to twenty years in men, sixteen to eighteen in women. During all this time, as we have seen, the periosteum has been responsible for increase in girth (Fig. 14).

Growing ends. Although each end of a long bone is thus a growing end, one epiphysis appears earlier than the other and fuses later, and is responsible for the main share in growth in length. In the arm these principal 'growing points' are at the shoulder end of the humerus, and the wrist end of radius and ulna; in the leg, at the knee end of all three bones. It is for this reason that amputation through the thigh gives rise to no trouble in a child, whereas in the corresponding amputation through the upper arm the humerus continues to grow from the shoulder and may

burst through the stump unless its epiphysis is deliberately
destroyed surgically. Complete destruction of an epiphysis
causes cessation of growth and corresponding stunting;
partial destruction by injury or disease causes a distortion
of growth in which the limb turns away from the still active
side.

A few bones, mainly those of the skull vault, mandible
and clavicle, are not pre-formed in cartilage as described
above but are ossified directly in primitive membranous
tissue, and these membranous bones have no epiphyses.

Remodelling. The articular cartilage, or joint surface, which
covers the free end of the epiphysis, persists throughout life.
It is that portion of the original cartilage system that has
not been substituted by bone formed in the secondary (epiphysial) centre of ossification. The shaft of the bone grows
in thickness by the formation of new bone within the deeper
layer of the periosteum. As bone is added externally, it is
removed internally, so that the shaft retains a tubular form,
is strong without being too massive and allows room for the
bone marrow. This process of external deposition and
internal reabsorption of bone is called 'remodelling'. Remodelling also occurs at the metaphyses (the metaphysis is
the 'neck' of the bone, just on the shaft side of the epiphysial
plate). The shaft attains, and retains, its gradual widening
out at each extremity, because at each metaphysis bone is
added internally and removed externally, the reverse of
what obtains at the centre of the shaft.

Growth. The foetus is, of course, growing continuously in
stature, but its *rate* of growth varies considerably. It is slow
during the first two months, while the main organs are being
laid down, and then reaches a peak at four to five months
and slows again in late pregnancy. The rate of growth falls
off rather rapidly after birth and then more slowly until
puberty, when a rapid growth spurt occurs—the adolescent
increment. All growth in height virtually comes to a stop

at around eighteen years in boys and sixteen years in girls; this is because epiphyseal fusion in the long bones has put an end to their growth. The spine, however, continues to grow very slightly until about thirty years of age. In middle and old age height begins to decline, mainly due to bending and degenerative changes in the spinal column. The changing *proportions* of the different components of the body are

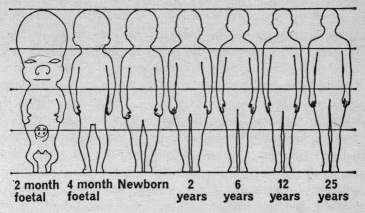

**2 month   4 month Newborn   2   6   12   25
foetal   foetal   years   years   years   years**

FIG. 15. To show the changing proportions of the body during growth and development. (From *A Companion to Medical Studies, Vol. 1.*)

shown in Fig. 15. At an early stage of foetal life, the head is disproportionately large compared with adult status because of the precocious development of the brain; at maturity, the lower limbs make the major contribution. Before puberty, the legs grow faster than the trunk, and boys are generally taller than girls because they have a longer prepubertal period. Most of the adolescent increment occurs in the trunk.

Factors controlling bone growth. As growth in height depends on growth in length of the long bones, which, in turn, depends on interstitial growth in the epiphysial cartilage plates, growth rate and final stature are determined by the rate of proliferation of cartilage cells in the growth plate and the duration of its active period. These are affected by a number of factors.

There is a strong *genetic* influence; identical twins are of similar height, and taller parents tend to have taller children. Prolonged *malnutrition* in childhood stunts growth. There is an intimate control of bone growth by certain *endocrine* glands. The somatotrophic or growth hormone secreted by the anterior lobe of the *pituitary* gland is a major factor in promoting growth from birth to adolescence by virtue of its effect on the proliferation of cartilage cells in the growth plates. An excess of secretion in childhood so stimulates epiphysial growth as to cause *gigantism*, the victim reaching 2·13 or 2·44 m (7 or 8 feet) in height. Its action in the adult is less dramatic, but even here there is gross hypertrophy of the skeleton and soft tissues, especially of the hands, feet, face and tongue—the condition known as *acromegaly*. Deficiency of growth hormone in early childhood is the cause of one form of dwarfism. The hormones of the *thyroid* gland are also important; in congenital deficiency (*cretinism*) there is stunting and idiocy, and the epiphyses are delayed in appearance and still unfused in middle age. Sex hormones, secreted both by the sex organs themselves and by the adrenal glands, cause the adolescent growth spurt.

Skeletal age. There is a great deal of variation in the calendar ages at which signs of maturity and sexual development become apparent—such features, for instance, as the onset of menstruation and the development of pubic hair. A boy or girl of chronological age of thirteen or fourteen may still be a child, or virtually an adult man or woman. True developmental age is best assessed from study of the dates of appearance of the secondary epiphysial centres of ossification in the long bones and the primary centres in the short bones, such as those of the wrist. In fact, by using an X-ray of the wrist and hand in comparison with a standard atlas of ossification dates, it is possible to arrive at a skeletal age that is a good index of maturity. Skeletal and calendar ages may differ by as much as two years.

The strong hard bones of the dried skeleton obscure from us their essential plasticity during life, for they respond to normal and pathological stresses, much as a tree responds slowly but surely to the deforming influence of wind and weather. The more work put on them, the more they hypertrophy, particularly in response to compression strain or muscular pull, and the internal arrangement of bone trabeculae is a purely mechanical solution to the engineering problems imposed on each bone. In view of this exact correspondence between final shape and functional needs, it is somewhat remarkable to find an inherent tendency to assume the exact adult shape, even in a bone removed from the growing embryo and left to develop alone in an artificial culture medium.

Joints and movements

Types of joint
The articulations between the bones vary considerably as regards mobility in different parts of the skeleton and some are quite immobile. The main types are:

(1) *Fibrous*. As a zigzag *suture* between the skull bones, whose interlocking irregularities are joined by a thin

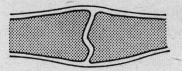

Fig. 16. Suture.

fibrous strand, often obliterated by ossification in old age; as a *syndesmosis*, which allows some stretching and twisting; as the distal tibiofibular joint; or as a broad sheet of *interosseous membrane* between paired bones—radius and ulna, tibia and fibula.

(2) *Cartilaginous*. Here the cartilage coverings of the ends
of the bones are united by a plate of thick fibrocartilage,
e.g. the cushion separating the two halves of the pelvis in
front at the pubic symphysis, or intervening as an intervertebral disc between the vertebrae constituting the
spinal column. Movement of such a joint is limited, but the
total of movements between adjacent vertebrae may be
quite considerable.

(3) *Synovial joints*, which allow free movement. The bone
ends are capped with smooth hyaline *articular cartilage*, and
the joint cavity is enclosed within the sleeve formed by the

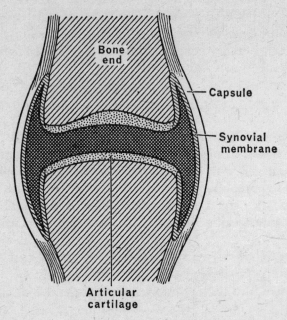

Fig. 17. A synovial joint.

fibrous joint *capsule*, a sleeve stretching from one bone to the
other and continuous with the periosteum. *Ligaments* of
strong, white fibrous tissue hold the bone ends in apposi-
tion. Foremost of these is the main capsular ligament, and,
in addition, there may be localized bands, or intrinsic

ligaments, running within the capsular substance and accessory or extrinsic ligaments traversing the joint or quite
outside the joint altogether. Tendons arising within a joint
obviously also function as extrinsic ligaments, e.g. the long
head of the biceps at the shoulder joint. In some cases, complete or incomplete cartilage plates, or discs, project from

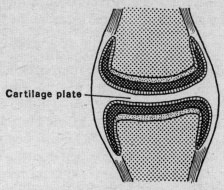

Cartilage plate

Fig. 18. Synovial joint partitioned by a cartilage plate.

the deep surface of the capsule into the joint cavity as partitions; these are found at either end of the clavicle, between
the lower jaw and the skull (temporomandibular joint) and
at the knee—the semilunar cartilages so often torn in foot-
ballers; they are also found at the wrist at the inferior radio-
ulnar joint.

Normally, the joint space is only potential, as the cartilage
surfaces and soft tissues lie everywhere in contact under a
negative pressure; it becomes a reality when the capsule is
distended by a fluid effusion due to injury or inflammation,
or if air is admitted by a wound or an operation. The joint
capsule is lined by a smooth, slippery *synovial membrane*, which
is reflected onto the bones but disappears at the periphery
of the cartilage surfaces. This highly vascular, fatty membrane, with its folds and fringes, packs the joint and secretes
the lubricant *synovial fluid*, which covers the joint surfaces in
a thin film. This is a clear or yellowish viscid fluid secreted
from the blood by the synovial membrane, which also

nourishes the articular cartilage, and the white cells it contains scavenge the debris from the moving surfaces. In the
area within the joint where the bones do not fit together
accurately, there may be *fatty pads* between capsule and
synovial membrane; these pads bulge into the dead spaces
and help to cushion movement. Any structure traversing
the joint, such as a ligament or tendon traversing the capsule,
carries an investing sheath of the synovial membrane.

Stability

The general integrity and coaptation of joints are maintained by a number of factors, the most important being the
strength of the ligaments and the tone of surrounding
muscles; these are interrelated, as muscle tone normally
prevents a severe strain on the ligaments being more than
momentary. Great mobility is essential at the shoulder, so
the capsule is lax, and the large humeral head fits poorly
into the shallow glenoid fossa of the scapula; this inherent
instability is overcome by having the tendons of the small
rotator muscles blend freely with the capsule so that the
anatomical deficiency is compensated by muscle guarding.
This contrasts with the arrangements at the hip, where
stability is of greater importance; here, the femoral head is
deeply buried in the socket of the acetabulum and an
intimate blending of the neighbouring muscles with the
capsule is unnecessary. Minor supportive factors are the
interlocking of the bone ends, rarely very secure in itself,
and the cohesion produced by outside atmospheric pressure.

The joint surfaces are least closely fitting during the main
range of movement and most closely packed at the extremes
of range. Thus there is a *close-packed* position when the joint
is well arranged to face stresses and strains, and a *rest*
position when it is not in use, with the surrounding muscles
relaxed. Although the ligaments are important strengthen-
ing structures, they have but little stretch and are not
adapted to resist sudden severe strain. Such movements are
usually resisted by the surrounding muscles, and muscles

can remain tense throughout a resisted movement or gradually pay out slack as their opponents contract in a manner impossible for the inelastic ligaments.

Varieties of synovial joints

Synovial joints are further subdivided by virtue of the shape of their bone ends and the movements possible between them.

The simple *plane* joint between the small, flat adjacent surfaces of carpal and tarsal bones allows minor gliding motion only. The *saddle* joint is self-explanatory, with reciprocating saddle-shaped surfaces, the best example being between the thumb metacarpal and the corresponding carpal bone. The *hinge* joint is a common mechanism, as in the elbow, ankle and fingers; here, one convex surface is gripped within a deep concavity, and the very strong collateral ligaments on each side allow motion round a transverse axis only. On the other hand, the capsule obviously needs to be very lax in front and behind to allow full flexion and extension. The *pivot* joint is again obvious, with a cylindrical bone end rotating on its long axis like a hinge within its socket—e.g. the head of the radius at the superior (proximal) radio-ulnar joint at the elbow in pronation and supination of the forearm.

The *ball-and-socket* joint of shoulder and hip has the most highly developed range of motion, with one spheroidal and one concave surface moving on each other round an infinite number of possible axes. Its main movements (see Figs. 19 and 20) are:

Flexion and *extension* in an anteroposterior plane; *abduction* and *adduction* in a mediolateral plane; *medial* and *lateral rotation* on the long axis of the limbs (not to be confused with pronation and supination of the forearm in the upper limb, where true rotation occurs only at the shoulder joint); and *circumduction*, in which the limb describes the boundary of a cone, in a combination of all the above.

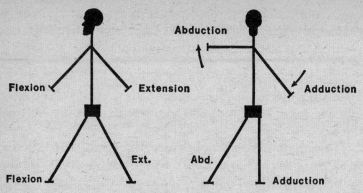

Fig. 19. Movements of the ball-and-socket joints of shoulder and hip.

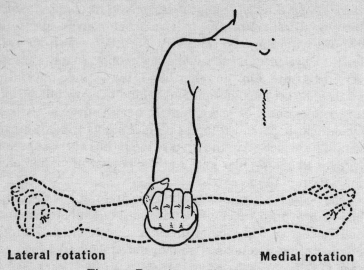

Fig. 20. Rotation at the shoulder.

The movements of joints are rarely as extensive as the shape of the bone ends would seem to allow, i.e. actual locking of the bones is only exceptionally a limiting factor. It is the soft parts that are usually responsible for checking the range, as in the contact of the front of arm and forearm in flexion of the elbow, and the tension of the hamstring muscles at the back of the thigh in limiting flexion of the hip.

Axial and appendicular skeleton: limb girdles

The skeleton as a whole is divided into the central *axial* portion of the head and trunk, and the peripheral *appendicular* portion of the limb bones. The axial skeleton is made

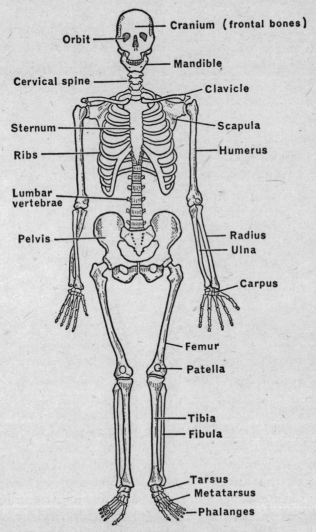

Fig. 21. The skeleton, anterior view.

up of the skull, mandible (lower jaw), spine, sternum (breast bone) and twelve pairs of ribs. In addition, there are the small hyoid bone in the upper part of the neck beneath the floor of the mouth and three tiny ossicles in each middle-ear cavity.

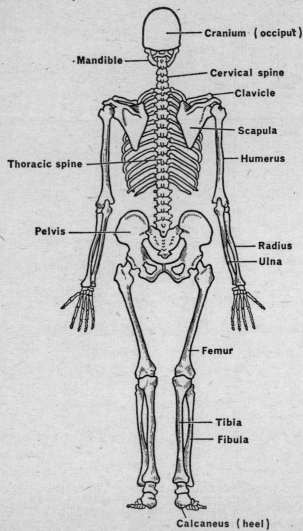

Fig. 22. The skeleton, posterior view.

Limb girdles

These are encircling arrangements of bones designed to connect the shoulder and hip regions to the central axial
skeleton in such a way as to provide a more or less firm basis
of attachment for the corresponding limb while allowing a
certain degree of mobility. The two girdles differ somewhat

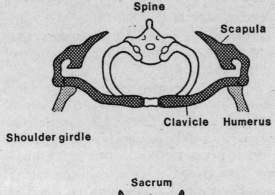

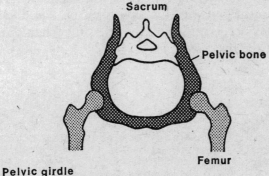

Fig. 23. The limb girdles.

in respect of these two functions. The arm does not have to
bear weight and must allow the freest use to be made of the
hand; the shoulder girdle is correspondingly unstable and
its main central connections are muscular rather than bony.
These conditions are reversed at the hip, where stability is
essential, and here the pelvic bones make up a stout complete bony ring, firmly attached to the spine behind (Fig. 23).

In the diagrammatic sections of the two girdles, the following points will be noted. They are both based on a circular model, but the shoulder girdle is very wide and is left open behind, where the scapulae are only attached to the spine by muscle so that they can move freely on the trunk. Each half of the shoulder girdle is made up of two bones: scapula and clavicle—the latter a relatively slender bone joining scapula to sternum and responsible for bearing a considerable part of the burden of the hanging arm. Halfway round the girdle on each side is the shallow glenoid fossa of the scapula, accommodating the humeral head at the shoulder joint.

In the hip girdle, the bony ring is complete in front and behind in a tight circle. Each half is composed of a single pelvic (innominate) bone carrying a very deep acetabular socket for the head of the femur at the hip joint. Posteriorly, the two pelvic bones are firmly attached to the spine—in this region the sacrum—at the sacro-iliac joints. The girdle is thus integrated with the axial skeleton and does not permit of any accessory leg movements other than those taking place at the hip itself.

5 Muscle

Apart from the amoeboid movements of the polymorph white blood cells and of certain connective tissue cells, and the lashing movements of hair-like protoplasmic extensions (cilia and flagellae) of certain layers of cells, such as those lining the air passages, all movement, both within the body and of the body, is the result of contraction in muscle cells. A pianist's fingering, a fencer's thrust, the heartbeat, the expulsion of urine, faeces or foetus, the rhythm of bowel movement, pupillary dilatation and contraction—all these are examples of such movements.

Muscular tissue is almost entirely composed of reddish muscle fibres arranged in bundles, but there is some scanty binding areolar connective tissue between the bundles. These fibres consist of highly specialized and greatly elongated cells characterized by their power of contraction on stimulation—an exaggeration of a property inherent in some measure in all protoplasm, even in simple unicellular creatures.

The muscles associated with the skeleton, and responsible for movements of the limbs and trunk, account for approximately half the body weight and contain half the body water. The functioning of these muscles is the chief factor governing energy production and expenditure in the body, for even slow walking increases basal oxygen consumption five or six times.

Muscular tissues vary in the speed, strength and duration of their contractions. Some muscles come into play as the result of conscious effort, others function automatically and unconsciously. Some will only contract if stimulated by impulses reaching them through their nerve supply, some

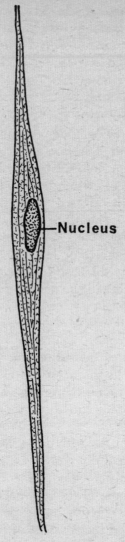

Fig. 24. A smooth muscle fibre.

have an inherent or intrinsic pattern of contraction and others respond to circulating hormones in the blood. These varying functions are associated with different types of muscle structure:

(1) *Smooth* or plain muscle (unstriated, involuntary or visceral) is the most primitive and least specialized type. It forms the contractile coat of blood vessels and the viscera—hollow internal organs like the bowel and bladder—structures that must work automatically, beyond the conscious control of the central nervous system and regulated by the semi-independent autonomic or vegetative nervous system. The degree of such control varies, and some smooth

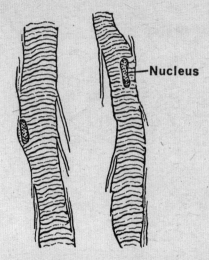

—Nucleus

Fig. 25. Striped muscle fibres. (After *Gray*.)

muscle, such as that of the bladder wall, has its own intrinsic rhythm which is only modified by nervous stimuli. All smooth muscle is stimulated to contract by being stretched. The typical smooth muscle cell is spindle-shaped, about 250 μm long and has a large oval nucleus at its middle.

(2) *Striated* muscle (skeletal, somatic, voluntary) has more intricate fibres, which appear cross-striated under the microscope. They are found in the muscles attached to the skeleton and are under the conscious control of the central nervous system. These striped cells can be very long and may

run for several centimetres from one attachment of the main muscle to the other. Each contains many hundreds of nuclei, placed to one side of the cell. And each cell receives near its midpoint the termination of a nerve fibre from the brain or spinal cord. Where this junction occurs, the muscle cell and the nerve fibre together produce a complicated structure—the myoneural junction or motor end-plate.

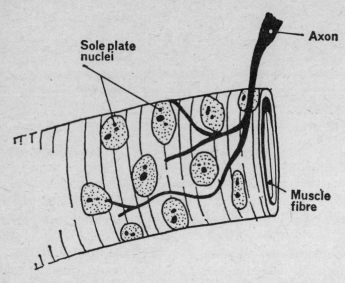

Fig. 26. Drawing of a silver stained motor end-plate from the rat. Note the terminal arborization of the axon and the accumulation of nuclei in this region. (From *A Companion to Medical Studies, Vol. 1.*)

(3) *Cardiac* muscle, the substance of the walls of the heart, occupies an intermediate position. Its fibres are cross-striated but their action is not under voluntary control, and the cells are branched so that the whole muscle consists of a network. There are less than a dozen nuclei to each cell, situated near its centre. Each of these cells has its own intrinsic rhythm and there is no clear demarcation between individual muscle cells.

Fig. 27. Cardiac muscle.

The muscle cell

The muscle cell may be considered as basically a typical cell that has become specialized to convert chemical energy into contractile force, and it is always elongated along the axis of contraction. It is surrounded by an excitable membrane known as the sarcolemma, while the cytoplasm is called sarcoplasm. This contains the normal components of cytoplasm, but the mitochondria are unusually large and numerous, there are very many glycogen granules and a special feature is the presence of contractile protein filaments, the myofilaments, which run along the length of the cell and are usually only visible under the electron microscope. When the filaments group together and become visible to light microscopy, they are known as myofibrils.

Myofilaments are of two types: thick and thin. The thin filaments consist of a protein, actin, in fibrillary form, which can also exist in globular form. The thick filaments consist of another protein, myosin. Each molecule has a rounded head and a long tail; the tails of the molecules run together to form the thick strands, and the heads project at the sides of the filaments. The two kinds of filaments lie side by side, and the contractile force is produced between opposing

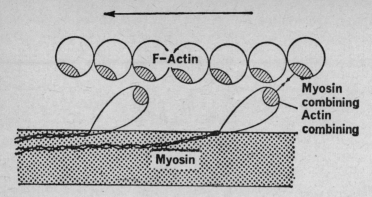

Fig. 28. The interaction of myosin and actin. (From *A Companion to Medical Studies, Vol. 1.*)

molecules, where actin globules and myosin heads come together like the opposite rows of a zip-fastener. As both filaments are polarized in the orientation of their molecules, contraction takes place in one direction only, i.e. a muscle can actively contract but cannot actively lengthen.

In *smooth muscle* the filaments run along the long axis of the cell; they are not regularly patterned in discrete lengths, so no cross-striations are visible under the microscope. There is no limit to the range of shortening, and smooth muscle, though it contracts rather slowly, can exert great force and maintain the same tension for long periods—a *tonic* activity, under the influence of the autonomic nervous system and of circulating hormones, that is very important in the functioning of the heart and blood vessels, bowel, bladder and air passages.

In *skeletal muscle* there is a regular arrangement in length of the two types of filament, producing the typical cross-striation associated with the grouping of filaments into myofibrils, an arrangement that allows rapid contraction. Thick and thin units are regularly packed into a composite contractile unit, the *sarcomere*. This has cross-connections at its middle, and the fibre polarity is reversed at this point so that contraction occurs from both ends of the sarcomere towards the centre. A single myofibril from a skeletal muscle may run for a few centimetres and contain 20 000–30 000 sarcomeres. A muscle fibre can shorten by over half its relaxed length, and this is about equal to the degree of contraction required of a skeletal muscle as a whole to produce a full range of movement at the joint over which it acts.

It is possible for additional smooth muscle cells to be produced in response to requirements throughout adult life—as in the wall of the uterus during pregnancy. This is not possible in skeletal muscle, and here any increase in bulk from exercise or training is due to increase in size, or *hypertrophy*, of individual muscle cells. Muscles hypertrophy from exertion and also from the stimulation of sex hormones at puberty, especially in males. They *atrophy*, or waste, from disuse, as when a limb is splinted or paralysed; interruption of the motor nerve supply produces particularly severe and rapid wasting.

We have already referred to the different types of muscle tissue: *striped* or voluntary; the *smooth*, unstriped or involuntary; and the *cardiac* muscle of the heart. Excluding the last, there is a general distinction to be made between the striped muscles attached to the skeleton of the trunk and limbs and concerned with voluntary movements, controlled by the will through nervous messages from the brain, and the smooth muscle forming the walls of such hollow internal organs as the bladder and bowel, and of the arteries and smaller blood vessels. The latter are required to function automatically and regularly without conscious attention, and are under the control of the semi-independent autonomic nervous system. Sometimes these two sets of muscles are referred to as the *somatic* muscles of the body wall and the *visceral* musculature of the internal organs.

Skeletal muscle

The voluntary muscles differ essentially from the involuntary in that they normally only contract in response to environmental changes, this response being mediated by the central nervous system. Fig. 29 shows how a stimulus to such a sensory surface as the skin is conveyed by a sensory nerve fibre to the central nervous system and is there relayed to a motor nerve cell, whose fibre runs out to end in

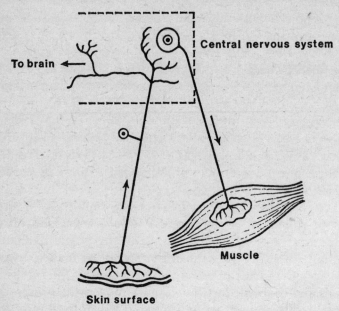

Fig. 29. A simple reflex arc. The stimulus conveyed from the skin is relayed through a cell of the central nervous system to elicit an automatic contraction of the appropriate muscle. At the same time information is passed to the brain for conscious control of the proceedings. (Adapted from Starling's *Principles of Human Physiology*.)

the muscle which it stimulates to contract. This is the primitive arrangement of the *reflex arc*, which is well established in quite simple animals, and has for its purpose the immediate and unthinking performance of such actions as withdrawal or blinking in protection from some noxious agent. In man, however, reflex actions have come increasingly under the control of the will, i.e. the message is simultaneously relayed up the spinal cord to the higher levels of the brain, which may so react as to modify the primitive response. The situation is comparable to that of a ship on the high seas that is being steered by a gyroscopic compass which is responding reflexly to any change in course but the information from which is being constantly noted by the captain, who may intervene at any time to effect a deliberate alteration in position. Meanwhile, deep in the bowels of the ship, the engines are pounding away, using up fuel, expelling wastes and pumping out bilge water, and the captain is quite unaware of these internal processes, though he may modify them indirectly.

Every skeletal muscle has motor nerve fibres connecting it with the central nervous system, but it is also provided with sensory fibres carrying information as to the degree of contraction and the tension within the muscle. This latter information is important, as muscles act in groups and in relation to each other so that they require central co-ordination based on the knowledge of how they are all behaving.

Individual muscle cells either run the entire length of the main muscle belly or end at intermediate points with an attachment to other muscle cells. When the first case applies, all the motor end-plates lie in a zone near the centre of the belly; in the second case, they are distributed throughout the entire substance of the muscle.

Each muscle cell has one end-plate associated with the termination of a nerve fibre, and each motor nerve cell in the nucleus of a cranial or spinal motor nerve sends out one much prolonged axon, or axis cylinder, within that nerve, which branches in the muscle to supply a number of muscle cells. The nerve cell and its axon exhibit an 'all-or-none' characteristic: they transmit either an impulse of a particular and constant magnitude or nothing at all. And, as the arrival of an impulse at a motor end-plate causes a similar 'all-or-none' depolarization of the related muscle cells, there is a similar response in that group of cells, i.e. an individual fibre contracts by some 50%–60% of its relaxed length or not at all. However, not all the fibres of a muscle necessarily contract at the same time, the proportion depending on the work to be done.

The group of muscle cells supplied by a single neurone is known as a *motor unit*, and the number of muscle cells in such a unit is related to the delicacy of the movement performed by that muscle, there being fewer cells per unit where fine movements are concerned. Thus, in the small muscles controlling the movements of the eyes and fingers, a single motor unit comprises perhaps only a dozen muscle

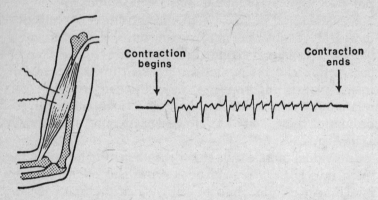

Fig. 30. Electromyogram.

cells, whereas in the coarser muscle of the limbs and trunk there may be several hundred cells per unit. An electrode placed over or within a muscle can detect the electrical activities of local motor units, and these can be amplified to produce audible evidence of 'firing' of these units—a slow or rapid machine-gun-like effect—or the visual record of an electromyogram.

The actual transmission of the nervous impulse to the muscle cell at the motor end-plate represents replacement of an electrical impulse by chemical stimulation. The impulse transmitted down the axon liberates acetylcholine at the end-plate, and this increases the permeability of the muscle membrane locally to sodium, potassium and other cations. An electrical potential is generated that spreads through the muscle fibre at the rate of several metres per second. It is this that can be recorded in the electromyogram, though the final record comes from many fibres comprising several motor units. Apart from some spontaneous random discharges, there is little electrical activity in a muscle at rest, but there is a burst of discharges during voluntary contraction.

The situation differs somewhat in smooth muscle. Whereas a skeletal muscle is completely paralysed and flaccid if its motor nerve is cut, smooth muscle may continue to function after interruption of its nerve supply, for the nerves modulate rather than initiate activity. Smooth muscle often has a double nerve supply, from both the central and the autonomic nervous systems, and one group of fibres releases acetylcholine and the other, noradrenaline. These exert opposing effects, one being excitatory and the other inhibitory; in addition, smooth muscle responds to the influence of circulating hormones.

Muscle contraction, tonus, posture

Muscle action moves the joint or joints lying between its origin and insertion, the former of these words meaning the fixed and the latter the moving attachment. This is clear enough when done against resistance, or against the mere weight of the limb. But even when a movement is assisted by gravity, as when the arm is dropped to the side, there is a controlled *relaxation*. In this case the arm-lifting muscles undergo relaxation, which is really a continuous readjustment of the degree of contraction; i.e. 'relaxation' is an active process. The ordinary contraction, when the origin and insertion are allowed to approximate freely and the muscle belly shortens adaptively, is *isotonic*, as the tension remains fairly constant throughout. If the attachments are kept separated by some resistance, the muscle cannot shorten and its tension rises greatly—an *isometric* contraction of constant length. And, since most actions are performed against some degree of resistance, contractions are usually a blend of isometric and isotonic.

Living muscle is never completely relaxed, except under deep anaesthesia, but is always in slight contraction or *tonus*. The state of the fibres is best expressed by comparison with a pianist continuously rippling over the notes of a piano; some notes are always sounding at a given moment, but they are always changing. Tonus is essential to the maintenance of posture, especially the erect posture of man, as it holds the body against the collapsing strain of its own weight, particularly the feet and legs, whose ligaments would be subject to enormous deforming stresses were it not for the constant guarding tone of the surrounding muscles. And the spinal column is not a straight rod but a series of curves from above downwards, whose general shape is maintained by the long spinal muscles forming the chords of these arcs. Essentially, postural reflexes in the erect position are devoted to preserving an upright, forward-looking head, and a complex reciprocal action of all the main muscle groups in calves, thighs, buttocks, spine and neck exists to secure this end. Hence the necessity for a *sensory* nerve supply from the muscles, telegraphing the exact state of tension from moment to moment to the central nervous system.

The biochemistry of muscle contraction

The *immediate* source of the energy used in contraction is the splitting up of 'high-energy' adenosine triphosphate, which is involved in the actin-myosin reaction. To take a simple example:

$$ATP + H_2O \rightleftharpoons ADP(\text{adenosine diphosphate}) + H_3PO_4 + 8 \text{ kcal.}$$

But this ATP has to be resynthesized, so energy is required from other sources. Under *anaerobic* conditions, when sufficient oxygen is not available locally, as at the beginning of violent exercise, or in a muscle held in prolonged static contraction which obliterates the local circulation, substances contained in the muscle fibres break down to provide the energy for this resynthesis. The most important of these is glycogen, its degradation product being lactic acid. If this accumulates locally in the muscle, as it may do during a forced static contraction, fatigue and inhibition akin to rigor develop. Normally, the excess lactic acid diffuses out into the circulation and is removed therefrom by oxidation; some is also taken up by the liver for reconversion to glycogen. Some of the lactic acid that remains in the muscle is restored to glycogen during the recovery period after contraction, but a considerable amount is burnt away to carbon dioxide and water by oxygen under *aerobic* conditions. The glycogen stores so depleted during heavy exercise cannot be fully restored at once; several days may be required.

In a steady state of exercise under *aerobic* conditions, both fat and carbohydrate brought to the muscles by the bloodstream provide energy for the resynthesis of ATP. The liver is an important source of the glucose used up, but even this source may become depleted, and in prolonged exertion there may develop hypoglycaemia (a concentration of glucose in the blood below normal) and fat becomes more important as a fuel. In prolonged exertion the muscle stores of glycogen repair the carbohydrate deficiency and the lactate level in the blood rises.

An adequate supply of oxygen is essential to contracting muscle, which may use up ten times as much as when it is in the resting state. The presence of lactic acid in the bloodstream during exertion stimulates the respiratory and cardiac centres to greater efforts, both to oxidize the circulating lactic acid and to deliver oxygen to the contracting muscles themselves. The oxygen is needed, to some extent, not so much to enable muscle to contract as to allow it to recover from the effects of contraction; a muscle stimulated to the point of fatigue in an inert gas such as nitrogen is incapable of such recovery.

The oxidative removal of excess lactic acid does not keep pace with its accumulation at the beginning of a continued effort, and its concentration in the blood rises. This is one of the main reasons for the breathlessness of exercise, an endeavour to increase the oxygen supply so as to get rid of lactic acid excess in the blood. When a steady state is achieved at which the oxygen intake can deal with the acid at the rate it is formed, the individual has acquired his 'second wind' and can continue comfortably for a long period. On the other hand, after a short period of very strenuous effort, one continues to be out of breath for several minutes afterwards as there is a dammed-up accumulation of lactic acid in the blood still to be oxidized, and the body is spoken of as having got into 'oxygen debt' by accepting more calls on its oxygen supply than it can meet at the time.

The situation is a complex one because the carbohydrate store in muscle is important as a short-term source of energy during the relatively anàerobic conditions obtaining at the beginning of exercise before the local and general circulation have had time to adapt, whereas the main source of energy during sustained activity is the free fatty acids of the blood—hence the importance of a high-fat diet for heavy workers under cold climatic conditions. In a short burst of exertion, such as a hundred metres' sprint, most of the energy comes from the breakdown of muscle glycogen and a large oxygen debt is incurred, i.e. the circulation has not provided oxygen and nutrients locally just when they were most needed. The adaptation of cardiac output, the redistribution of blood to favour the muscles and the increased intake of oxygen by respiration are not immediate. There is also the problem that muscles used rhythmically, as in running, only have a very intermittent blood supply because, as mentioned above, the contractions themselves empty the blood vessels; so, here again, there is partial oxygen deprivation and dependence on the glycolysis of a diminishing local glycogen store. A step to overcoming this difficulty is the presence in muscle of a pigment, myoglobin, akin to the haemoglobin of the blood, which can liberate oxygen locally by dissociation. This system is well developed in some animals, e.g. in the 'red' pectoral muscles of birds.

Oxygen uptake from respiration increases with the rate of exercise, and there is a maximal uptake level which varies with size, age, sex and state of training. The maximum uptake for average young men is around 3·5 litres per minute, for young women 2–3 l/min and for trained athletes as high as 6 l/min. The oxygen content of the atmosphere, the inflating capacity of the lungs and the work of the heart are all general limiting factors in the availability of oxygen. Locally, the muscles themselves may use the oxygen supplied more or less efficiently—the arm muscles utilize oxygen less efficiently than the leg muscles. Other factors are the efficiency of the circulation and the haemoglobin content of the blood. If less haemoglobin is available to transport the oxygen, as in anaemia, or if the atmospheric oxygen is inadequate to saturate the haemoglobin that is available, as at high altitudes, muscle efficiency will be impaired until there has been an adaptive increase in the number of red corpuscles.

A good deal of *heat* is produced during muscular contraction, partly from the oxidative processes involved and partly from the mechanical work itself. At best, only some 25% of our total energy output can be translated into mechanical energy that can influence the environment as work; the rest produces heat. Most of the increased heat generated by exertion is dissipated by the evaporation of sweat. But the body temperature does actually rise during the first hour of exercise; not all the extra heat is lost and thereafter the raised temperature is maintained at a definite level. Bodily mechanics are more efficient at a higher temperature than at normal, chemical processes are accelerated and resistances decreased; hence the importance of a warming-up period for athletes.

Smooth muscle

It may be useful to reiterate here some of the earlier remarks on smooth muscle, as this differs considerably from
the voluntary muscle we have been considering, both in
inherent properties and in nervous control. It has the
property of inherent and independent rhythmic contractions, and is under the control of a different nervous system,
which permeates the muscle tissue itself with nerve plexuses
and clumps of nerve cells (ganglia) exerting a local control
over contraction. Moreover, while the motor nerves reaching voluntary muscle from the central nervous system have
only one action—to make it contract—the autonomic fibres
supplying smooth muscle are of two kinds, sympathetic and
parasympathetic, which stimulate it to relax and contract
respectively. And, as we have seen, these actions are reproduced powerfully by the effects on the muscle of certain
chemical agents: adrenaline, which mimics sympathetic
action, and acetylcholine, which mimics the parasympathetic. Smooth muscle is much more sluggish in its response than voluntary muscle. It is particularly sensitive to
stretching, and this is important since the hollow organs
like the bowel and the bladder, which contain air and
urine, are made of smooth muscle that needs to respond to
distension or relaxation of the organ. A *slowly* increasing distension, as when the bladder steadily fills with urine, causes
the muscular coat to relax proportionately so that the pressure of the fluid does not rise until the last stages of great
distension; at this point the rising pressure stimulates the
muscle to rhythmic waves of contraction, which end up in a
mass contraction and expulsion of the contents of the organ.
On the other hand, the *rapid* distension of a hollow organ
provokes an immediate and intense spasmodic response.

The entrances and exits to hollow organs are controlled
by specialized contracting circular bands of smooth muscle,
or *sphincters*, e.g. the cardiac and pyloric sphincters at each
end of the stomach; these are normally in a state of con-

traction until relaxation is necessary to allow the organ to fill or empty. It is obvious that the nervous and chemical control of a sphincter must be exactly the converse of that of the organ as a whole, since it must contract while the organ relaxes and vice versa. So we find that the parasympathetic and sympathetic nerves, and the chemical agents that correspond to them, exert exactly the opposite actions on the muscular wall of an organ and the sphincter muscle that guards its exit.

Gross anatomy of muscles

We have already noted the three main types of muscle tissue: voluntary, involuntary and cardiac. Here we deal only with the voluntary or skeletal muscle, the flesh of the body, making up some 40% of its weight and divided into the *axial* muscles connected with trunk, head and neck, and the *appendicular* muscles of the limbs. The separate fibres are bound in bundles by connective tissue, and these in turn are bound into larger bundles to compose the individual muscle, with its sheath derived from the deep fascia.

Attachments
Muscles are usually attached to bone or cartilage, sometimes to ligaments or skin, and these attachments are either directly by muscle fibres or by an intervening tendon, a sinew or 'leader', a tough, inelastic cord-like structure of white fibrous tissue. Most muscles have a tendon at one or both ends; and wide flat muscles have broad, sheet-like tendon expansions or aponeuroses, e.g. those of the flank abdominal muscles which extend from lower rib to pelvis. When a muscle contracts one attachment remains fixed—the *origin*—and the other—the *insertion*—is moved towards it. Generally, in the limbs the origin is proximal and the insertion distal, while in the trunk the origin is medial and the insertion lateral. But origin and insertion are sometimes interchangeable with circumstances. Thus the normal

action of the pectoralis major, inserted into the humerus, is to approximate (adduct) the arm to the side of the chest. But if the arm is fixed by gripping a table, the same muscle can be used to pull on the ribs at its origin, expanding the thoracic cavity as an accessory muscle of respiration, an action commonly performed in asthmatic attacks when breathing becomes difficult.

Attachments are usually just distal to the joint moved, e.g. the biceps into the upper radius just beyond the elbow, an arrangement rather inefficient mechanically but allowing great speed of action. A muscle may have two or even three heads of origin (*cf.* the biceps and triceps in the arm), but its insertion is nearly always single. Tendinous attachments to bone produce a local ridged elevation of bony substance, while direct muscular insertion leaves the bone quite smooth. Occasionally, muscles are attached to each other, usually a symmetrical arrangement between a pair on opposite sides of the midline; thus the fibres of the mylohoids under the chin criss-cross at a line of junction (called the median raphe), while the aponeurotic fibres of the abdominal muscles interlace in herring-bone fashion at the midline of the abdominal wall. A few muscles have two separate bellies connected by an intermediate tendon, the object being a redirection of pull secured by the binding down of the tendon to a neighbouring bone by a band of deep fascia. Not all skeletal muscles act on joints, for some excite movements of soft tissues only, e.g. those that move the eyeball and the soft palate.

The *form* of a muscle depends on the arrangement of its fibres. When these have a direct pull, the muscle may be fusiform (with a definite belly tapering above and below), strap-like, quadrilateral or triangular (Fig. 31). But more power is obtainable when the fibres are inserted indirectly into tendinous prolongations within the muscle substance, for then a greater concentration of short efficient fibres is possible, as in the unipennate, bipennate and multipennate arrangements shown.

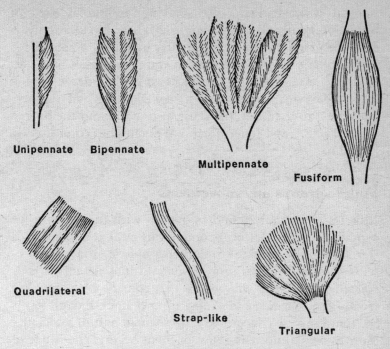

Fig. 31. Variations in the form of muscles and the arrangement of their fibres.

A strap-like muscle may be segmented transversely by tendinous intersections, e.g. the rectus abdominis of the abdominal wall (page 200). And in situations like the hand and foot, where bulky muscles would obstruct movement, the fusiform bellies of the muscles producing movements of the fingers and toes are contained in the forearm or leg, and their motors are continued as long, slender tendons into the extremities.

Muscles are *named* mainly after their attachments and actions—e.g. the flexor pollicis longus, the long flexor of the thumb—occasionally by shape—e.g. pronator quadratus, the square muscle pronating the forearm—or by some other feature—e.g. the quadriceps with four heads of origin. Muscles show considerable variation in their exact ana-

tomical arrangements, often having additional slips of
origin or being absent altogether. Their blood supply is
extremely rich, owing to the greatly increased demands for
oxygen made during contraction, and they receive branches
from one or more nerves carrying two kinds of nerve fibre—
motor, carrying the stimuli to contraction, and sensory,
conveying to the brain the sense of position and the degree
of contraction and tone, which are essential to co-ordinated
function.

Muscle groups and movements

Muscles are usually found in groups with the same nerve
supply, acting to perform common or related actions. It is
this movement, not the individual muscle, that is represented in the brain and no individual muscle can be contracted by an act of will, nor can any muscle act alone. We
try to lift our arm, *not* to contract our deltoid, though this is
the main muscle concerned, and a whole host of associated
muscles come into play, most of them beyond the fringe of
consciousness. In its relation to a particular movement an
individual muscle may fill one of the following roles:

(1) the *prime mover*, directly responsible for the actual
movement;

(2) the *antagonist*, which is capable of causing the reverse
motion and therefore has to relax reciprocally as the prime
mover contracts, actively paying out just enough slack to let
the latter do its job;

(3) the *fixators*, associated muscles steadying the base, or
fulcrum, against which the prime mover acts;

(4) the *synergists*, which act by controlling an intermediate
joint so that the prime mover may act with maximal
efficiency; thus the muscles closing the fingers can only produce a strong grip if other, synergist, muscles simultaneously bend the wrist backwards.

And, of course, in connection with different movements the
same muscle will play different parts.

Fine movements, such as writing, are carried through with both prime movers and antagonists in some degree of contraction throughout, the actual motions reflecting the changing balance of the opposing forces; whereas in rapid forceful movements, such as striking a nail with a hammer, both prime movers and antagonists relax once the stroke has been initiated and the movement is continued by sheer momentum.

Muscles may be artificially excited to contract by such insults as the application of heat, mechanical pinching, irritation with chemicals, or an electric current; and the last is of particular interest to the experimental physiologist because the nature of the normal nervous stimulation of muscle is itself electrical. A constant galvanic current excites a single contraction at the beginning and end of its flow, i.e. at make and break; a faradic current does likewise, but the current is of much shorter duration and there is a minimum length of time of current-flow below which there is no response. This *threshold value* is only a fraction of a second in quickly responsive warm-blooded animals, but may be several seconds in sluggish cold-blooded creatures. The effectiveness of response of a muscle to electrical stimulation rises with the strength of the current, and there is a minimum strength below which stimulation is ineffective. This increase of response is due to a larger number of individual muscle fibres taking part in the contraction. But as far as each fibre is concerned, each contraction in which it takes part is an all-or-none effort; it contracts completely or not at all, and the greatest contraction a muscle can make is when all its component fibres are contracting. After contraction there is a short *refractory period* of a minute fraction of a second during which the muscle is inexcitable; this is a period of recovery and preparation for the next contraction, and is, of course, an essential part of the cycle of cardiac muscle contraction, as we have seen.

If a muscle is repeatedly stimulated, it does not continue contracting indefinitely but becomes *fatigued*; it does less work and eventually ceases to respond. This state of affairs is due to the using-up of all the food substances providing energy for the contraction, e.g. glycogen, and the accumulation of waste products such as lactic acid. In combating fatigue a plentiful supply of blood and oxygen is essential. But fatigue in man is not just a simple matter relating to the muscle itself; it is a much more complex one, involving the tiring out of the controlling nerve cells and fibres.

The tense generalized muscular contraction that occurs after death in *rigor mortis* must not be confused with ordinary contraction in life; it is associated with the accumulation of lactic acid and an irreversible change in the muscle proteins. But it is accelerated by previous fatigue, so that its onset in animals that have been hunted to death is very rapid.

6 Standard positions, terms and references; nomenclature; symmetry and segmentation; body coverings and body systems

Positions, terms and references

It is essential to understand that, for the purposes of anatomical description and the localization of parts, the body is always considered as being in a particular conventional position. This is as important as in map-reading, where two people can understand each other's references because of a prior agreement to stick to the same methods of plotting and the same arrangement of North and South. In this standard position the body is imagined erect, with the arms by the side and the thumbs out so that the palms of the hands face forwards. The front of the body and limbs is the *anterior* surface, the back is *posterior*, so that if the relation of any two structures is being considered one may be described as anterior or posterior to the other insofar as it is nearer the anterior or posterior surface. Anterior is sometimes also called *ventral* and posterior *dorsal*, a derivation from the condition in animals where the belly faces the ground.

The position of structures is also related to the median plane of the body (AA in Fig. 32). One point is said to be *medial* to another which is placed further from the midline, the second point being *lateral* to the first. And points nearer the head end are *superior* to those *inferior* ones nearer the feet; *cranial* is sometimes used for superior and *caudal* for inferior, corresponding to the head and tail of animals.

Internal and *external* are descriptive of the boundary walls of body cavities or hollow organs; thus the ribs have an

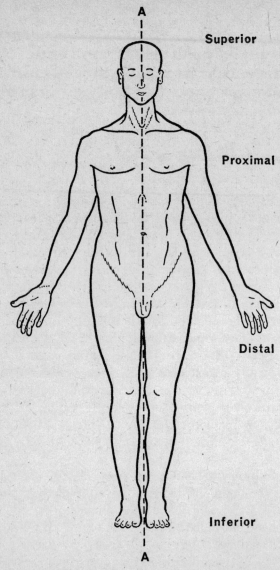

Superior

Proximal

Distal

Inferior

Fig. 32. The conventional view of the body from in front, showing its anterior surface. AA is the median plane.

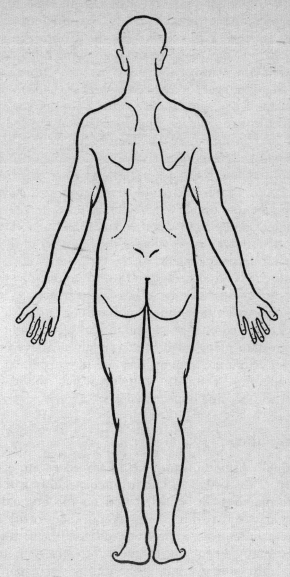

Fig. 33. The posterior surface of the body.

external surface directed outwards and an internal surface facing towards the thoracic cavity. *Superficial* and *deep* are obvious references to relative distances from the surface, e.g. the skin is superficial to the underlying muscles.

Certain other terms are used in connection with the limbs. Here, anterior may become *palmar* (hand) or *plantar* (sole of the foot), with *dorsal* for the posterior surface, the back of the hand or the upper surface of the foot. Also, lateral and medial may be referred to in the limbs by names derived from the corresponding paired bones of forearm and lower leg, i.e. *radial* and *ulnar* or *fibular* and *tibial*. Points nearer the shoulder or groin are *proximal* to those *distal* ones situated nearer the fingers or toes. *Peripheral* corresponds roughly to distal but is usually employed for the more outlying distribution of the branches of the circulatory and nervous systems.

Finally, it is often necessary to refer to sections through the body; these may be *horizontal* (transverse), *sagittal* (along or parallel to the median plane) or *coronal* (along or parallel to the coronal plane (BB in Fig. 34) at right angles to the median plane). All the terms given above must be mastered before proceeding; they are the cardinal points of the anatomical compass, without which the exploring student will become hopelessly lost.

Nomenclature

It is equally important to have an agreed system of names for parts of the body; a standard nomenclature is therefore used throughout the world, based on the original Latin terminology and revised at intervals. These revisions are, however, infrequent and, since different generations of students may be brought up under different systems, there is inevitably some overlapping. Surgeons particularly tend to retain the older names, and in some cases two or more terms may be colloquially used for the same structure.

In this book no attempt has been made to adhere to any

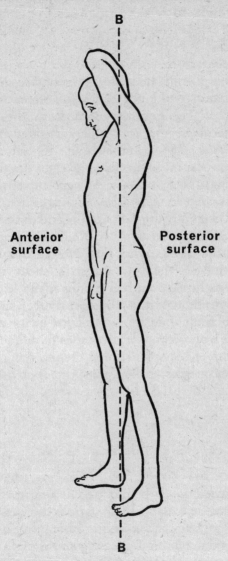

Fig. 34. The body in side view. BB is the coronal plane.

particular scheme and the names will be introduced naturally in the course of description.

Symmetry and segmentation

The two halves of the body on each side of the median plane are very largely similar, with corresponding right and left limbs, right and left kidney, and so on, i.e. many structures are _symmetrical_. This symmetry is far from perfect, however, as some internal organs are mainly one-sided (the liver) or entirely so (the spleen), and even the opposite limbs or the two halves of the brain are never exactly the same.

The human body repeats, in very modified form, the primitive arrangement of _segmentation_; this is widespread in the animal kingdom and well exemplified in the earthworm, which is made up of a number of identical segments, each containing the same organs and to a certain extent independent, though the gut traverses them in common. Integrated as he is, man still retains many features of this segmentation, though it is more clearly seen in the embryo than in the adult state. It is displayed in the arrangement of the vertebral column, in the series of paired ribs and in the segments of the spinal cord, each of which gives off a pair of spinal nerves to the appropriate body segment, distorted though these have become by reason of the emergence of the limbs.

The body covering; skin and fasciae

Skin

The skin covers the body and is continuous with the linings of the internal canals at their various orifices. It is highly elastic and mobile on the deeper structures, except where tightly bound down on the scalp, ears, palms and soles. On its surface open the hair follicles, sweat and sebaceous glands, and its brown melanin pigment is most developed on exposed sites and at the genital and nipple areas. It contains the peripheral endings of sensory nerves, acts as an excretory agent through its glands and helps to regulate body temperature through the rate of water loss by evaporation. It is also protective, and is modified at certain sites to form the hair and nails.

Microscopically, the skin has two main zones:

(1) The superficial *epidermis*, which varies in thickness—greatest in the palms and soles—and is permanently creased

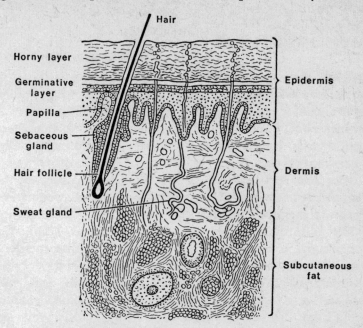

Fig. 35. Section through the skin to show its layers.

opposite the flexures of joints. It has an outer *horny layer* of dead or dying flattened cells being continually shed as scales by friction and is continually renewed by the growth of the *germinative layer*.

In more detail (Fig. 36) five layers may be recognized. Within the basal layer, the melanocytes form a specialized group of cells whose function is the synthesis of melanin pigment. In dark-skinned individuals pigment is present in all

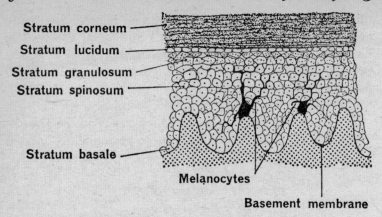

Stratum corneum

Stratum lucidum

Stratum granulosum

Stratum spinosum

Stratum basale

Melanocytes

Basement membrane

Fig. 36. The detailed structure of epidermis. (From *A Companion to Medical Studies, Vol. 1.*)

the basal cells. The keratin of the horny layer is characteristic of skin and derived structures, such as hair, nails, claws and hoofs.

(2) The deeper *dermis*, or true skin, which is a layer of tough elastic and fibrous connective tissue intervening between the epidermis and the subcutaneous fat; it is this layer in animals that is converted into leather. Unlike the epidermis, it is highly vascular and projects into the deep aspect of the latter in little bays or papillae, which contain the terminal capillary loops, and the end-bulbs of the sensory nerves, which carry impressions of touch, pain, heat and cold from the overlying surface. Although the hair follicles, sweat and sebaceous glands actually lie in the dermis, they have developed by ingrowth from the epidermis so that the hair shafts and the ducts of the sweat glands traverse the latter to reach the surface. The sebaceous glands open alongside the hairs to ensure their lubrication.

The *sweat glands* occur over the whole skin surface and are most numerous on the palms and soles; there are something like three to four million in all. They are simple, coiled, tubular glands abundantly supplied with blood vessels. *Apocrine* glands are modified sweat glands, localized to the armpit, anogenital area and breast, that are related to sexual phenomena; they begin to function only at puberty and their secretion produces the characteristic body odour. The individual *hair* consists of a long, dead, keratinized

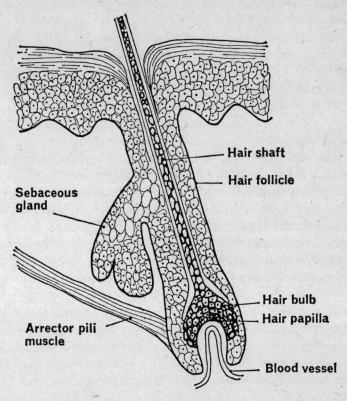

Fig. 37. Diagram of hair follicle and related sebaceous gland. (From *A Companion to Medical Studies, Vol. 1.*)

shaft and the basal growth segment or *bulb*, invaginated on its deep aspect by the highly vascularized hair *papilla* of specialized connective tissue. Strands of involuntary muscle inserted into the follicles are responsible for the erection of hairs, or 'goose flesh', that occurs during cold or emotional stress. Hair growth is intermittent. After a growth phase, the lower part of the follicle degenerates, the hair shaft is loosened and then shed as it is pushed out by the growth of a new hair, neighbouring follicles being at different stages of this cycle.

The *sebaceous glands* are found everywhere, except on the palms and soles of the hands and feet. They are lobulated structures found in the angle between a hair follicle and a strand of smooth muscle, which aids the expulsion of their secretion. This, the *sebum*, lubricates the skin and helps to protect it against the environment; it is a complex mixture of fatty acids, glycerides and cholesterol. Sebaceous secretion increases rapidly at puberty under the influence of the sex hormones.

The *nail* is a translucent keratin plate embedded at the sides and base in folds of skin—the nail folds. This plate is free at the tip but elsewhere is firmly attached to the underlying epidermis. The nail bed, from which growth occurs, is the epidermis underlying the basal nail fold. At the half-moon, or *lunula*, this epidermal layer is thick and pale. Nail growth is quicker at the fingers than at the toes, the times for complete replacement being some six and twelve to eighteen months respectively.

The protective function of the skin depends mainly on the barrier effect of the intact epidermis. Substances may pass into the body either directly via the epidermis or through the openings of the hair/sebaceous system. Penetration depends on the water and lipid solubility of the agent concerned. The skin is almost impermeable to water, but fat-soluble substances such as alcohol enter because they dissolve the lipid of the cell walls; absorption is aided by the increase of skin temperature and blood supply, or if the skin is damaged by injury or inflammation. The intact skin is virtually impermeable to electrolytes and only slightly permeable to gases; in animals such as amphibia the skin is an important respiratory organ, but in man cutaneous respiration only constitutes about 0·5% of total (lung) respiration.

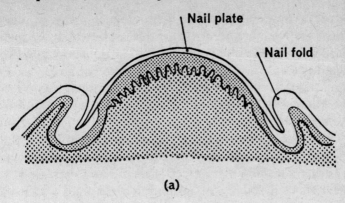

(a)

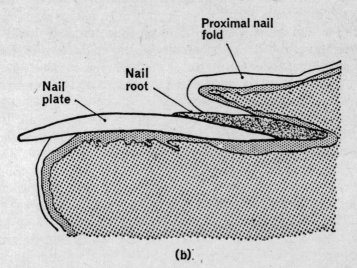

(b)

Fig. 38. Diagram of nail, (a) longitudinal, (b) transverse. (From *A Companion to Medical Studies, Vol. 1.*)

Fasciae

Intervening between the skin and the muscles that clothe the bones of the skeleton are two important layers of connective tissue, the superficial and the deep fascia.

The *superficial fascia*, the ordinary subcutaneous fat, is a continuous sheet over the whole body and is everywhere fatty, save at the eyelids and over the male genitals. In one or two sites it contains muscle fibres, such as the muscles of facial expression and the dartos muscle which corrugates the scrotum. Fat is more developed in the abdomen, breast and buttocks, and is thicker in women than in men. It insulates the body, so retaining warmth; and it also contains the cutaneous nerves and vessels on their way to and from the skin.

Where the superficial fascia is abundant, as in the thigh, the skin can move freely over the deeper structures; where it is virtually absent, as over the nose and ear, the skin is firmly tacked down. The insulating action means that those with a thick layer of fascia survive longer than others when exposed to cold. The thicker deposit in women is responsible for their rounded body contours and, indeed, the distribution of this fat is a secondary sexual characteristic; the female breast develops in this layer, and women have more fat deposited at the buttocks and thighs, *cf.* the admired steatopygy of some primitive races. In the male, fat tends to accumulate in later life in the upper abdominal wall. In certain sites—the pads of the tips of the fingers and toes, the heel-pad and also the buttocks—the fat is honeycombed by fibrous strands in loculated fashion; this is a cushioning device against pressure in standing and sitting, and when handling tools.

The elastic tissue of the dermis is so arranged as to produce definite planes of cleavage, varying in direction in different parts of the body. Thus, in the anterior abdominal wall, the natural skin cleavage is roughly transverse so that an operative incision in this line is under little tension and gapes only slightly, while a longitudinal incision at right angles to this is pulled open by skin tension and is much more difficult to close.

Internal to the superficial fascia is the *deep fascia*, a tough, membranous sheet of white fibrous connective tissue that covers and partitions the various muscle groups, and lies in close relation to bones and ligaments, gaining an attachment to the subcutaneous bony prominences. From its deep surface, other sheets, or *septa*, extend inwards between the muscle groups, and form sheaths for nerves and vessels and compartments for the viscera. This fascia varies greatly in strength and thickness at different sites; thus it is filmy over the small muscles of the hand, virtually absent over the face, but extremely thick and strong in the lower part of the back.

The general arrangement of fasciae is seen most clearly in the cross-section of a limb (Fig. 69). Here we see, in succession from without inwards: the skin, the superficial fascia and the deep fascia, enclosing the muscle masses of the limb in a continuous restraining envelope and sending down between them intermuscular septa that separate and demarcate the various groups and reach the bone to blend with the periosteum. This arrangement is important as regards the return of blood and lymphatic fluid from the limbs. The heart pumps arterial blood into these compartments; the return of fluids towards the chest is due to the pumping action of muscular contraction within the non-yielding fascial envelope which raises pressure within each compartment. As both veins and lymphatics are provided with one-way valves, this pushes their fluid content towards the heart, and if muscular contraction is eliminated by paralysis or immobilization in a plaster cast tissue fluids tend to accumulate in the limbs, causing the swelling known as *oedema*. In various situations the deep fascia is modified to form restraining bands, or *retinacula*, to hold down tendons and prevent them from bowstringing when their muscles contract; as *synovial sheaths* to facilitate the smooth gliding of tendons, as in the fingers and toes; and as *bursae*—simple, closed synovial sacs over points of friction or compression, as between skin and bony points (knee-cap and elbow), and between tendon and bone.

The deep fascia is not readily penetrated by fluid, and collections of fluid, such as blood or pus, take the line of least resistance and spread along the guiding fascial planes, and may come to the surface quite remotely from the site of injury or inflammation; thus the first evidence of tuberculosis of the lumbar spine may be an abscess 'pointing' in the groin. Some glands are invested with a firm fascial layer which becomes stretched and painful when the gland is inflamed, as in mumps or inflammation of the testicle.

Body wall and body cavities; somatic and visceral structures

The body wall, or *parietes*, encloses the great cavities, abdomen and thorax, which are also named after their

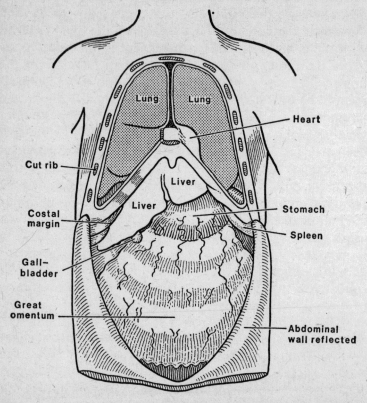

Fig. 39. The viscera exposed by removal of the anterior walls of abdomen and thorax.

lining membranes as the peritoneal and pleural spaces. Essentially, the body wall is composed of the skeleton, with its attached muscles and connective tissues, and the overlying skin and fat—the parietal or somatic structures.

The *cavities*, with their smooth serous linings, contain the internal, visceral or splanchnic organs: the lungs and heart in the chest, intestines and other organs in the abdomen. The viscera have developed in the embryo from the posterior body wall and still retain this attachment in the adult, the lungs by their roots, the intestine by its double-layered, fat-laden supporting mesentery. And we have already noted that the parietal skeletal muscles are under the control of the will, while those of the viscera function unconsciously and are largely independent of the central nervous system.

The body systems

These are:

The skeletal system—the bones. ⎫ Together constituting
The joints or articulations. ⎬ the locomotor
The muscles. ⎭ system.
The respiratory system. ⎫ Together constituting
The digestive system. ⎬ the visceral
The urogenital system. ⎭ organs.
The vascular system—heart, blood vessels, lymphatics.
The nervous system, and the sense organs.

The digestive system

Food passing through the mouth enters an expanded cavity behind, the *pharynx*, which is also common to the air passage at this level. It then travels down the gullet (*oesophagus*) to the *stomach*, thence to the coils of *small intestine* and then into the large intestine or *colon*. The waste matter in the food finally reaches the lowest part of the large bowel, or *rectum*,

and is expelled through the short *anal canal*. At various
points along the digestive tract, certain glands are situated

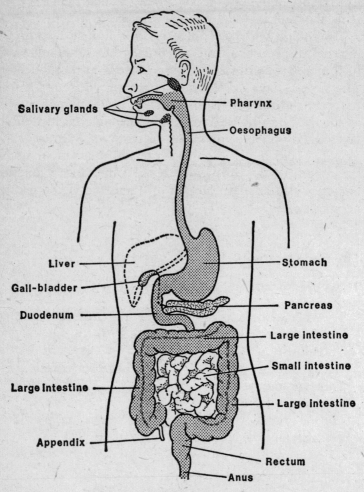

Salivary glands

Pharynx

Oesophagus

Liver

Gall-bladder

Duodenum

Large intestine

Appendix

Stomach

Pancreas

Large intestine

Small intestine

Large intestine

Rectum

Anus

Fig. 40. The digestive system.

and discharge their secretions into it. These are: the three
pairs of *salivary glands* around the mouth; the *pancreas*,
below the stomach; and the *liver*, which overlies the
stomach.

The respiratory system

Air inhaled through the nose or mouth enters the pharynx behind it and then travels down through the air passage proper. The first part of this is the *larynx*, which is also the

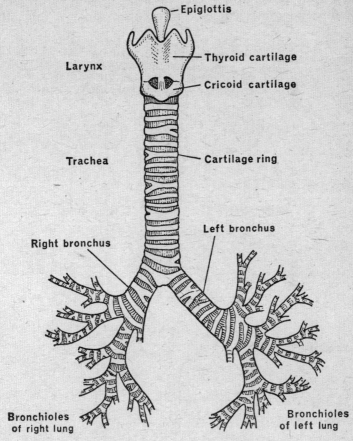

Fig. 41. The respiratory passages.

organ of voice; and this leads on to the windpipe (*trachea*), which divides in the upper part of the chest into a right and left *bronchus* for the right and left lung. Each bronchus sub-divides within the lung to form numerous branching *bronchioles* which end in clusters of tiny *air sacs*; and it is in the walls of the latter that the actual interchange occurs between the gases dissolved in the blood and those of the inhaled air.

Urinary system
The urine is secreted in the two *kidneys*, lying at the back of the abdominal cavity. From each kidney a tube, the

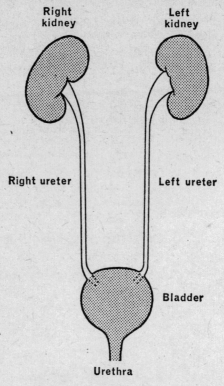

Fig. 42. The urinary system.

ureter, carries the urine down to the bladder in the pelvis, the space between the hip bones, and it is discharged from there into the urethra. In the female this is a short channel, soon opening externally; in the male it is a long, curved pathway traversing the prostate gland and then running through the penis, which is also used for performing the reproductive act.

The associated genital or sex organs will be considered later (page 477).

Vascular system

The circulatory system is essentially a closed circle, round which the blood is propelled by the contractions of the muscular heart. Blood is driven into the arteries, thick elastic tubes that aid by their recoil the distribution of blood to all parts. The arteries divide into smaller branches in their course to the organs and limbs, and finally break up into a meshwork of fine capillaries, microscopic thin-walled vessels that permeate every tissue of the body except the cornea of the eye and the outer layer of the skin.

The blood contained in the capillaries discharges oxygen and food materials to the tissue cells, and takes up carbon dioxide and wastes in return. The network reforms to form small veins, which become large venous trunks as they travel centrally towards the heart. These are thin-walled, have no pulse and contain valves to prevent any backward flow of blood.

Now there are, in fact, two separate circulations: a *systemic*, concerned with the body as a whole and driven by the left side of the heart, and a *pulmonary*, concerned with the passage of blood through the lungs and driven by the right side of the heart.

The heart has a right and left side, which are quite shut off from one another; and each side has an upper chamber, or atrium,* receiving blood from the great veins, and a lower chamber or ventricle discharging blood into the great arteries. Stale venous blood from the body enters the right atrium, passes to the right ventricle and is expelled through the pulmonary artery to traverse the capillaries of the lungs. Here, it becomes aerated, receiving fresh oxygen from the air in the air sacs and giving up carbon dioxide to be breathed out or expired. The fresh blood returns from the lungs in the pulmonary veins to the left atrium of the

* Sometimes called the auricle.

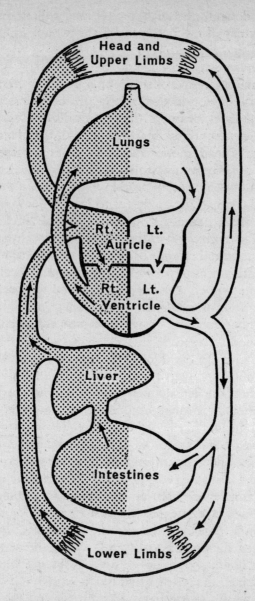

Fig. 43. A diagram of the circulation of the blood.

heart, then down to the left ventricle and is discharged into the great artery of the body, the aorta, which supplies the head, trunk and limbs through its branches.

In the body tissues the blood is rendered dark and venous, and is ultimately collected up into great veins—the superior vena cava, draining the head and arms, and the inferior vena cava, draining the trunk and legs. Note that, whereas the arteries of the body contain bright red blood and the veins dark blood, the reverse is inevitably the case for the pulmonary arteries and veins, for the lungs are concerned with reversing the chemical states of the blood.

There is a special arrangement of the abdominal vessels that must also be noted. Whereas the veins leaving most structures pass directly to the heart, those from the stomach and intestines enter another organ, the liver, where they break up into a second set of capillaries, so that the blood is filtered through the liver before reaching the heart via the veins of the liver itself. This is to secure that the liver utilizes and stores the dissolved food substances carried by the blood from the bowel; the arrangement is known as the *portal* circulation.

Lymphatic system

This is a sort of accessory to the main vascular system. Not all the fluid portion of the blood that exudes into the tissues from the capillaries returns to these vessels, with the result that there is an accumulation of tissue fluids. This excess is removed by a separate set of fine channels, the lymphatics, which begin as clefts and crevices between the cells and join up to form a plexus draining the various organs.

These vessels pass up the limbs and trunk, and are interrupted at certain points by a number of gland-stations or filters, which lie at the elbow and knee, armpit and groin, and in the trunk along the great blood vessels.

The lymphatics of the trunk join to form a wider vessel known as the *thoracic duct*, of matchstick thickness, which runs up in the chest to the left side of the neck; there it

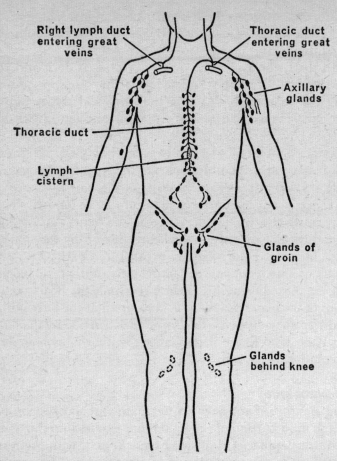

Fig. 44. The lymphatic system. (The glands of the head and neck are not shown.)

gathers the lymphatics of the left arm and the left side of head and neck, and discharges into the great veins. On the right side, the vessels of the arm, head and neck discharge directly into the veins.

One of the main functions of this system is the absorption of digested fat via the lymphatics of the bowel, and the lymph glands deal with any infection brought to them by the lymphatics.

7 Regional anatomy: the arm

Bones of the upper limb

Shoulder

This region is made up of the rounded *shoulder-cap*, the prominence formed by the head of the humerus and the overhanging acromion process of the scapula. It also includes the *scapular* region behind, over the shoulder blade, the *pectoral* region or front of the upper part of the chest below the clavicle, and the *axilla* or armpit intervening between the two. We have seen that the shoulder girdle is made up of scapula and clavicle, articulating at the acromio-clavicular joint; and the medial end of the clavicle articulates with the sternum at the sternoclavicular joint.

The *scapula* itself consists of the main *body* or blade, a thin, triangular plate or bone carrying certain local elevations or processes. The body has superior, medial (vertebral) and lateral (axillary) *borders*, with a superior and inferior *angle* at either end of the medial border. But where the corresponding lateral angle would be expected, at the junction of lateral and superior borders, there is the expanded thickened mass of the *head* of the scapula, hollowed out on its lateral aspect to form the shallow *glenoid fossa*. The deep (anterior) surface of the bone is slightly concave and is applied to the backs of the upper ribs; the superficial (posterior) surface carries a prominent ridge, the *spinous process*, which runs up and out from its root on the medial border to end as a broad lozenge-shaped expansion, the *acromion process*, overlying the shoulder joint. Lastly, there is the short, stubby *coracoid process*, shaped like a crook, arising from the superior border and also overhanging the joint but at a lower level and in front. It can be seen in Fig. 46 how

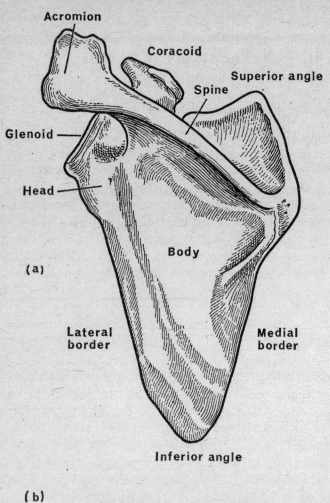

(a)

(b)

Fig. 45. (a) Left scapula, posterior aspect. (b) Left clavicle, from above.

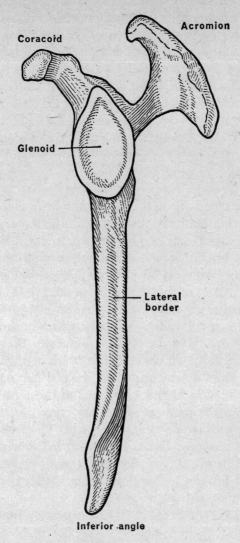

Fig. 46. Left scapula, lateral view.

the acromion and coracoid together form a protective arch over the joint—the so-called secondary socket—an additional safeguard against upward displacement of the humerus.

The *clavicle* is a long, slender, curved rod connecting the acromion to the upper portion of the sternum, the manubrium. It lies more or less horizontally, forming the lower boundary of the neck at each side, the floor of a space called the posterior triangle (see also Fig. 149). Its two ends are somewhat expanded and the shaft has a rather flattened S-shaped curve, so that the medial half has a forward-bulging convexity, which is reversed in the lateral portion of the bone as it sweeps away and back to reach the scapula.

Arm

The *humerus*, the bone of the upper arm, articulates with the scapula at the shoulder joint and with the forearm bones at the elbow; it is a long bone with a shaft and two expanded ends. The upper or proximal end carries the smooth, rounded *head*, which is directed medially and upwards. On the lateral aspect of this region, opposite the head, are two prominences, the *greater* and *lesser tuberosities*, giving attachment to the small rotator muscles which intimately surround the joint. It is the greater tuberosity that forms the point of the shoulder beneath the overhanging acromion. The tuberosities are separated by the *bicipital groove*, which carries the tendon of the long head of the biceps from its origin within the joint, at the upper border of the glenoid fossa, on its passage down into the arm. The true *anatomical neck* of the humerus is the narrow strip immediately encircling the head; the junction of the shaft proper with the whole mass formed by head and tuberosities is the *surgical neck*, a common site for fractures.

The *shaft* is cylindrical in its upper portion but the lower third is triangular in section. Halfway down, its outer aspect is marked by the rough and elevated *deltoid tubercle*, the insertion of the great deltoid muscle which abducts the humerus from the side. And curving round in spiral fashion from the back of the bone, just distal to the tubercle, is the groove for the radial nerve as it passes round to the front of the arm.

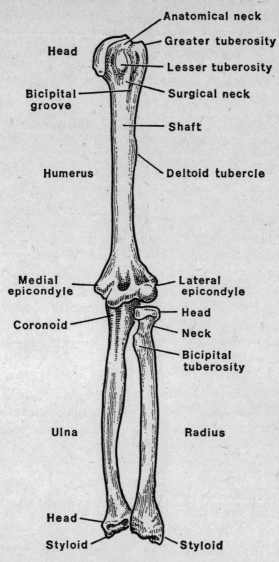

Fig. 47. Bones of the left upper limb, anterior aspect.

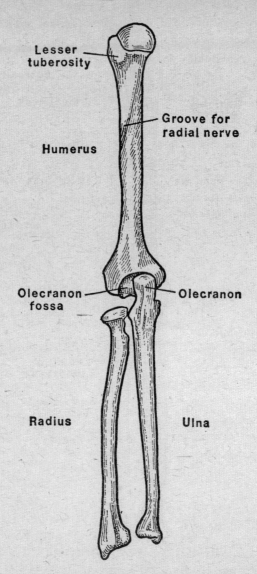

Lesser
tuberosity

Groove for
radial nerve

Humerus

Olecranon
fossa

Olecranon

Radius

Ulna

Fig. 48. Bones of the left upper limb, posterior aspect.

At the expanded lower end, the two prominences over-hanging the elbow joint on each side are the medial and lateral *epicondyles*, the former being the more developed as it is the origin of the strong flexor muscles of wrist and fingers.

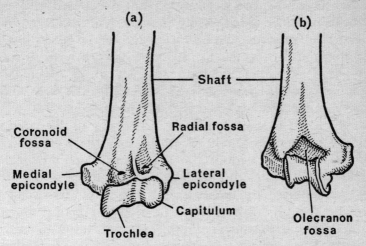

Fig. 49. Lower end of left humerus, (a) anterior, (b) posterior.

The polished, rounded projections downward from the lower end of the shaft—the lateral *capitulum* and medial *trochlea*—are the parts of the humerus concerned in the elbow joint; and it will be seen that there are hollowed-out depressions in the bone just above them, the *radial* and *coronoid fossa* in front and the *olecranon fossa* posteriorly, which accommodate the bones of the forearm in the extremes of flexion and extension of the joint.

Forearm

The *radius* and *ulna* are the paired bones of the forearm, articulating with each other at either extremity to form the superior (proximal) and inferior (distal) radio-ulnar joints; the lower end of the radius (but not of the ulna) forms the wrist joint with the carpal bones.

The *ulna* is the larger and lies on the medial side; its shaft has a sharp subcutaneous border which can be felt under

the skin throughout the back of the forearm. The upper end
carries the olecranon process behind, the point of the elbow,
which fits into the olecranon fossa of the humerus, and the
coronoid process in front, corresponding to the coronoid
fossa. These two processes are separated by the C-shaped
trochlear notch, which closely embraces the trochlear process of the humerus; and just below this on the outer aspect
of the ulna is a notch to accommodate the head of the
radius. The head of the ulna is at its lower end, a rounded
knob seen at the inner side of the back of the wrist and
carrying the small, pointed styloid process.

The *radius* is shorter than the ulna, lying on the lateral
side of the forearm, and its expanded ends are the other way
round, the larger end distal and the small head proximal.
The head is a smooth disc, hollowed-out above to receive
the capitulum of the humerus, and it fits within the circular
embrace of the ring formed by the radial notch of the ulna
and the annular ligament of the elbow joint. It is within this
ring that the head rotates during the movements of pronation
and supination (Fig. 50(a)). The neck is a slight constriction

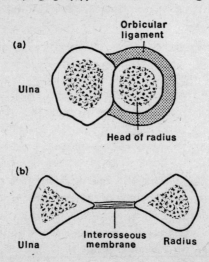

Fig. 50. Cross-section of radius and ulna, (a) at elbow, (b) at
midforearm.

immediately below the head, and just below this on the inner side of the upper shaft is the bicipital tuberosity for the insertion of the biceps. The lower end of the radius is very broad and carries a styloid process which can be felt to lie rather lower than the ulnar styloid; the back of the bone here is grooved by the passage of the extensor tendons of the wrist and fingers.

Both forearm bones have a sharp interosseous border facing each other and connected in life by the broad sheet of interosseous membrane, which lines the whole length of the forearm and separates the anterior flexor compartment from the posterior extensor compartment of the limb (Fig. 50(b)).

Wrist and hand

The wrist or *carpus* is composed of eight carpal bones, small, irregular cuboidal or rounded structures, arranged in two rows of four, a proximal and a distal. They are, named from medial to lateral sides: proximal row—pisiform, triquetrum, lunate, navicular; distal row—hamate, capitate, multangulum minor, multangulum major. They interlock and are bound together by strong interosseous ligaments so that, although the movement at the joint between any pair is small, the sum total of composite range gives the wrist its flexibility.

The skeleton of the *hand* itself is made up of the five *metacarpal* bones; each is a good example of a short long bone, the miniature of a bone like the humerus, with a shaft and expanded ends. The proximal *base* of each articulates with a carpal bone at the carpometacarpal joint, and the rounded distal *head*, which forms the prominence of the knuckle, makes up the metacarpophalangeal joint with the proximal phalanx of the corresponding finger.

The fingers contain three *phalanges*, articulating at the two (proximal and distal) interphalangeal joints. These phalanges—proximal, intermediate and terminal—are also short long bones; each has a proximal base and a distal end,

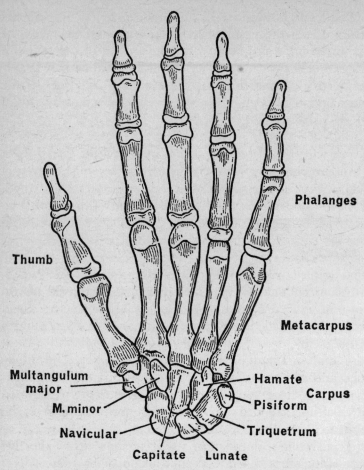

Fig. 51. Bones of left hand, anterior aspect.

formed by two rounded condyles making a hinge joint with the base of the phalanx in front. And at the end of the finger the tip of the terminal phalanx ends in an expanded bony tuft which supports the nail and nail bed. It should be noted that the phalanges are rather concave on their anterior (palmar) surfaces; this is because they form the floor of a tunnel roofed by fibrous tissue, through which the flexor tendons of the fingers glide in their synovial sheaths.

The *thumb* is considerably specialized so that its skeleton differs considerably from that of the other digits. The metacarpal is short and slight; it does not lie parallel to the others but is set freely away from the hand so that it can be opposed to the fingers. And there are only two broad phalanges.

The joints of the arm and their movements

Clavicular joints

The *acromioclavicular* and *sternoclavicular* joints at either end of the clavicle are simple plane joints, allowing only a limited range of gliding movement. They each contain in their synovial cavity a projecting fibrocartilage disc or meniscus. And they are important mainly as accessories to the complete range of shoulder movement, which they complement. The inner is much more stable than the outer, which is easily dislocated by a fall on the arm, the outer end of the clavicle riding up over the upper surface of the acromion.

Shoulder joint

This, as we have seen, is designed for mobility at the expense of stability and the articulating surfaces are far from congruous. The large rounded humeral head, about half a spherical surface, does not fit well into the shallow glenoid fossa of the scapula, only one-ninth of a sphere. But in life the socket is deepened by a fibrocartilaginous rim, the *glenoid labrum*. The joint capsule is attached round the margins of the glenoid and the humeral head as a rather lax sleeve, hanging down in a fold on the inferior aspect of the joint; this laxity is necessary if enough slack is to be present to allow the arm to be lifted freely from the side.

The essential instability is compensated in various ways. Several thickened bands running in the substance of the capsule form accessory ligaments; and the tendon of the long head of the biceps originates within the joint from the

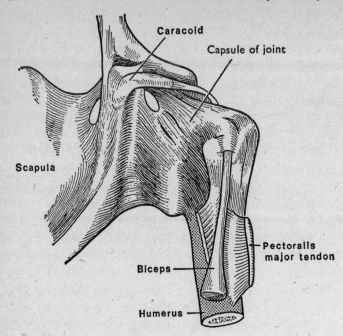

Fig. 52. Left shoulder joint, seen from in front. Note the loose capsule and the emerging tendon of the long head of biceps.

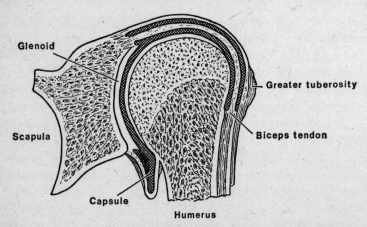

Fig. 53. The shoulder joint, longitudinal section. Note the dependent fold of capsule below, and the biceps tendon traversing the joint cavity.

upper pole of the glenoid fossa, traversing the cavity through a synovial sheath to emerge in the upper part of the bicipital groove between the two tuberosities. In its passage it functions as a ligament. But the main supporting factor is the intimate blending with the capsule of the tendons of the four small rotator muscles arising from the scapula as they pass to their insertion on the tuberosities. These muscles are the supraspinatus, infraspinatus, teres minor and subscapularis. And these muscles which actually move the joint are therefore, at the same time, guarding against displacement by tightening up the capsule and holding the humeral head firmly against the glenoid. Even so, dislocation is common, the head passing off the glenoid to lie anteriorly under the coracoid process. This is the common subcoracoid dislocation; but the head may be displaced vertically downward below the glenoid, or backward beneath the spine of the scapula.

The *movements of the shoulder* are:

(1) *Abduction* and *adduction*. In *abduction*, the arm is lifted away from the side in the coronal plane by the deltoid muscle; the reverse movement of *adduction* is effected by gravity, or actively by the pectoralis major muscle in front of the chest, and by the latissmus dorsi and teres major behind (Fig. 19, page 64).

(2) *Flexion* and *extension*, which occur in the sagittal plane, through a combination of muscles (Fig. 19).

(3) *Rotation*, produced by the small rotators, arising from the scapula (Fig. 20, page 64); it is lateral or medial.

(4) *Circumduction*—a combination of all the above movements, a swinging motion of the outstretched arm.

What makes these movements more complex—particularly abduction—is that they do not occur alone but in association with movement of the scapula, for this bone can be rotated on the trunk by the large, flat, triangular trapezius muscle of the back arising from the spine. The arm can be abducted through 180°, but only half this is true shoulder motion, the rest being produced by scapular

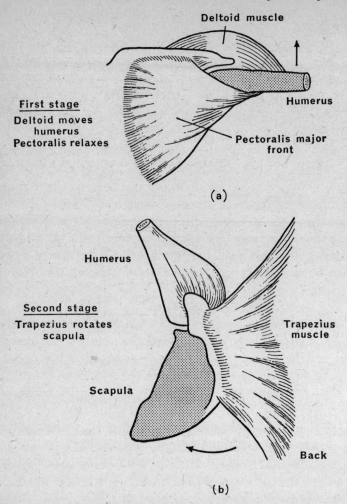

Fig. 54. Stages in abduction of the left shoulder joint, (a) anterior, (b) posterior. These stages are not as distinct as the figures suggest, for humerus and scapula both move from the outset, the former more at the beginning and the latter more at the end of movement.

rotation. Both movements occur in simultaneous co-ordination from the outset, and at the same time the outer end of the clavicle rises and there is gliding movement at both clavicular joints (Fig. 54).

Elbow joint

This comprises three articulations: the true elbow hinge joint between the trochlear process of the humerus and the corresponding notch of the ulna; the shallow articulation of the radial head and the humeral capitulum; and, inferiorly, the proximal radio-ulnar joint, whose synovial cavity is continuous with that of the others.

As in all hinge joints, the collateral ligaments at the sides are very strong, and these, with the deep interlocking of

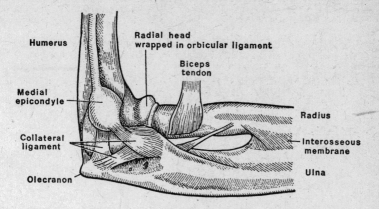

Fig. 55. Medial aspect of left elbow joint. (After *Gray*.)

humeral and ulnar surfaces, allow only flexion and extension to occur. To facilitate these the anterior and posterior portions of the capsule are so lax as to be easily stretched in the full extremes of range. When the elbow is fully extended, arm and forearm are *not* in line but at an angle of 5–10°, the carrying angle—larger in women than in men—which disappears as the joint is flexed. *Flexion* is produced by the biceps and brachialis muscles, the superficial and deep muscles of the anterior compartment of the arm; it is checked by the contact of the soft tissues of arm and forearm before the coronoid process of the ulna has settled very far into the corresponding fossa of the humerus. *Extension*, which usually goes a few degrees beyond the straight posi-

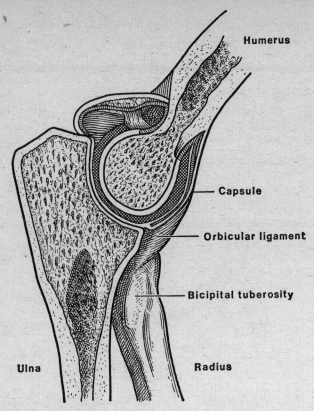

Fig. 56. Elbow joint, longitudinal section. (After *Gray.*)

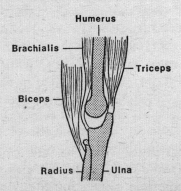

Fig. 57. Muscles responsible for flexion and extension of the elbow.

tion, is produced by the triceps at the back of the arm and checked by the fitting of the olecranon process into its humeral fossa. The triceps tendon blends closely with, and strengthens, the back of the elbow joint capsule.

We have seen that the *proximal radio-ulnar joint* is formed by the round head of the radius rotating like a pivot in the circular embrace formed by the radial notch of the ulna and the annular ligament of the joint (Figs 50, 55). Although its movements are distinct, its synovial cavity is freely continuous with that of the elbow joint proper and infectious processes spread easily from one to the other.

Pronation and supination are the names of the rotary movements of the forearm responsible for twisting the palm of the hand forwards and backwards. In this movement the

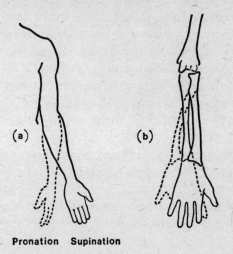

Pronation   Supination

Fig. 58. Pronation and supination. Note in (a) the 'carrying angle' in the supine position, and in (b) the movement of the radius.

ulna remains stationary while the radius rotates round it, moving at both radio-ulnar joints and along the axis of the interosseous membrane connecting the two bones. At the inferior radio-ulnar joint the relations of the two bones are just the opposite of the condition at their upper ends, the small, rounded ulnar head fitting into an ulnar notch on the broad radial expansion. In addition, the bones are connected by a triangular cartilage plate with its apex fixed to the ulnar styloid process. The apex of this triangle is the lower end of the axis of pronation-supination, the upper being the centre of the head of the radius; and the disc completely shuts off the lower end of the ulna from the wrist joint proper, the radiocarpal articulation.

Joints of wrist and hand

The *wrist joint* is formed between the proximal row of carpal bones (navicular, lunate, triquetrum) and the composite surface made up by the distal end of the radius, plus its cartilage plate fixed to the ulna. It is a modified hinge joint, with strong ulnar and radial collateral ligaments whose main movements are flexion (*palmarflexion*) and extension (*dorsiflexion*). But, in addition, there is movement of the whole wrist and hand to either side, *abduction* and *adduction*—ulnar and radial deviation—and the normal position for a strong grip is in slight adduction, not in exact line with the forearm.

There are numerous small *carpal joints* between the individual bones of the carpus which are bound together by short interosseous ligaments, the whole forming a complicated synovial cavity continuous with the wrist-joint proximally and the *carpometacarpal joints* distally, where the metacarpal bases articulate with the distal row of carpal bones. Little motion occurs at these latter joints, except at the specialized saddle joint between the thumb metacarpal and the multangulum major. Here motion is very free and its principal components are:

(1) *adduction* to the side of the hand and *abduction* away from it, in the plane of the palm;

(2) movement in a plane at right angles to the palm, *palmar abduction* and *adduction*;

(3) the characteristic human movement of *opposition*, in which the thumb is carried across the palm and its tip opposed to that of another digit in grasping. Note that this is a complex motion, in which the metacarpal rotates on its long axis so that the thumbnail faces forwards instead of outwards (Fig. 59).

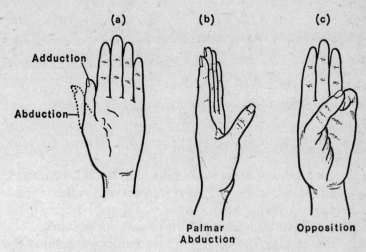

Fig. 59. Movements of the thumb.

The *metacarpophalangeal joints* between the heads of the metacarpals and the bases of the phalanges are again modified hinge joints, with strong collateral ligaments on each side and only a loose dorsal and palmar capsule. The main movements are flexion into the palm and extension backwards, and usually there is some 10–15° of hyperextension behind the plane of the palm. In addition, there are side-to-side movements, known as adduction and abduction, based on the relation to the axis through the middle finger, abduction being away from this centre and adduction towards it. The middle finger itself is abducted whether it moves laterally or medially.

The small interphalangeal joints allow flexion and extension only.

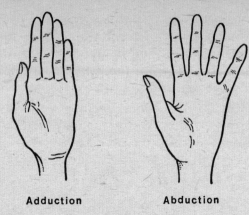

Adduction   Abduction

Fig. 60. Abduction and adduction of the fingers.

Surface anatomy of the arm

Now that the bony framework of the limb has been reviewed, some aspects of the surface anatomy may be considered. Inspecting and palpating the *anterior aspect* of the arm, the following features are noted from above downwards.

At the *shoulder* the clavicle is seen under the skin; the acromion process and the upper end of the humerus give the general rounded contour, and the coracoid process can be felt by deep pressure in the hollow under the outer end of the clavicle. The great bulge of the deltoid muscle is obvious, clothing the lateral aspect of the joint, and the pectoral muscles appear when the arm is adducted against resistance.

In the *arm* there is the bulge of the biceps.

At the *elbow* the prominences of the medial and lateral epicondyles are visible and palpable on either side, particularly the former, and the biceps belly fades away into its tendon, which can be felt as it crosses the elbow hollow, the *antecubital fossa*.

In the *forearm* the general muscular bulge of the flexors of the wrist and fingers fades into the visible or palpable tendons of these muscles just above the wrist, and on each side of the latter joint are felt the styloid processes of radius and ulna, the former at a slightly lower level.

In the *hand* the anterior (palmar) aspect consists of the central palm proper, with a muscular eminence on either side. The larger lateral *thenar* eminence, the ball of the thumb, is made up of the short muscles of that digit, and the shallower *hypothenar* eminence on the medial side covers the corresponding muscles of the little finger. The three main palmar creases are fairly constant, and it should be noted that the line of the webs between the fingers as seen from the

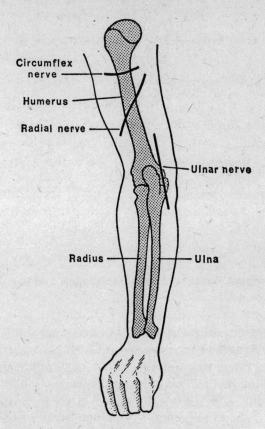

Fig. 61. Surface markings of important vessels and nerves of the left arm, posterior aspect.

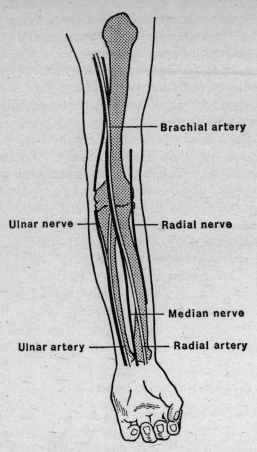

Fig. 62. Surface markings of important vessels and nerves of the left arm, anterior aspect.

front is not opposite the metacarpophalangeal joints, **i.e.** the knuckles behind, but lies a good inch more distally.

Running down the *posterior aspect* of the limb in the same way, we note at the *shoulder* the subcutaneous spine of the scapula ending in the overhanging acromion process and again the bulge of the deltoid. The muscle mass at the back of the *arm* is the triceps, and the same bony points are noted at the elbow, with the addition of the olecranon process of the ulna forming the point of the elbow midway between the epicondyles. In the *forearm* the bellies of various extensor muscles of wrist and fingers can be seen or felt, and at the *wrist* the head of the ulna is prominent on the medial side of the dorsum. On the back of the *hand* the knuckles mark the metacarpal heads, and the extensor tendons are seen when their muscles contract.

The surface markings of certain underlying nerves and vessels are indicated in Figs 61 and 62. The only nerves that are really superficial and easily felt are the ulnar nerve in its groove behind the medial humeral epicondyle at the elbow, which can be rolled to give the familiar tingling in forearm and little finger, and the median nerve at the front of the wrist. Both are very vulnerable to penetrating injury at these sites. The main arteries are also fairly deep except at the wrist, where the radial and ulnar vessels can be felt pulsating on either side under the skin; and it is often possible to see the beating of the superficial arterial arch in the palm.

The subcutaneous structures of the arm

Fig. 63 shows the flayed arm, divested of skin and superficial fascia. It demonstrates how the deep fascia forms a continuous sheath for the deep structures of the limb, and it can be seen that a network of veins and cutaneous nerves ramifies between skin and deep fascia, branches which all ultimately penetrate the fascia to join their parent vessels or nerves. The venous network in hand and forearm is complex but is gathered up in two main channels in the arm: the *cephalic vein*, running up the lateral side of the biceps to the shoulder, and the *basilic* vein on the medial side, piercing the fascia halfway up the arm to run alongside the brachial artery.

There is a subcutaneous *bursa* over the olecranon process separating it from the skin.

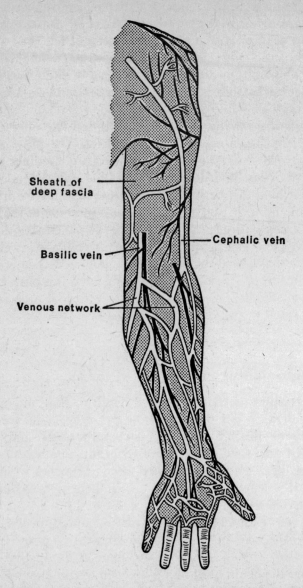

Fig. 63. The flayed arm, showing the subcutaneous structures, anterior aspect.

The muscle groups of the arm

Shoulder

The *deltoid* muscle on the outer side overlaps the front and back of the limb, lying in the same plane as the *pectoralis major*, which passes from the clavicle and upper ribs to its insertion in the upper humerus. The deltoid abducts the arm and the pectoralis adducts it to the side. Adduction is also performed by the *teres major* and *latissmus dorsi* posteriorly, the latter being one of the great muscles of the back that arise from the spinal column (Fig. 134, page 238). Lying very deeply, and arising from the scapula, are the short rotators of the humerus, inserted into the tuberosities under cover of the deltoid; their actions in rotating the bone laterally or medially are sufficiently clearly demonstrated in Figs 66–68, page 140.

Upper arm

This is seen in cross-section to be divided into anterior and posterior muscular compartments by the humerus itself, with a lateral and medial intermuscular septum of deep fascia attached to each side of the bone (Fig. 69). The anterior (*flexor*) compartment contains those muscles flexing the elbow, the biceps superficially and the brachialis nearer the bone. The *biceps* has two heads of origin, a long tendon arising within the shoulder joint and a short one from the tip of the coracoid process; they join to form the main belly, from which the tendon of insertion issues just above the elbow to reach the bicipital tuberosity of the radius. In addition to its flexing action, the biceps is the most powerful supinator muscle of the forearm owing to the fact that it is inserted near the back of the radius and can rotate it on its long axis; it is essential in powerful supinating actions such as the use of a corkscrew or screwdriver. The underlying *brachialis* arises from, and clothes, the front of the humerus; it is a simple elbow flexor, inserted into the coronoid process of the ulna.

At the back of the arm, in the *extensor* compartment, the *triceps* arises by three muscular heads, two from the humerus and one from the scapula below the glenoid. The tendon of

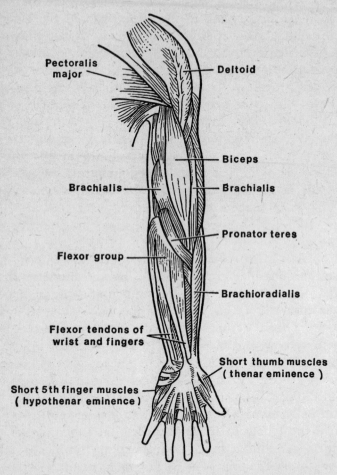

Pectoralis major

Deltoid

Biceps

Brachialis

Brachialis

Pronator teres

Flexor group

Brachioradialis

Flexor tendons of wrist and fingers

Short thumb muscles (thenar eminence)

Short 5th finger muscles (hypothenar eminence)

Fig. 64. Muscles of the left upper limb, anterior aspect.

insertion goes into the olecranon process of the ulna, but some fibres are also attached to the back of the joint capsule and, by pulling it out of the way, save it from being nipped as the joint is extended.

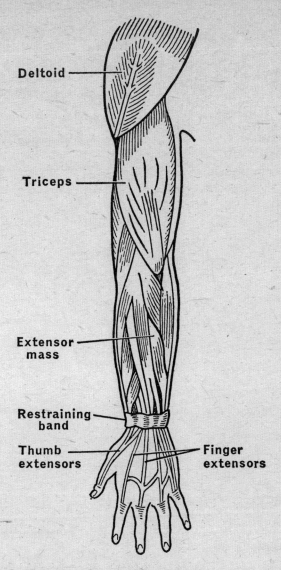

Fig. 65. Muscles of the left upper limb, posterior aspect.

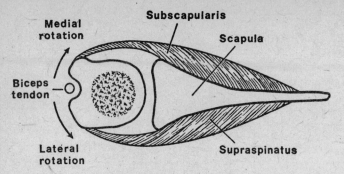

Fig. 66. Cross-section of scapula and head of humerus to show action of short rotator muscles.

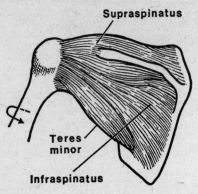

Fig. 67. Outward rotation by the group of short rotator muscles.

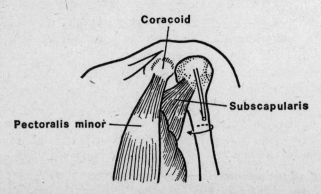

Fig. 68. Inward rotation by the subscapularis muscle.

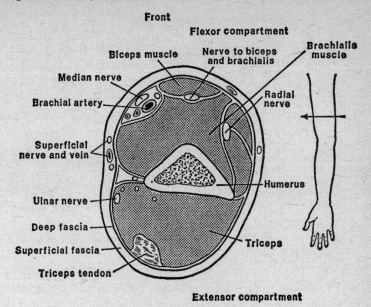

Fig. 69. Cross-section of the upper arm near the elbow.

Forearm

This is similarly divided into flexor and extensor compartments by means of the interosseous membrane between radius and ulna, and each compartment contains a complex group of muscles with an associated nerve trunk.

In the *flexor* compartment are a superficial and a deep group of muscles lying in two planes, with the median nerve, which supplies them, running between the layers. The *superficial* group arises from a common flexor origin on the medial humeral epicondyle. It includes the pronator teres, the main pronator muscle of the forearm, inserted into the middle of the radius; the superficial flexors of the fingers; and the flexors of the wrist joint. The *deep* group arises from the forearm bones and interosseous membrane, and includes the deep flexors of the fingers and the long flexor of the thumb. And the lateral muscular boundary of the front of the forearm is the brachioradialis, which passes

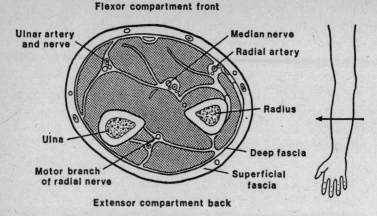

Fig. 70. Cross-section of right forearm, viewed from above.

from the lateral humeral epicondyle to be inserted into the lower radius, acting alternately as pronator and supinator. All these flexor bellies become tendinous 2·5 or 5 cm above the wrist, and those tendons destined for the fingers

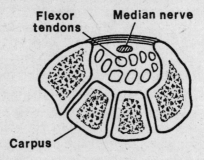

Fig. 71. Cross-section of wrist to show the flexor tendons in the carpal tunnel.

pass into the hand in company with the median nerve under a transverse ligament spanning the arch of the carpal bones, an arrangement known as the carpal tunnel (Fig. 71). It will be seen that each finger, but not the thumb, has two long flexor tendons, a superficial and a deep, and the arrangement of these in the hand will be studied later.

The *extensor* forearm compartment is again arranged in superficial and deep layers of muscles; and, just as the superficial flexor group arises from the front of the medial epicondyle, so the superficial extensor group arises from the back of the lateral epicondyle. And, just as the principal nerve of the flexor compartment is the median, so the main nerve at the back is the radial. The main groups of extensor muscles are:

(1) the extensors of the wrist joint;

(2) the extensors of the fingers (only one tendon to each finger, except the index and fifth, which have two);

(3) the extensors of the joints of the thumb.

All these tendons are bound down by a restraining band of deep fascia at the wrist to prevent them from bowstringing backwards when their muscles contract.

The blood vessels of the arm

The main artery to the upper limb begins in the neck as the *subclavian*. On the right this is a branch from the great innominate artery. On the left, it arises directly from the arch of the aorta in the chest (Plates 9, 10).

As the vessel passes out through the axilla to reach the arm, it is known as the *axillary* artery and in the upper arm proper is termed the *brachial*. Here it lies on the medial side of the biceps with its companion vein, in close relation to the median and ulnar nerves. At the elbow the brachial divides into radial and ulnar branches, which course down the corresponding sides of the forearm; at the wrist the ulnar artery continues into the palm, while the radial turns round on to the back of the carpus. In the palm there are two *arterial arches*, or arcades, formed between branches of the two vessels; from these arches the digital arteries to the fingers are given off. A less important *dorsal arterial arch* is formed on the back of the carpus by the radial vessel alone, with twigs to the fingers (Plate 2).

The *veins* are less clear-cut in their peripheral arrange-

ment than the arteries, forming a diffuse subcutaneous network in hand and forearm, but the main lateral (*cephalic*)
and medial (*basilic*) channels emerge at the elbow. The
cephalic runs up to the shoulder before it dives under the
clavicle to join the subclavian vein; the basilic vein runs
alongside the brachial artery to the axilla, where it continues as the axillary vein and ultimately as the subclavian
in the neck (Fig. 63).

The fine *lymphatic* vessels are arranged in a superficial
network corresponding to the venous pattern, and the main
lymphatic trunks of the limb are gathered up to run with
the brachial artery, entering the main group of lymph
glands of the arm in the axillary space. There is a small
gland-station halfway up the arm, just in front of the medial
epicondyle, the epitrochlear gland (Fig. 44, page 112).

The nerves of the arm (Figs 61, 62; Plates 1, 10(b))

The nerves of the limb are the ultimate branches of a complex *brachial plexus* of spinal nerve roots situated in the lower
neck, behind the clavicle, and in the axilla. From this plexus
emerge the main nerve trunks of the arm—median, radial
and ulnar—which are grouped closely around the axillary
vessels in the lower part of the axilla.

The *radial* nerve soon runs round to the back of the upper
arm, passing spirally round the humerus close to the bone
in its radial groove to re-emerge anteriorly just above the
elbow in the fold between the biceps and the brachioradialis muscles. It then winds back round the neck of the
radius to become the main nerve of the extensor compartment of the forearm. It is mainly a motor nerve, supplying
the triceps, brachioradialis, and extensors of wrist, thumb
and fingers; but it has a small sensory supply to the skin on
the back of the hand between thumb and index, and to the
outer side of the forearm, and this superficial branch runs
with the radial artery in the flexor compartment.

The *median* nerve passes down the arm in company with

the brachial artery and lies in the forearm between the superficial and deep planes of flexor muscles in the anterior compartment. It supplies most of the forearm flexor group and enters the palm by passing under the transverse carpal ligament with the flexor tendons of the fingers. In the hand it gives an important branch to the small thumb muscles in the thenar eminence, and sensory branches, which accompany the digital arteries, to the fingers, supplying the skin of the thumb, index, middle and lateral half of the ring finger on their anterior aspects (Plate 2).

The *ulnar* nerve also runs with the brachial artery in the arm, but it enters the forearm separately by passing behind the medial epicondyle in the ulnar groove. It runs on the inner side of the flexor compartment with the ulnar artery, enters the hand superficial to the transverse carpal ligament and supplies all the small intrinsic hand muscles responsible for the fine co-ordinated movements of the fingers. It also supplies sensation to the fifth and inner half of the ring fingers, on front and back.

The hand

The main *palmar space* of the hand is bounded by the thenar and hypothenar eminences on each side; its floor is the skin of the palm and its roof the metacarpal bones. Through it pass the flexor tendons on their way to the fingers and in it lie the arterial arches of the ulnar artery with their digital branches to the fingers, and the accompanying digital branches of the median and ulnar nerves. Lying between the metacarpals and arising from them are the *interossei*, the small intrinsic muscles of the hand. Their tendons wind round the metacarpal necks to be inserted into the extensor tendon expansion on the back of the proximal phalanges, and they abduct and adduct the fingers, as well as performing certain characteristically human movements described below.

The *dorsum* of the hand is considerably simpler, a shallow

space between the skin and the backs of the metacarpals
traversed by the extensor tendons of the fingers.

In the *fingers* the arrangement of the flexor tendons is
complex. Each digit, save the thumb, has two flexors,

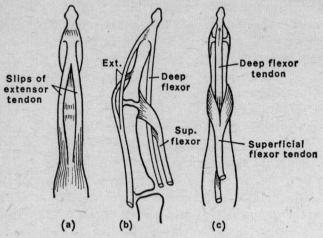

Fig. 72. The interosseous muscles of the hand. (a) Back of finger;
(b) side view; (c) front of finger.

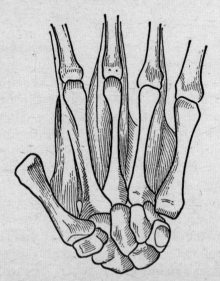

Fig. 73. Tendon arrangements in an individual finger.

superficial and deep, which are inserted into the phalanges as shown in Fig. 73, the deep tendon on its way to the terminal phalanx splitting the other in two. The tendons are facilitated in their gliding by being enclosed in a smooth, double-layered *synovial sheath*, and each sheath in turn is

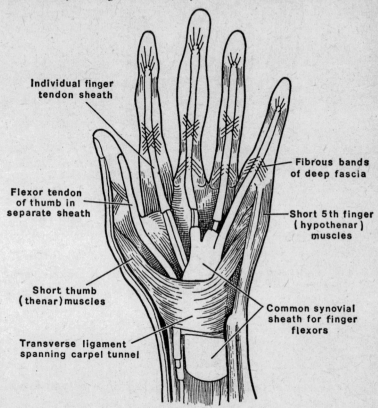

Individual finger
tendon sheath

Fibrous bands
of deep fascia

Flexor tendon
of thumb in
separate sheath

Short 5th finger
(hypothenar)
muscles

Short thumb
(thenar) muscles

Common synovial
sheath for finger
flexors

Transverse ligament
spanning carpel tunnel

Fig. 74. Flexor tendons and their sheaths in wrist and hand. (After *Gray*.)

enclosed in a fibrous tunnel which is attached to the phalanges on each side. The general arrangements of these sheaths is shown in Fig. 74, and it will be seen that each is separate except that of the little finger, which runs down into the palm to expand into a general sheath for all the tendons passing through the carpal tunnel. Because of this

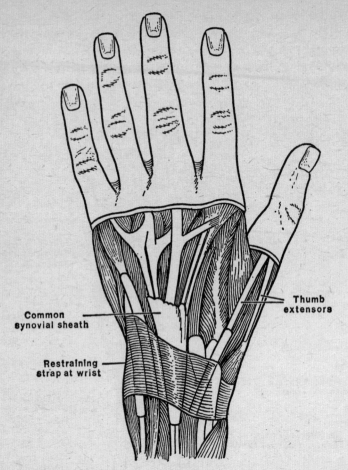

Thumb
extensors

Common
synovial sheath

Restraining
strap at wrist

Fig. 75. Extensor tendons and their sheaths in wrist and hand.
(After *Gray*.)

connection, infections of the little finger are more dangerous
than those of other digits, as they may spread along the
sheath into the hand and forearm instead of remaining confined to the finger.

On the dorsum, the arrangement of extensor tendons is
much simpler; there is a common synovial sheath as they
pass over the wrist, but the individual tendons of the fingers

are without sheaths. Fig. 73 indicates how the tendon of each finger splits up as it is inserted; note the broad extensor expansion which strengthens the back of the capsule of the metacarpophalangeal joint, the expansion that is also the insertion of the intrinsic muscles.

The main features of the *nails* are sufficiently indicated in Fig. 38 (page 101) and Fig. 76.

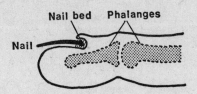

Fig. 76. The fingernail grows forward from the nail bed at its root.

Function of the hand

In the lower animals the digits are used purely for the crudest of gripping purposes, the important muscles being the long flexors and extensors, which alternately contract and relax their hold on an object (Fig. 77(a), (b)).

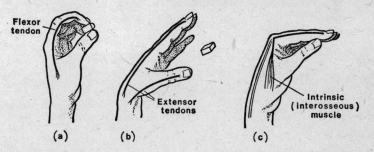

Fig. 77. Movements of the fingers: (a) and (b) crude movements; (c) fine use of the fingers.

Man retains this power but has also developed a fine co-ordinated action of the fingers themselves as distinct from the forearm muscles. This activity has been developed by the evolution of the local intrinsic muscles, confined to the hand, that are required for writing and the use of delicate tools. Their essential action is the adoption of the writing position in the fingers and of the opposition of the thumb to them in a fine grip. This the small muscles achieve by flexing the metacarpophalangeal joints and then extending the interphalangeal joints by pulling on the distal stub of the long extensor tendons (Fig. 77(c)).

8   Regional anatomy: the leg

Bones of the lower limb

Innominate bone

The general arrangement of the pelvic girdle has already been indicated at page 67. Each half of the pelvic ring is made up of one innominate or hip bone. The two bones articulate with each other in the midline anteriorly at the pubic symphysis, the bony prominence felt just above the external genitalia. Behind, each innominate bone articulates with the sacral portion of the spine at the sacroiliac joint, a strong irregular junction allowing very little motion.

The innominate bone itself combines the functions of a proximal support for the limbs and an attachment for the limb muscles with those of a bony container for the pelvic organs; the latter aspect is considered in Chapter 9. Here, we need only note that it consists of three main portions—ilium, ischium and pubis—all centred on the acetabulum, the socket for the femoral head.

The *ilium* is a broad, flaring sheet of bone whose upper border is the iliac crest, which is felt in putting the hand on the hip; and this crest has an anterior and a posterior spine at either end.

The *pubis* has a superior and an inferior strut or ramus, the two rami enclosing a large gap in the bone, the obturator foramen; on the superior ramus is a little knob, the pubic tubercle, which can be felt at the inner end of the fold of the groin.

The *ischium* is the dependent portion of the pelvis carrying the broad ischial tuberosity which takes the body weight in sitting.

The *acetabular socket* of the hip joint is a deep cavity in the

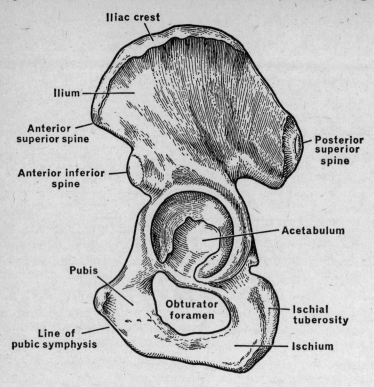

Fig. 78. Left innominate bone, outer aspect.

centre of the external aspect of the bone, with an encircling, upstanding rim shaped like a horse-shoe, deficient at the acetabular notch in front and below.

In life there are important ligamentous and muscular attachments to the innominate bone. The anterior spine of the ilium and the pubic tubercle are spanned in a straight line by the *inguinal ligament* of the groin, an important structure separating thigh from abdomen under which the great vessels and nerves pass into the limb from the trunk. The obturator foramen is bridged across by the *obturator membrane*, pierced by nerve twigs and small vessels, and the acetabular notch is crossed by the *transverse ligament*. The broad outer surface of the ilium is the origin of the gluteal

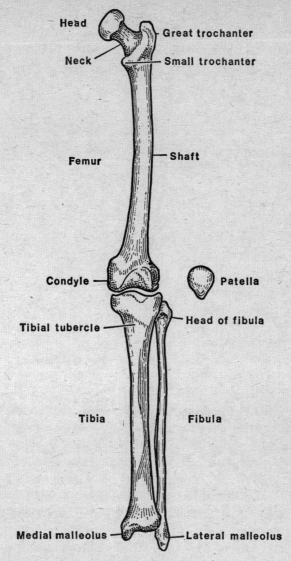

Head

Great trochanter

Neck

Small trochanter

Femur

Shaft

Condyle

Patella

Tibial tubercle

Head of fibula

Tibia

Fibula

Medial malleolus

Lateral malleolus

Fig. 79. Bones of the left upper limb, anterior aspect.

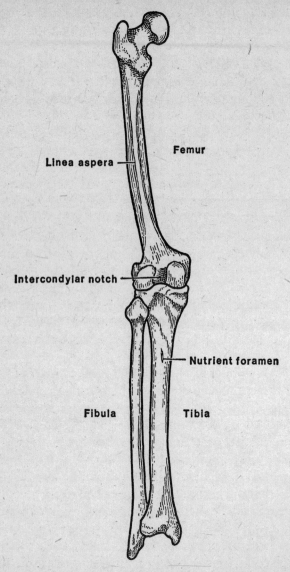

Fig. 80. Bones of the left lower limb, posterior aspect.

muscles of the buttock; the ischium, of the hamstring muscles of the back of the thigh; and the pubis, of the adductor muscles which bring the leg in towards the midline.

Femur

This bone has a long, cylindrical shaft with a normal slight outward and forward bowing, which is grossly exaggerated by the effect of body weight in diseases that soften the skeleton. At its upper end the well-rounded *head*, some two-thirds of a complete sphere, is set well off the shaft at a fairly constant angle of some 130° by the long stout *neck*. And at the base of the neck are the two *trochanters*, both rather more obvious on the posterior aspect of the bone, the great trochanter laterally and the small trochanter medially. The neck acts as a lever for the muscles attached at its base, so that the relative restriction of motion, which is the price of security at the hip as compared with the shoulder, is compensated by an increase in the mechanical advantage of its surrounding muscles. The neck itself is at a constant mechanical disadvantage under the stress of body weight, and its angle with the shaft is diminished by softening disease or by fractures.

The shaft is thickly clothed by muscles, many of which gain an attachment to the great bony ridge running down the length of its posterior aspect—the *linea aspera*. Not far below the neck, the posterior surface is marked by the gluteal tuberosity, the insertion of the great gluteus maximus muscle of the buttock. A few inches above the knee, the shaft expands in triangular fashion, ending in two massive condyles, lateral and medial, separated by a deep intercondylar notch; the medial is the more prominent and is set at a slight angle to the line of the shaft, as seen from below (Fig. 81). The lower end of the femur has a smooth surface above the condyles in front for articulation with the patella, and a broad trigone or popliteal surface posteriorly facing back into the *popliteal fossa*, the hollow at the back of the knee. There is a small epicondyle on the free surface of each

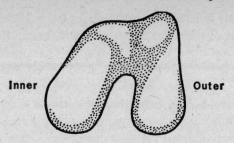

Fig. 81. Condyles of left femur, end-on view.

condyle, and the adductor tubercle at the summit of the
medial condyle is the lowest attachment of the adductor
group of muscles.

The *patella* or knee-cap is a sesamoid bone, i.e. one
situated entirely within tendon, in this case the tendon of the
great quadriceps muscle which lies in front of the thigh and
extends the knee joint. The bone is roughly triangular—
apex downwards—and has an upper and lower pole, and
anterior and posterior surfaces. The posterior surface
articulates with the smooth surface of the femur just above
the origin of the condyles. The quadriceps tendon is inserted into the upper pole, and the muscle pull is transmitted to the patellar ligament, which connects the lower
pole to the tubercle of the tibia.

Tibia and fibula

These are the paired bones of the leg, on the medial and
lateral sides of the limb respectively. They articulate with
each other at the superior (proximal) and inferior (distal)
tibiofibular joints; and the tibia articulates with the femur
at the knee and with the talus bone of the tarsus at the
ankle. The fibula also participates in the ankle, but is excluded from the knee joint. Each bone has an opposing
interosseous border, sharp margins connected by the interosseus membrane in life.

The *tibia* has a long shaft, triangular in cross-section, with
medial, lateral and posterior surfaces, and medial, lateral

and anterior borders. The anterior border is the sharp subcutaneous crest that can be felt as the shin from knee to
ankle, and the flat medial surface is directly beneath the
skin on the inner side of the leg. The upper end of the bone
is expanded as two broad, flat masses, the medial and
lateral condyles, which articulate with those of the femur,
the semilunar cartilages intervening (see page 162). Looking down on this upper aspect of the bone, the smooth
condylar surfaces are seen to be separated by a rough intercondylar area carrying the projecting tibial spine. At the
upper end of the tibial crest, just below the condyles, is the
prominent tibial tubercle, the ultimate insertion of the
extensor apparatus of the knee joint. The lower end of the
tibia is broadened to a much smaller degree than the upper
and has a downwardly projecting process, the medial
malleolus, overhanging the inner aspect of the ankle.

The slender *fibula* transmits little, if any, of the body
weight. Its shaft is polygonal in section, with numerous ridges
for muscle attachments, and the head at the upper end
articulates with the side of the outer tibial condyle. The
head carries at its apex a pointed styloid process, the insertion of the biceps muscle of the thigh and the attachment of
the lateral ligament of the knee. The lower end of the bone
forms the lateral malleolus, which overhangs the outer side
of the ankle at a considerably lower level than the medial
malleolus of the tibia.

Foot

The bones of the foot fall into three groups, corresponding
with those of the hand: the *tarsals*, the *metatarsals* and the
phalanges of the toes. But in general, and particularly in the
tarsus, they are much stronger and thicker than those of the
upper limb. The *talus* lies immediately beneath the long
bones, articulating with them at the ankle, and itself rests
on the *calcaneus* or heel bone. But the talus so inclines
downwards and medially to the inner side of the foot that
the anterior ends of talus and calcaneus lie almost side by

side, forming the proximal row of the tarsus. The distal row consists of three *cuneiforms* on the inner side of the foot and the *cuboid* laterally, and sandwiched between the two rows on the inner side of the foot is the *navicular*, separating talus from cuneiforms.

These bones need not be discussed in great detail. The *talus* has a *body*, with a rounded superior surface facing

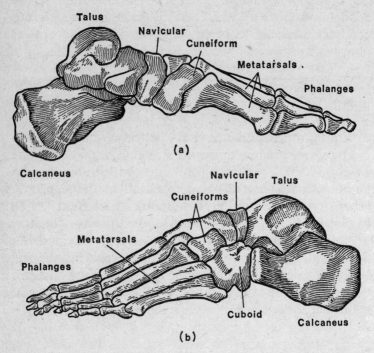

Fig. 82. The bones of the left foot, (a) inner side, (b) outer side.

upwards at the ankle articulation, and a *neck* carrying the globular *head*, which fits into the socket of the navicular. The *calcaneus* has a broad, projecting posterior *tuberosity*, the prominence of the heel, set at an angle to the main body of the bone. It articulates anteriorly with the back of the cuboid. Talus and calcaneus are connected by a very strong interosseous ligament.

The *cuboid* and *cuneiforms* are irregular bony masses articulating with the bases of the metatarsals. And the latter are essentially similar to the metacarpals of the hand except that, because of the relative lack of mobility of the great toe, the first metatarsal lies closely parallel to the others. It is considerably stouter than the rest and its head is supported underneath by a pair of tiny sesamoid bones. Note that the long axis of the foot, to which adduction and abduction of the toes are referred, is the second metatarsal, not the middle bone as in the hand.

Joints of the lower limb

Hip joint

The deeply embedded head of the femur forms the hip joint by articulation with the acetabulum, the socket in the lateral surface of the innominate bone. As in the case of the shoulder, this socket is deepened by a fibrocartilaginous rim or *labrum* round its periphery, and the acetabulum is so deep that its most central portion is actually out of contact with the femoral head and is not lined with cartilage but filled in by a fatty pad. In the centre of the spherical femoral head is a small crater-like depression, the *fovea centralis*, opposed to this non-articulating region of the acetabulum. And from the fovea a round cord, the *ligamentum teres*, runs to be attached to the margins of the acetabular notch; it carries blood vessels which help to nourish the head of the femur.

The capsule of the hip is extremely strong and thick, a blending of bands from each portion of the innominate bone—iliofemoral, pubofemoral and ischiofemoral ligaments. It is attached proximally around the acetabular margin and distally it reaches right down to the base of the femoral neck in the trochanteric region, i.e. the whole of the neck is inside the capsule. The synovial membrane lining the deep aspect of the capsule is reflected upwards at the

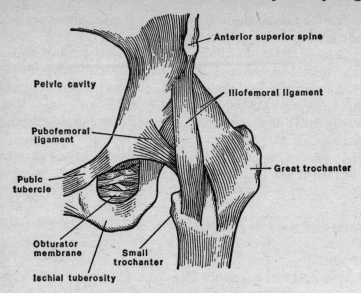

Fig. 83. Left hip joint, anterior aspect.

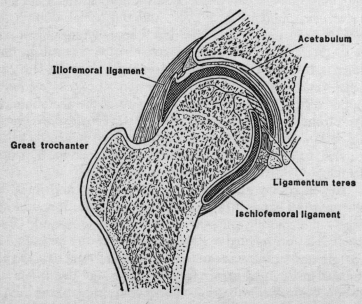

Fig. 84. Hip joint, longitudinal section. (After *Gray*.)

base of the neck, clothing the latter as far as the margins of
the head.

Movements at the hip have already been outlined at page
63 and correspond to those of the shoulder. They are:

(1) *abduction* and *adduction* in the coronal plane away
from and towards the midline;

(2) *flexion* and *extension* in the sagittal plane, in front of
and behind the trunk;

(3) *lateral* and *medial rotation*, a rolling movement of the
thigh on its long axis.

The joint is liberally supplied with nerves, some of which
also serve the knee, and it is for this reason that pain from
hip disease is often referred to the knee and thought to be
arising from that joint. The common 'fracture of the hip'
in the elderly, still too often a crippling affection, is a fracture through the femoral neck, and its bad reputation for
healing is due to the poor blood supply reaching the frac-
tured surfaces via the synovial reflections and the ligamentum teres.

Knee joint
This joint is very complex, with several communicating
compartments; note that the fibula plays no part in the
articulation, unlike the radius at the elbow. The main parts
are:

(1) the *tibiofemoral* joint between the tibial and femoral
condyle of each side, a semilunar cartilage intervening;

(2) the secondary *patellofemoral* articulation between back
of patella and trochlear surface of femur.

The knee is essentially a hinge joint, with flexion and
extension as its principal movements, the strong collateral
ligaments resisting any sideways strain. But, in addition,
there is a small element of rotation, a screwing motion that
locks the joint home as it is being finally straightened out,
so that in the fully extended position in standing it is very
stable compared with the tendency to 'give' when flexed
only a few degrees. One important feature is the manner in

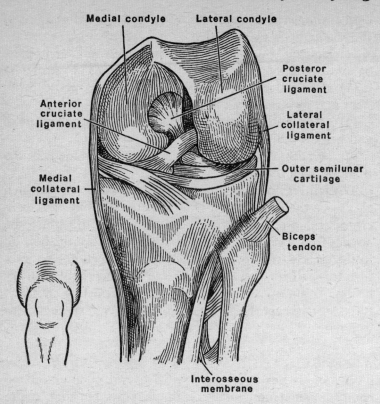

Fig. 85. Left knee joint, anterior aspect. The joint is bent at a right angle, and the front of the capsule has been removed. (After *Gray*.)

which the loose and extensive anterior portion of the capsule is strengthened by the expansion of the tendon of insertion of the quadriceps muscle of the thigh as it passes down in front of the joint of the tibial tubercle. There is an intimate blending of tendon and capsule, with the result that the tonus of the muscle is responsible for guarding the joint against strain, keeping the capsule taut and preventing the accumulation of any effusion in the joint which would distend the capsule. To a much lesser extent the hamstring muscles posteriorly serve the same purpose.

This muscle-joint interrelation is of the greatest importance, as muscle wasting leads to instability; and, *vice versa*, any internal derangement of the knee, such as a torn semi-

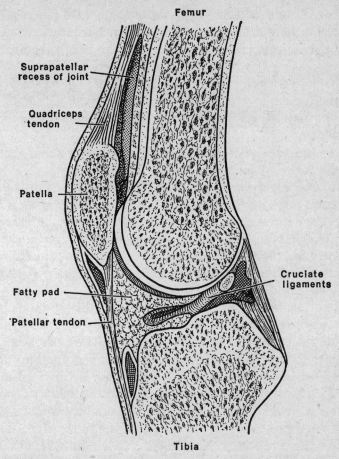

Femur

Suprapatellar
recess of joint

Quadriceps
tendon

Patella

Fatty pad

Patellar tendon

Cruciate
ligaments

Tibia

Fig. 86. The knee joint, longitudinal section. (After *Gray.*)

lunar cartilage, causes a reflex wasting of the thigh which predisposes to still further derangement. This vicious circle must always be broken by exercises to re-educate the quadriceps in bracing the joint straight before any improvement can be expected.

The collateral ligaments and the posterior capsule are fairly taut; but the anterior capsule is loose and extensive, especially where it spreads up, around and above the patella as the suprapatellar recess in front of the lower part of the femur, a pouch which is necessarily baggy to allow freedom in full flexion of the joint. When the knee is fully extended, the patella articulates with the femur above the condyles; in increasing flexion it lies against portions of both condyles, and corresponding facets are marked out on the back of the bone.

The *synovial cavity* of the knee is obviously complex, and includes:

(1) the *patellofemoral space*, with the suprapatellar recess in the highest part of the anterior compartment;

(2) the main *anterior joint cavity* between tibia and femur on each side;

(3) the *intercondylar portion*, the tunnel between the femoral condyles, traversed by the X-like crossing of the cruciate ligaments which tie each femoral condyle to the tibia (Fig. 86);

(4) the *postcondylar space* on each side, a loose synovial pouch behind each femoral condyle.

The *semilunar cartilages* are concentric discs intervening between the tibial and femoral condyles, and fixed to the deep aspect of the capsule (Fig. 87). They are wedge-shaped in cross-section, projecting for 1–2 cm into the centre of the joint, with a thin free edge. Their front and back ends are called the anterior and posterior horns; those of the more elliptical medial cartilage embrace those of the outer margins to the upper surface of the tibia, with which they move and rotate. Nevertheless, they also have, via the capsule, an attachment to the femur, and this double fixation is responsible for the ease with which a cartilage, usually the medial, is torn or dislocated when the joint is under stress and its bony components are being distracted.

The *tibiofibular* joints include:

(1) the superior tibiofibular articulation, a simple plane circular lateral disc. The two cartilages cushion the contact

of the bony surfaces and are firmly bound down at their joint formed by the fibular head abutting against the side of the lateral tibial condyle;

(2) the connection of the shafts by means of the interosseous membrane;

(3) the inferior tibiofibular joint, a firm fibrous union.

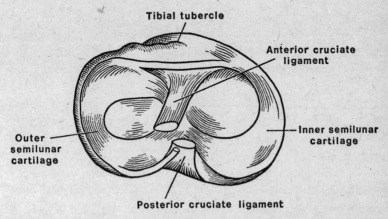

Fig. 87. Semilunar cartilages and cruciate ligaments of the knee joint. The view is of the upper surface of the left tibia, the femur having been removed.

Little or no motion occurs at any of these; the fibula hardly moves on the tibia and is almost entirely bypassed in the transmission of body weight, which is taken mainly in the line of the tibia.

Ankle joint

This is formed between the upper surface of the talus and the lower ends of tibia and fibula; note that the latter bone is not excluded from the ankle as the ulna is from the wrist. The malleoli, overhanging the talus on each side, form a deep mortice in which that small bone is deeply wedged; there are strong collateral ligaments, and the only movements are those of plantarflexion (flexion) and dorsiflexion (extension), with a neutral position at right angles.

The *tarsal joints* form a complex system of interconnecting articulations, which is best considered from the point of view of the movements that actually take place.

Movements of the foot

The foot as a whole may be *inverted* or *everted*, i.e. the sole turned inwards or outwards, and this movement is best understood by considering it as a bony block moving *en*

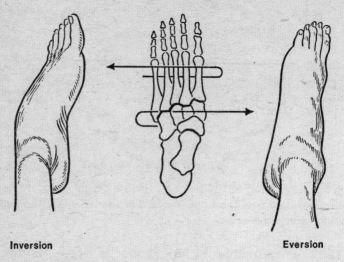

Inversion   **Eversion**

Fig. 88. Inversion and eversion of the foot. The arrows merely indicate the direction of motion; both movements occur at the same system of joints.

masse at the subtaloid joint (the joint between talus and calcaneus) and the talonavicular ball-and-socket, with the talus acting as a stationary pivot (Fig. 88).

In addition, the forefoot—metatarsals and toes—may be *adducted* or *abducted*, i.e. deviated medially or laterally, but keeping the sole parallel to the ground; this occurs at the midtarsal joint, which traverses the foot and is made up of the talonavicular and the calcaneocuboid articulations (Fig. 89).

In actual practice, it is not possible to separate all these movements; inversion of the sole is always accompanied by some adduction of the forefoot and eversion by some abduction, the movements of the subtaloid and the midtarsal joints being reciprocal. And it is difficult at first to

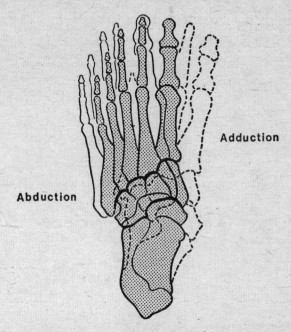

Fig. 89. Abduction and adduction of the forefoot. The shaded position is neutral, the outlines indicate the extremes of range.

realize that none of these rotary movements is occurring at the ankle; inversion in particular appears to be taking place at ankle level. In fact, they all occur well below the level of the ankle, which is capable only of simple up and down movement.

The remaining tarsal joints are small plane articulations of an accessory nature; and the joints of the toes are similar to those of the fingers, with the slight difference that the central axis of reference for abduction and adduction is

the second metatarsal and toe, not the third as in the
hand.

The arches of the foot
These are longitudinal and transverse. The *longitudinal
arches* lie in the long axis of the foot, with a higher medial
long arch and a lower lateral long arch, both having a common posterior pillar in the calcaneus. But they diverge

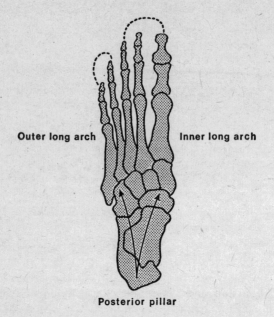

Fig. 90. Longitudinal arches of the foot.

anteriorly. The line of the medial arch is calcaneus-talus-
navicular-cuneiforms-inner three metatarsals. The line of
the outer is calcaneus-cuboid-outer two metatarsals, this
arch being much the flatter. The talus thus lies at the summit of the long arches of the foot, so that the body weight
transmitted to it down the leg is constantly tending to
flatten them out—and would do so were it not for the
supporting ligaments and muscles.

The *transverse arch* is the natural downward side-to-side concavity seen in cross-section and is most marked in the region of the bases of the metatarsals.

The long arches are not present at birth but develop with activity in the first eighteen months; the transverse arch is already present in the foetus. A great deal has been written in the past about the importance of the arches to the efficient functioning of the foot that is no longer altogether accepted. 'Flat foot' in itself is of little importance; what

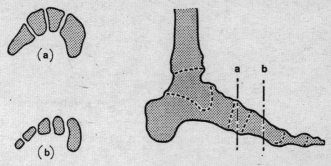

Fig. 91. Cross-section of the transverse arch of the foot, (a) in the tarsal, (b) in the metatarsal region.

matters is that the foot should be supple and capable of assuming the arched position voluntarily when relieved of body weight. This is the state of affairs in children and ballet dancers, who have excellent function although their feet may be very flat. It is rigidity that is painful and disabling, whether in the arched or the flat position, and some of the most troublesome feet are the very highly arched variety that are just beginning to break down.

What are the supports of these arches? They are two: the binding ligaments, and the tone of certain muscles of the calf and the sole; the latter is the more important. To a certain extent also the actual shape of the bones contributes to the maintenance of a normal arch. The *ligaments* of greatest importance are, of course, those on the concave underside of the foot, the ties of the arches—either short bonds between individual bones or longer structures, running from one pillar of the arch to the other.

The tendon of the *tibialis anterior* muscle runs down from the front of the leg to pull on the medial cuneiform and thus actively maintain the long arch; the tendon of the *tibialis*

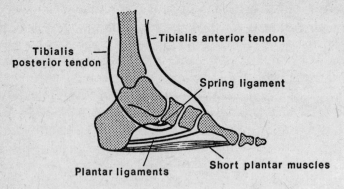

Fig. 92. Ligaments and tendons supporting the longitudinal arches of the foot.

posterior of the calf curves round behind the ankle and sustains the head of the talus from below as it passes to its insertion in the navicular. And the *peroneus longus* tendon from the outer side of the leg crosses the sole of the foot transversely from lateral to medial side, bracing the transverse arch.

Finally, the muscular mass of the sole, the short plantar muscles that are attached behind to the underside of the calcaneus and spread forwards to the toes, exert a constant tonic support for both long arches.

It is important to note that the ligaments are never intended to take the entire strain of body weight, except perhaps momentarily; they are themselves guarded by the tone of those calf muscles whose tendons have been mentioned above as supports for the arches. If this tonus is weakened by fatigue or paralysis, the ligaments are overstretched and an acute foot strain results, the first stage of a chronic flat foot.

Gait and the mechanism of the foot

The arches are spring-like structures, yielding slightly under body weight in walking and then exhibiting an elastic recoil; indeed, the principal ligament, the short plantar band between calcaneus and navicular which directly supports the head of the talus, is known as the *spring ligament*. In taking a step, the weight is first planted on the heel, then transmitted rapidly along the outer border of the foot and finally across the transverse arch to be borne by the first metatarsal head; and flexion of the metatarsophalangeal joints gives a propulsive kick-off onto the other foot. The remarks on page 149 about primitive and fine movements of the fingers also apply, though with less force, to the toes. In walking on rough pebbly ground, particularly with naked feet, the long flexors of the toes act as powerful gripping agents to steady the balance. But on smooth pavements and in stout shoes it is necessary to bring the toes flat down on the bearing surface to give an efficient thrust, and this is accomplished by the intrinsic muscles. Should this intrinsic control be lost, the unopposed long flexors and extensors produce clawing of the toes, which are hyperextended at the metatarsophalangeal

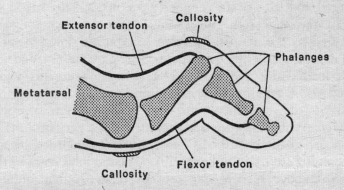

Fig. 93. The clawed toe resulting from loss of intrinsic control.

and flexed at the interphalangeal joints. This exposes the
metatarsal heads in the sole; walking gives the painful
sensation of treading on stones, and callosities form both in
the sole and on the dorsum of the toes over the prominences
(Fig. 93).

Surface anatomy of the leg

Bearing in mind the underlying bony structures, we can now
review the surface anatomy of the lower limb.

Anterior aspect

The deep fold of the groin marks the transition between
abdominal wall and thigh; pressure in this fold reveals the
resistance of the underlying inguinal ligament, which can be
traced laterally to the anterior superior spine of the ilium
and medially to the small, knobby pubic tubercle; the
latter is often obscured in men by the overlying spermatic
cord running down to the testicle.

Immediately distal to the ligament is the *inguinal region*,
which the main vessels and nerves of the limb have entered
from the abdomen; the pulsation of the femoral artery can
be felt just below the midpoint of the ligament, and the
femoral nerve rolled under the finger a little more laterally.
In this region also are numerous lymph glands, often enlarged and felt as soft, rounded structures (Plate 7(b)).

In the *thigh* itself the main anterior muscle bulge is
formed by the quadriceps femoris, tapering down towards
its insertion into the upper pole of the patella. The mass on
the medial side is the adductor group of muscles, whose
tendons of origin can be felt running up to their attachment to the pubic bone, just lateral to the external genital
organs. When the knee is braced straight, a very firm resistance is felt under the skin on the outer aspect of the thigh,
extending from hip to knee; this is the iliotibial band of the
deep fascia, which is kept taut by certain of the buttock
muscles and helps in maintenance of the erect posture.

At the front of the *knee* the patella is seen, and in line with it, about 5 cm below, is the tibial tubercle; the patellar tendon connecting the two and transmitting the quadriceps pull becomes obvious when the knee is extended, and a hollow on either side of the tendon indicates where the anterior joint compartment lies immediately beneath the skin. The tibial condyles are easily seen or felt and the head of the fibula on the outer aspect; just below the head, the peroneal nerve winds round the neck of the fibula, over which it can be rolled, giving rise to a tingling sensation in the outer side of the calf.

The *tibial crest*, its anterior border or shin, is subcutaneous from tibial tubercle to ankle, curving slightly inwards as it passes down, with the flat subcutaneous medial surface on its inner side. On the outer side of the crest is the bulge of the muscles in the anterior compartment of the leg, and more laterally still the peroneal muscles obscure the underlying fibula, which emerges under the skin a few centimetres above the ankle.

At the *ankle* the two malleoli stand out on either side, the medial malleolus at a higher level; and 2·5 cm behind the latter can be felt the pulsation of the posterior tibial artery.

The *navicular* often makes a prominence on the medial border of the foot, as does the base of the fifth metatarsal on the outer side, and the joint between the medial cuneiform and the base of the first metatarsal is often a troublesome dorsal prominence under the shoe.

Posterior aspect

The great muscle masses at the back of the limb are, from above downward: the gluteal muscles of the buttock, the hamstrings in the thigh and the calf muscles of the leg.

The *buttock* has a rounded *gluteal* fold below, marking its junction with the thigh, and as this is followed round to its lateral extremity there is felt under the skin the resistance of the great trochanter. Deep in the lowest point of the buttock is felt the apex of the ischial tuberosity.

At the back of the *knee* the tendons of the hamstrings diverge; they can be seen or felt as the biceps passes down and out to the head of the fibula, the semimembranosus and semitendinosus down and into the tibia. These tendons make the upper boundaries of the diamond-shaped space behind the knee joint, the *popliteal fossa*, in which lie the great vessels, which have travelled round from the front of the thigh in its lower third, as well as the sciatic nerve and its divisions. It may be possible to feel these structures by deep palpation, though the course of the sciatic nerve in the thigh itself is entirely protected by the overlying muscle.

The bulk of the *calf* is made up of the triceps, i.e. the gastrocnemius and soleus muscles, which taper down as they joint to form the Achilles tendon, or heel cord, passing down the back of the ankle to the tuberosity of the calcaneus.

The surface markings of the main vessels and nerves are indicated in Plates 3 and 4. Note the spiral course of the *femoral artery* and *vein* round the inside of the femur to reach the back of the limb in the popliteal fossa; and the straight course of the *sciatic nerve* down the middle of the back of the thigh to divide in two a few centimetres above the knee, the lateral peroneal division winding round the neck of the fibula to reach the front of the leg.

Subcutaneous structures of the leg

Fig. 94 displays the flayed limb, with the superficial structures that intervene between the skin and the envelope of deep fascia. The *venous network* is gathered up into two main channels. The *great saphenous vein* runs upwards in front of the medial malleolus, along the medial border of the leg and thigh, and finally turns round to the front of the inguinal region, where it passes through an oval window in the deep fascia to join the main *femoral vein*. The *small saphenous vein* is formed on the lateral side of the foot, passes upwards behind the lateral malleolus and travels up the midline of

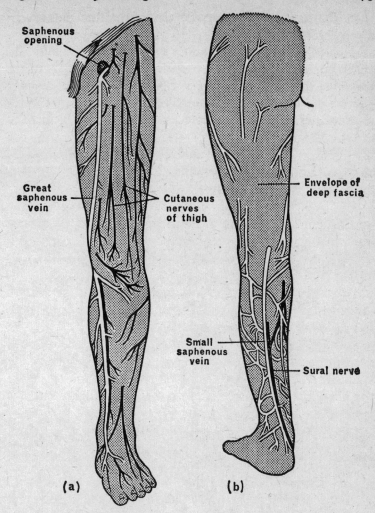

Fig. 94. The flayed lower limb, (a) anterior, (b) posterior.

the back of the calf as far as the knee, where it pierces the deep fascia to join the *popliteal vein*.

Both veins are accompanied by lymphatic vessels; there are a few small popliteal lymph glands at the termination of the small saphenous vein and numerous large inguinal glands surrounding the upper part of the main saphenous trunk at the groin.

The *cutaneous nerves* of the limb are many. The front of the thigh is supplied by lateral, intermediate and medial branches from the *femoral* nerve; the back by a small sciatic branch of the main *sciatic* trunk. The *saphenous* branch of the femoral nerve runs the whole length of the limb, in com-

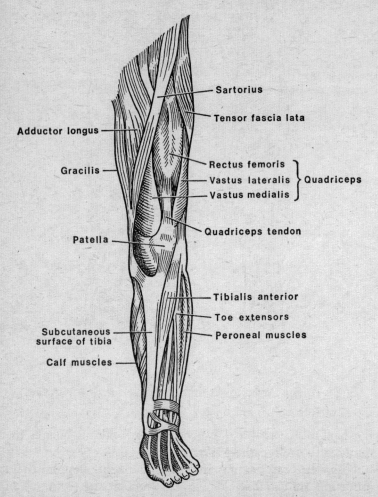

Fig. 95. Muscles of the left lower limb, anterior aspect.

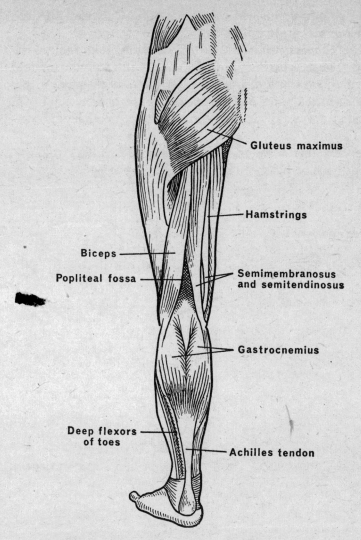

Fig. 96. Muscles of the left lower limb, posterior aspect.

pany with the great saphenous vein, and the *sural* branch of the sciatic accompanies the small saphenous vein to the outer side of the foot.

The main subcutaneous *bursae* in the leg are:

(1) A bursa between skin and ischial tuberosity to relieve pressure in sitting.

(2) A bursa over the lateral aspect of the great trochanter.

(3) The prepatellar bursa in front of the patellar tendon and the lower half of the patella. This is not a part of the knee joint; it takes a considerable share of the weight in kneeling and its distension is the cause of ordinary 'housemaid's knee'.

(4) A bursa between the skin and the Achilles tendon—when the shoe has been rubbing against the heel.

The muscles of the upper part of the leg

These consist of the following main groups:

(1) *Muscles connecting trunk and femur:* iliacus and psoas (Fig. 97). These muscles are responsible·for flexing the hip

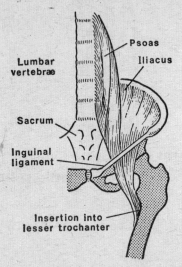

Fig. 97. Iliopsoas group of muscles, left side.

joint. The *iliacus* is a broad sheet arising from the inner (pelvic) surface of the iliac portion of the innominate bone, while the *psoas* is a long belly at the back of the abdominal cavity alongside the lumbar vertebrae, from which it arises.

The two muscles form a conjoined *iliopsoas tendon*, which enters the thigh by passing under the inguinal ligament lateral to the femoral vessels and so to its insertion at the lesser trochanter.

(2) *Muscles connecting pelvis and femur.* There are two main groups here: the gluteals and the adductors.

The *gluteal* muscles of the buttock arise from the outer surface of the ilium and from the back of the sacrum. The most superficial, covering all the others, is the great

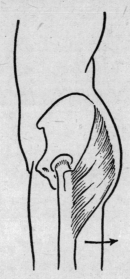

Fig. 98. Action of gluteus maximus.

gluteus maximus, the rump, which is inserted partly into the gluteal tuberosity of the femur and partly into the iliotibial band of deep fascia. It is one of the most important postural muscles, for it extends the hip, carries the leg backwards in walking and braces the whole limb by tightening up the deep fascia.

Under cover of it are the smaller *gluteus medius* and *minimus*, which are the main abductors of the hip away from the midline. These are again of the highest postural importance in that they make it possible to stand on one leg by

pulling the pelvis over in line with the weight-bearing
limb and they resist the tendency of the trunk to fall to the
other side. They thus make walking possible, for this is no
more than an alternate standing on either leg (Fig. 99).

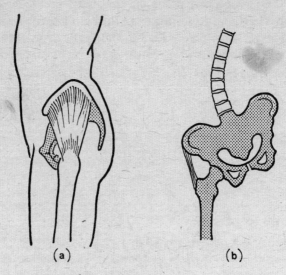

Fig. 99. (a) The gluteus medius. (b) The stabilizing action of the
gluteus medius in standing on one leg.

The *adductors*—longus, brevis and magnus—lie on the
inner side of the thigh. They arise by tendons from the pubic
tubercle, pubis and ischium, running downwards and out
to be inserted into the shaft of the femur as far down as the
adductor tubercle. And their action is clearly to adduct the
femur to the midline. The three muscles are arranged in
layers, from before backwards as named, and are supplied
by the *obturator nerve*, which enters this region of the thigh
from within the pelvis after piercing the obturator membrane. One strap-like member of this group, the *gracilis*,
reaches as far as the tibia.

(3) *The short rotators of the hip*, analogous to those of the
shoulder, are situated very deeply in the buttock, in close
relation to the sciatic nerve, and are seen only after re-

moval of all the overlying gluteal muscles. They arise from the sacrum and obturator membrane, insert into the base of the femoral neck and laterally rotate the hip.

(4) The *quadriceps femoris* is the great mass on the front of the thigh responsible for extension of the knee. Three heads arise from the femur—the *vastus lateralis, intermedius*

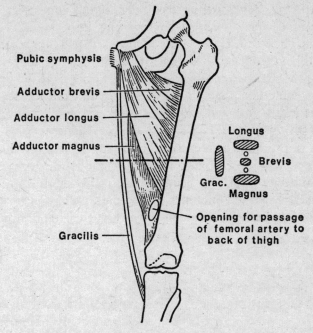

Pubic symphysis

Adductor brevis

Adductor longus

Adductor magnus

Longus

Brevis

Grac.

Magnus

Gracilis

Opening for passage of femoral artery to back of thigh

Fig. 100. Adductor muscles of the left thigh. Their disposition is also shown in cross-section, with the two branches of the obturator nerve sandwiched between the layers.

and *medialis*—and the fourth is the *rectus femoris* from the ilium just above the acetabulum. We have seen that there is a common quadriceps tendon emerging in the lower thigh and inserted into the upper pole of the patella; its expansions on either side of the patella blend with, and greatly strengthen, the capsule of the knee joint.

(5) The *sartorius* is a long strap-like muscle arising from

the anterior superior spine of the ilium and traversing the length of the front of the thigh very superficially in a downwards and inwards direction to its insertion in the upper tibia on its subcutaneous aspect. It helps to flex both hip and knee, and to produce the movement of sitting with crossed legs—the 'tailor's position', after which it is named.

(6) The *hamstrings* are the bulk of the back of the thigh, a powerful group arising from the ischial tuberosity and destined to flex the knee by inserting into the tibia and fibula. These muscles are the *semimembranosus*, *semitendinosus* and the *biceps* which diverge in their passage down the thigh; the first two pass inwards to the upper tibia, while the biceps, which acquires an additional head from the back of the femur, is inserted on the outside of the knee into the head of the fibula. This Λ-shaped arrangement of the tendons produces the upper boundary of the ◇-shaped popliteal fossa at the back of the knee, the lower sides of the diamond being the two heads of the gastrocnemius muscle. The sciatic nerve, which lies in the thigh under cover of the hamstrings, emerges between the fork of their tendons to lie

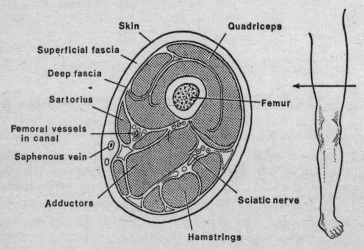

Fig. 101. Cross-section of the left thigh, viewed from below, to show the muscle masses.

comparatively superficially at the apex of the popliteal region. The space is roofed over by deep fascia and has the popliteal surface of the femur as its bony floor.

The disposition of all these muscle groups in the thigh is clearly shown in cross-section (Figs 101, 102). The intermuscular septa of deep fascia mark off three compartments, each containing a muscle group with its supplying nerve,

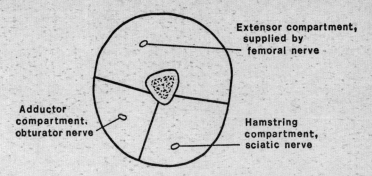

Fig. 102. Key to Fig. 101. The three compartments of the thigh, with their nerves.

which also innervates the overlying skin. The anterior or extensor compartment contains the *quadriceps*, supplied by the femoral nerve; the medial contains the *adductors*, supplied by the obturator nerve; and the posterior the *hamstrings*, supplied by the sciatic.

Note how the sartorius in its diagonal course is so applied to the side of the vastus medialis as to produce a tunnel-like space; this is *Hunter's canal*, and through it pass the femoral vessels on their journey round the femur to emerge in the popliteal fossa behind, where they are then called the popliteal artery and vein.

The muscles of the lower leg

Cross-section of the leg reveals the main compartments and muscle groups as follows. The interosseous membrane between the bones separates a main *anterior compartment*, containing the muscles that extend the ankle and toes, from the *posterior compartment* for the calf muscles; and there is a small separate *lateral compartment* on the outer side of the fibula for

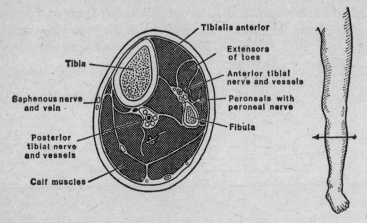

Fig. 103. Cross-section of left lower leg, seen from below, showing the main muscle masses.

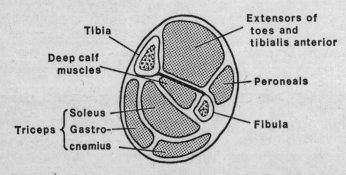

Fig. 104. Key to Fig. 103. Compartments of the lower leg.

the peroneal muscles which evert the foot. Finally, the main posterior compartment is subdivided into a *superficial* portion for the gastrocnemius and soleus (together forming the triceps), and a *deep* portion for the long flexors of the toes and the tibialis posterior, which inverts the foot.

The anterior group: tibialis anterior, extensor hallucis longus, extensor digitorum longus

These are all extensor muscles, supplied by the anterior tibial division of the peroneal nerve. Their tendons run down in front of the ankle, where they are bound down by the retinacular bands of deep fascia. The tibialis anterior, on the medial side, is inserted into the first metatarso-cuneiform junction, where it helps to maintain the inner long arch (Fig. 92, page 170) and dorsiflexes the foot. The extensors of the toes and of the great toe (hallux) traverse the dorsum of the foot to be inserted in the same way as the corresponding tendons in the hand—with this difference, that in the foot there is a duplicate set of extensor tendons arising from a short extensor muscle situated on the dorsum of the foot itself.

The *peroneal* muscles, longus and brevis, evert the foot and act incidentally as dorsiflexors of the ankle; they are supplied by the peroneal nerve. P. brevis has a short course to the base of the fifth metatarsal (the styloid process of that bone), whereas p. longus dives into the sole at that point and runs transversely to reach the base of the first metatarsal, so helping to maintain the transverse arch. Both tendons pass round the back of the lateral malleolus to reach the foot, and they have a constant tendency to slip forward round the edge of that bone which is resisted by a special peroneal retinaculum.

The posterior group

(1) *Superficial layer: triceps.* The triceps is the great calf muscle acting to flex (plantarflex) the foot through its insertion, via the Achilles tendon, into the tuberosity of the calcaneus. It comprises the superficial *gastrocnemius*, with a head arising from the back of each femoral condyle and forming the lower V-shaped boundary of the popliteal fossa, and the underlying *soleus* arising from the back of the upper

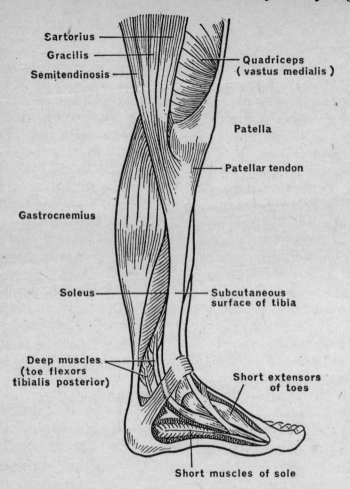

Fig. 105. Muscles of left lower leg and foot, inner aspect.

tibia and fibula. These form the Achilles tendon halfway down the calf, and the tendon is separated from the upper part of the back of the calcaneus by a deep bursa.

(2) *Deep layer: tibialis posterior, long flexors of the toes*. These arise from the back of the tibia and fibula and the connecting interosseous membrane. The emerging tendons enter the sole of the foot by passing behind the ankle joint and the

medial malleolus. The long flexors are attached to the toes, as in the case of the similar flexor digitorum profundus in the arm. But whereas the fingers have two sets of flexor tendons, both arising in the forearm, those of the calf correspond only to the profundus, and the duplicate tendons are

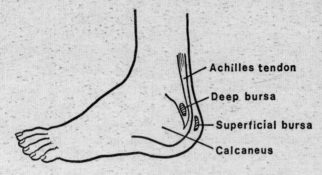

Fig. 106. Achilles tendon and related bursae.

supplied by a short flexor muscle which is confined to the sole of the foot. The tibialis posterior is inserted into the navicular and sustains the inner long arch, taking some weight off the spring ligament by supporting the head of the talus; it is a powerful invertor of the foot as well as a plantarflexor of the ankle.

The foot
The general arrangement of the foot and toes, the flexor and extensor tendons, and their synovial sheaths, etc., resembles that of the hand but with local differences, due to the complete duplication of the long flexor and extensor tendons by short extensors and flexors situated in the dorsum and sole of the foot itself, and arising from the upper and undersurfaces of the calcaneus.

The deep plantar fascia of the sole is very dense and covers a complex pattern of small plantar muscles arranged in layers.

The blood vessels of the leg (Plates 3, 4, 7(b))

The *femoral artery* and *vein* are the downward continuation of the external iliac vessels from within the pelvis. They enter the thigh by passing beneath the inguinal ligament, carrying with them a femoral sheath prolonged from the deep fascia of the abdomen, and lie in the groin with the vein medial to the artery. Their course in the thigh lies in three parts: in the upper third, in the inguinal region; in the middle third, in Hunter's canal, under the sartorius muscle; and in the lower third they pass round the back of the femur close to the bone to enter the popliteal fossa as the *popliteal* vessels. In the thigh the artery has a number of branches, which encircle the femur and supply the muscle masses, and it divides in the lower part of the popliteal fossa into an anterior and posterior tibial section. The *posterior tibial* continues the main line of the vessel down into the calf, where it lies close to the back of the interosseus membrane, enters the sole behind the medial malleolus in company with the flexor tendons and supplies digital arteries to the toes via an arch similar to the one we have seen in the hand. The *anterior tibial* passes forward from the popliteal fossa through the interosseous membrane into the anterior compartment of the lower leg and runs down the front of the membrane, emerging on the dorsum of the foot with the extensor tendons to give branches to the toes.

The main branches of the venous trunk are similar, save that the venous pattern to the toes is much more superficial and inconstant in arrangement.

The nerves of the leg (Plates 4, 7(b))

The *femoral* and *sciatic* nerves are the two great anterior and posterior trunks, and these, with the *obturator* and certain other twigs, are derived from the *lumbosacral plexus* of spinal nerve roots situated within the abdomen and pelvis.

The *femoral nerve* enters the thigh beneath the inguinal

ligament to the lateral side of the femoral vessels and very soon breaks up into the following branches:

(1) the lateral, intermediate and medial cutaneous nerves of the thigh;

(2) muscular branches to the quadriceps femoris;

(3) twigs to the hip and knee joints.

The *obturator nerve* enters the medial adductor compartment of the thigh from within the pelvis by piercing the obturator membrane; it supplies the adductors, the overlying skin and the knee joint.

The *sciatic nerve* is formed within the pelvis, on the deep aspect of the sacrum; emerges into the buttock, where it lies deeply among the small rotator muscles under cover of the glutei; and runs vertically down the limb to the popliteal fossa, where it divides into medial (*tibial*) and lateral (*peroneal*) branches. In the upper part of the thigh it lies midway between ischial tuberosity and great trochanter, is covered over by the hamstrings in the midthigh and is revealed more superficially where the hamstring tendons diverge to their insertions.

The *tibial* nerve continues the straight course of the sciatic from the upper angle of the popliteal fossa to the lower, lying fairly superficially under the deep fascia and crossing diagonally over the deeper popliteal vessels; it supplies the calf muscles, forming a common neurovascular bundle with the posterior tibial artery, with which it enters the foot to supply the intrinsic muscles and the sensory branches to the toes.

The *peroneal* branch follows the biceps tendon closely to the head of the fibula, winds superficially round the lateral aspect of the neck of that bone, gives a superficial branch to the peroneal muscles and the skin over the lateral side of the calf, and continues as the anterior tibial nerve in company with the anterior tibial artery in the extensor compartment. Here it supplies the extensor muscles, enters the dorsum of the foot and gives digital branches to the dorsum of the toes.

9 Regional anatomy: the abdomen

Abdominal cavity and boundaries

The abdominal cavity is the largest of the body spaces and even more extensive than is at first obvious, for its roof, the diaphragm, reaches high into the bony thorax under cover of the lower ribs; in fact, the ribs enclose a part of the

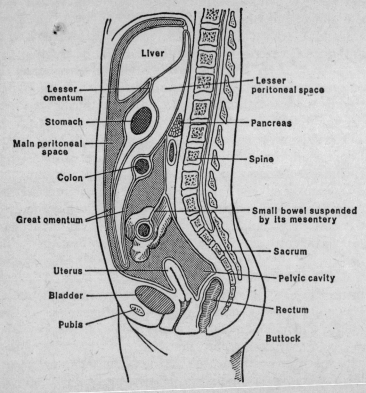

Fig. 107. Longitudinal section of the (female) abdomen, showing the distinction between pelvic cavity and abdomen proper.

abdominal cavity almost equal to that of the whole thoracic cavity itself.

The right and left domes of the diaphragm separate the base of the right and left lungs above from the corresponding lobes of the liver below, with the heart and pericardium sitting on the flat middle part of the partition. In full expiration—i.e. on breathing out as far as possible—the right dome of the diaphragm is at the level of the fifth rib, just below the nipple, and the left is 2·5 cm lower.

It is necessary to distinguish between the *abdomen proper*— which contains the main digestive organs, bowel, liver, pancreas, spleen, the kidneys, adrenals and great vessels— and the *pelvic cavity* below, enclosed by the innominate bones and containing the termination of the bowel, the bladder and genital organs. The two cavities are freely continuous at the pelvic inlet, which forms, as it were, a deficiency in the floor of the main abdominal cavity. And a longitudinal section through the trunk shows how the main axes of the two cavities are set at such an angle that there is a considerable backward inclination of the pelvis, whose organs in consequence are relatively separate from those in the abdomen proper.

Unlike the cranial and thoracic cavities, the abdomen is relatively unprotected by bony framework, though this is compensated for to some extent by a strong muscular belly-wall anteriorly and in the flanks.

The boundaries of the abdomen are:

(1) *behind*, the lumbar vertebrae of the spinal column, clothed by the psoas and quadratus lumborum muscles, which help to form the actual posterior wall of the cavity;

(2) *in front* and *at the sides*, the muscles of the flank and anterior abdominal wall;

(3) *above*, by the diaphragm;

(4) *below*, on each side, the iliac fossae of the innominate bone, clothed with the iliacus muscles, support part of the abdominal contents.

The cavity is lined by a serous membrane, the peritoneum (page 199), which also clothes most of the contained organs, the viscera; these peritoneal surfaces are normally in close contact and the cavity is thus only potential, unless air is admitted by operation or injury, when the organs fall

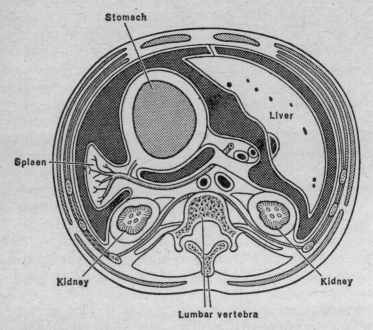

Fig. 108. Cross-section of the abdominal cavity. Note the recess on either side of the vertebral column in which each kidney lies. The large black space is the greater sac of the peritoneal cavity; the lesser sac is the small space immediately behind the stomach.

away from each other. Note also, in cross-section, how the vertebral column encroaches forward so that it can be easily felt through the anterior abdominal wall, leaving great bays on either side of the posterior abdominal wall in which lie the kidneys.

There is considerable respiratory excursion in the abdomen; on inspiration the diaphragm is considerably lowered as the lungs expand, with corresponding effects on the abdominal organs, the liver descending 5–8 cm with a deep breath. This is less conspicuous in the lower-placed organs; and the cavity constantly varies in size with contraction and relaxation of its muscular walls and the degree of distension of the hollow viscera.

Surface anatomy

Anterior abdominal wall

In the surface anatomy of this region, the main points are as follows. On each side superiorly the *costal margins*, the lower borders of the ribs, form a Λ-shaped angle whose acuteness varies with the general build of the individual. At the apex of this angle, and rather depressed below the surface, is the tip or *xiphoid process* of the sternum. The *umbilicus*, or navel, is in the midline, on a level with the fourth lumbar vertebra behind; it is the scar marking the former attachment of the umbilical cord joining foetus to placenta. In the lowest part of the midline anteriorly is the *pubic symphysis*, formed by the junction of the two halves of the pelvis. Xiphisternum, umbilicus and symphysis are all connected by a tough strip of deep fascia, the *linea alba*, a very firm central attachment for the muscles, whose aponeuroses intersect at this point. When these muscles are contracted, the linea alba is seen as a central depression between the prominent *rectus* muscle on either side; and the belly of the latter is crossed by two or three transverse intersections.

Inferiorly, the symphysis can be traced out to the pubic crest on each side, and then the *inguinal ligament*, spanning across from pubic tubercle to anterior superior iliac spine, marks the junction between abdomen and thigh; from the iliac spine the *iliac crest* can be followed round to the back.

The surface of the anterior abdominal wall is divided into conventional zones for convenience in the reference and location of the organs. These are shown in Fig. 109. In the upper zone lies the epigastric region centrally, with the hypochondriac regions on either side; in the middle zone the umbilical region lies centrally, flanked by the lumbar regions; and in the lower zone the hypogastric region lies centrally, and has the iliac regions to right and left. A conventional representation of the surface projection of some

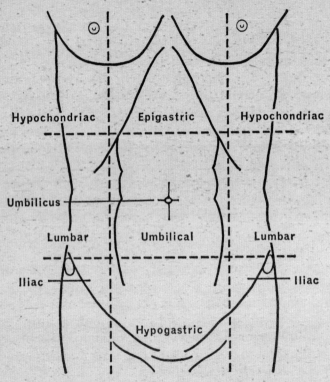

Fig. 109. The zones of the anterior abdominal wall.

of the internal organs on this pattern is given in Plate 5, but it must be remembered that there are very considerable variations in their positions with change of posture and in different individuals.

Posterior abdominal wall

In the midline of the back the spinous processes of the lumbar vertebrae are felt between the thorax and the sacrum; and on each side of the spine is the belly of the great *erector spinae* muscle, which is constantly helping to maintain the erect position. The short twelfth rib on each side marks the lowest limit of the bony thoracic cage above, while below the iliac crests can be traced back towards the sacrum, where they end in the posterior superior iliac spines, marked by a dimple in the overlying skin. Thus on each side of the spine, beyond the lateral border of the rector spinae muscle, is an unprotected portion of the abdominal wall, the loin or lumbar region, lying between twelfth rib and iliac crest. However, the flank musculature is extremely strong and resistant. Plate 5 indicates the relation of some of the internal organs to the surface of the back, and it should be noted that the colon and kidneys, because they are *behind* the peritoneum, are relatively fixed—unlike the mobile liver and small bowel, which are slung freely in the peritoneal cavity.

Abdominal muscles

These are more easily drawn than described and are best considered in the following groups:

(1) the muscles of the posterior abdominal wall: *psoas*, *quadratus lumborum*;

(2) the muscles of the flanks: *internal* and *external obliques*, *transversus abdominis*;

(3) the muscles of the anterior abdominal wall: *rectus abdominis*, *pyramidalis*.

The cross-section in Fig. 110 gives a general idea of the interrelation of these groups.

Muscles of posterior abdominal wall

The *psoas* has already been encountered in connection with the hip joint (page 178); a long belly beside the lumbar vertebrae, it finally skirts the pelvic brim to enter the thigh. The *quadratus lumborum* lies immediately lateral to it, a flat quadrilateral muscular plate extending vertically between twelfth rib and iliac crest.

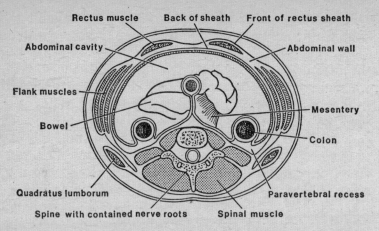

Fig. 110. Cross-section of abdomen below the level of the kidneys. Note the suspension of the small bowel by its mesentery from the posterior abdominal wall; also the relations of the muscle groups.

Flank muscles

These make up a layered arrangement of muscles in which the *internal oblique* is sandwiched between the *external oblique* superficially and the *transversus abdominis* deeply; their fibres cross each other in different directions, thus adding to their protective power. They all arise from the lower ribs above, from the iliac crest below and, via a dense sheet of lumbar fascia, from the tips of the lumbar transverse processes behind. The external oblique fibres run downwards and in—the direction of a man putting his hand in his pocket; the internal oblique fibres are exactly at right angles to this and those of transversus run straight round in the horizontal plane.

Near their origins, above, below and behind, these muscles are very fleshy, but as they curve round the flank to reach the anterior abdominal wall they merge into great aponeurotic sheets extending between costal margin and iliac crests. These aponeuroses are intimately related to each other and to the rectus abdominis muscle, as indicated below; they intersect at the linea alba in the midline.

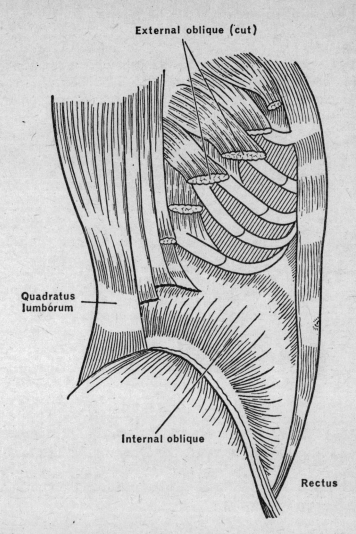

External oblique (cut)

Quadratus
lumbórum

Internal oblique

Rectus

Fig. 111. The flank abdominal muscles of the right side, intermediate layer. The external oblique has been removed, and the
internal oblique is seen extending between the quadratus lumborum behind and the rectus abdominis in front.

The aponeuroses, the rectus sheath, nerves and vessels of the abdominal wall

We have seen that, at a point about halfway between the flank and the linea alba, the flank muscles are continued anteriorly as aponeurotic sheets. As these sheets approach the lateral border of the rectus muscle, the external oblique

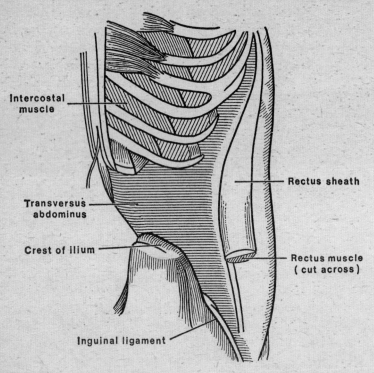

Fig. 112. The deepest layer of the flank muscles, transversus abdominis. (After *Gray*.)

aponeurosis passes in front of that muscle and the transversus aponeurosis behind, while that of the internal oblique splits in two, one layer joining each of the others (Fig. 110, page 196). This arrangement provides a continuous rectus sheath, whose anterior and posterior walls rejoin on the inner side of the rectus to form the linea alba.

The nerves and vessels of the anterior abdominal wall are the lower six of the twelve pairs of thoracic vessels and nerves; this is because, as the lower ribs overhang the abdominal cavity and as these ribs are incomplete anteriorly, the neurovascular bundles running in the rib spaces emerge between the flank muscles of the abdominal wall. Here, they run in the plane between the transversus and internal oblique until the lateral border of the rectus is reached, when they enter the sheath behind the muscle and finally turn forward at its inner border, piercing the linea alba to reach the skin. In their circular course, superficial branches have been given off in the flanks and twigs to the muscle layers.

Finally, within the rectus sheath, a vertical arrangement of arteries and veins is formed behind the rectus muscle by anastomosis between the superior epigastric vessels coming down from the thorax and the inferior epigastrics running up from the external iliac trunks below (Fig. 113).

Peritoneum, mesenteries, omenta

The abdominal cavity is a closed sac, lined continuously with the peritoneal serous membrane; that part of the membrane clothing the deep aspect of the abdominal wall is the *parietal* peritoneum, which is reflected over the contained viscera as the *visceral* peritoneum. The parietal layer is attached to the deep surface of the anterior and posterior abdominal walls, to the undersurface of the diaphragm and to the upper surface of the pelvic floor. Its smooth surface, lubricated by a small amount of serous fluid, allows the structures to glide freely, and a loose layer of extraperitoneal connective tissue intervenes between the parietal peritoneum and the abdominal wall, a layer in which any infection—derived, for instance, from a perforating wound of the bowel—spreads with great ease and virulence.

The main peritoneal cavity of abdomen and pelvis is known as the *greater sac*, in contradistinction to a smaller recess, the *lesser sac*, which lies immediately behind the stomach and is almost cut off from the general cavity. These spaces, and the relations of the various organs to the peritoneum, are rather complex and best understood by

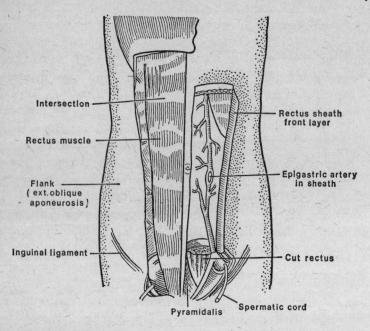

Fig. 113. The left rectus sheath opened and the muscle removed to show the epigastric vessels.

studying Figs 107 and 108 (pages 190 and 192), bearing in mind that most of the organs lying freely in the abdominal cavity have developed in the embryo from the posterior wall, to which they still retain an attachment. The main points are these (Fig. 114):

(1) Certain organs are entirely retroperitoneal and therefore fixed, e.g. the kidneys and pancreas; they lie on the posterior abdominal wall, with a peritoneal covering, if any, on their anterior surfaces only.

(2) Other organs, the ascending and descending portions of the colon and the rectum—all parts of the large bowel—

have a peritoneal covering in front and at the sides; they are still retroperitoneal but with rather more mobility.

(3) The small intestine, and the transverse and pelvic portions of the colon, hang freely in the cavity; to do so, they have pulled out a double-layered sheet of peritoneum from the posterior abdominal wall which is known as a

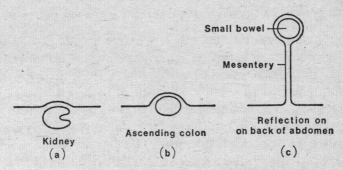

Fig. 114. Different degrees of peritonization of abdominal organs. (a) A retroperitoneal organ. (b) Organ with peritoneum on three sides. (c) Freely suspended organ with mesentery.

mesentery. This is laden with fat, contains lymph glands and vessels, and the blood vessels reach the bowel by running between its layers from the great vessels on the posterior abdominal wall. Small bowel, transverse and sigmoid colon each have their own mesentery; but the main one is the great sheet supporting the small bowel, known simply as *the* mesentery, whose line of attachment behind crosses the lumbar spine obliquely from left to right from above downwards (Fig. 110, page 196).

(4) The stomach has two special mesenteries known as *omenta*. The *great omentum* hangs down from the lower border of the stomach as an apron-like fold in front of the small bowel and then turns up to embrace the transverse colon before its final reflexion onto the posterior abdominal wall, i.e. the mesentery of the transverse colon is really continuous with the great omentum of the stomach (Figs 107, page 190, and 39, page 104). The *lesser omentum* connects the

upper border of the stomach to the liver, and its layers split
to enclose that organ before their final reflexion on the
undersurface of the diaphragm as the suspensory ligaments
of the liver. It can be seen that the lesser omentum forms the
anterior boundary of the lesser sac (Fig. 107).

The general disposition of the abdominal organs

Fig. 39 on page 104 shows the structures revealed when the
anterior abdominal wall is removed. The *liver* occupies the
upper right portion of the cavity, emerging a little below the
costal margin and occupying a little of the left side, with its
superficial intermediate portion lying in the epigastric region immediately below the xiphisternum. Beneath its lower
border, on the right side, projects the *gall bladder*. The
stomach is largely under cover of the left ribs, but a part of
its anterior surface is seen in the angle between the lower
border of the liver and the left costal margin. The *great
omentum* descends from the lower border of the stomach like
an apron, covering and obscuring the *transverse colon*, which
lies immediately below the stomach, and the numerous coils
of *small bowel*. These last are the main contents of the cavity,
insinuating themselves into its recesses and dropping down
into the pelvis. The *spleen* is barely seen as it is tucked up
entirely under cover of the left costal margin. The more
superficial viscera will now be considered in turn.

Stomach

The stomach is the widest part of the digestive tract, connecting the oesophagus (gullet) with the duodenum, the
beginning of the small intestine. The oesophagus joins the
stomach very shortly after entering the abdomen through
the diaphragm, and the duodenum leaves the organ at the
pyloric orifice. The stomach lies mainly in the epigastric and
left hypochondriac regions, but there is considerable variation with posture, digestive state and emotion, and part may
lie in the umbilical region or even descend to the pelvic

brim. It is a rather J-shaped structure, with anterior and posterior surfaces parallel to the abdominal wall, and upper and lower borders, known as the lesser and greater curvatures respectively. It can be seen in Fig. 115 how the lesser

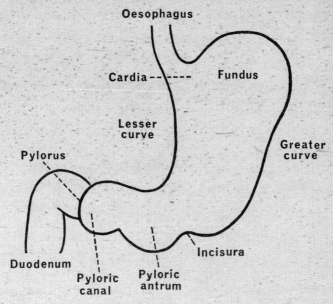

Fig. 115. The stomach, anterior surface.

curve forms a continuous sweep between the oesophageal entry at the cardia at its left extremity and the departure of the duodenum at the pylorus at the right-hand end. The main subdivisions are:

(1) the dome-shaped *fundus*, the receptive portion, often distended with air

(2) the main *body*, concerned with digestion, and marked off by a definite notch from

(3) the *pyloric* or expulsive portion, narrowing down as the pyloric antrum to the pyloric canal itself.

Both cardiac and pyloric orifices are surrounded by what are known as *sphincters*, muscular rings that are normally in contraction, keeping the openings closed until temporary relaxation is required for the entry or exit of food. This sphincteric control of the exits of hollow muscular organs is a common mechanism in the body, e.g. at the neck of the bladder and the opening of the anus.

In structure the stomach is made up of several layers (Fig. 116). There is the smooth outer *serous* or peritoneal coat; the

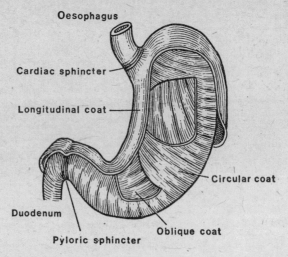

Fig. 116. The muscle coats and sphincters of the stomach.

thick, intermediate *muscular* wall, which contains circular, longitudinal and oblique fibres; and the internal lining *mucous membrane*, a velvety layer of intricate folds whose glands secrete the gastric juice.

The *anterior surface* of the stomach is under cover of the left lobe of the liver and the left costal margin, with a small intermediate portion which is applied to the back of the anterior abdominal wall. The *posterior surface* lies on a definite 'stomach bed', shown in Fig. 117, which is formed by several organs of the posterior abdominal wall—pancreas, left kidney and suprarenal, and spleen. And between the posterior surface and its bed is, of course, the lesser sac. The *fundus* lies immediately in contact with the under-surface of the left dome of the diaphragm.

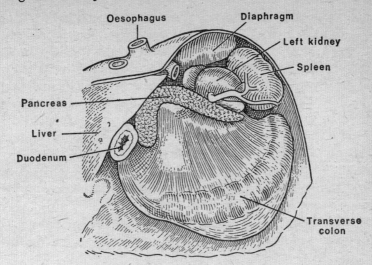

Fig. 117. The stomach bed. The left lobe of the liver has been cut away, and the stomach itself removed, to show the organs on which it lies. (After *Gray*.)

We have already seen that the two layers of peritoneum clothing the surface of the stomach meet at its borders to form the omenta; the lesser omentum connects the lesser curvature to the liver and the greater omentum hangs down from the greater curvature as a large apron.

Small intestine

This is the portion of the digestive tract between the pyloric orifice of the stomach and the large bowel. Some 6 m long, it is divided into:

(1) the *duodenum*, a short coil of 25 cm immediately continuous with the stomach and bound down tightly to the posterior abdominal wall;

(2) the *small intestine proper*, suspended in coils from the posterior wall by its mesentery.

The *duodenum* is rather C-shaped, embracing in its concavity the head of the pancreas. A short first part runs horizontally from the pylorus under cover of the liver and

gall bladder; the vertical second part descends in front of
the right kidney; and the third part crosses transversely in
front of the lumbar vertebrae, from which it is separated
only by the great vessels—the aorta and inferior vena cava.
It finally joins the small intestine at the duodenojejunal
flexure. The pancreatic duct and the bile ducts from the
liver have a common opening into its second part (Plate 6).

The *small intestine* proper runs from the duodenojejunal
flexure to the ileocolic valve, which marks the junction of the
small intestine with the caecum of the large bowel. Its coils
are completely surrounded by a peritoneal coat, except for
a narrow strip, the mesenteric border, where the two layers
of the mesentery diverge to enclose them. In essential structure the small bowel resembles the other parts of the in-

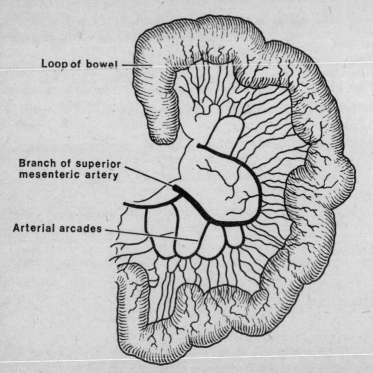

Fig. 118. A loop of small bowel, with its blood vessels.

testine, possessing an outer *serous* coat, a *muscular* wall and a lining *mucosa*. But characteristic are the circular folds projecting into the lumen; the enormous number of minute fringes or *villi* of the mucosa which give it a velvety appearance like the pile of a carpet, a device for increasing the absorptive area within a limited space; and the scattered patches of lymphoid tissue. Networks of blood and lymph vessels and of nerves form plexuses between the layers, all eventually forming larger trunks which run into the mesentery. The blood vessels of the bowel enter the root of the mesentery as the superior mesenteric branches of the aorta and vena cava, and form a complicated pattern of arches or arcades between its layers, from which the ultimate twigs are given off which encircle the bowel itself.

The *jejunum* is the upper two-fifths of the small bowel, the *ileum* the lower three-fifths. There is no sharp distinction, but the ileum is thinner, narrower, less vascular and contains more lymphoid tissue, indications that digestion and secretion are preponderant at the upper end of the bowel, absorption at the lower. The coils of the small gut as a whole are enclosed within a space bounded by the limbs of the colon. In general, the jejunum is above and to the left, while the ileum is central and inferior, some of its loops usually spilling over into the pelvis before rising again as the terminal ileum to join the large bowel.

The mesentery
The leaves of the mesentery enclose between them:
(1) considerable fat and connective tissue;
(2) many lymph glands receiving the bowel lymphatics or *lacteal* vessels, so called because of the milky fat-laden fluid in them;
(3) the mesenteric vessels;
(4) nerves supplying the intestine.

It is stimulation of the latter that gives rise to shock and collapse when the mesentery is injured or twisted. The posterior attachment or 'root' of the mesentery is an oblique line of some 15 cm running down the posterior abdominal wall from the left of the second lumbar vertebra to the region of the right sacro-iliac joint below, and crossing over the duodenum and great vessels. The great disparity between this short origin and its extensive attachment to the bowel causes the membrane to be thrown into fan-shaped folds.

Large bowel

The large intestine runs from the end of the ileum to the external orifice of the anus in the perineal region. Only some 1·5–1·8 m feet long, it includes the following portions:

(1) the *caecum*, with its *vermiform appendix*;

(2) the *ascending*, *transverse* and *descending* limbs of the colon;

(3) the *pelvic colon*;

(4) the *rectum* and *anal canal*.

Of these, only the transverse and pelvic segments of colon have mesenteries and are freely mobile; the remainder are retroperitoneal, while the lower rectum and anus are entirely below the level of the peritoneum in the depths of the pelvis.

The *caecum* lies in the right iliac fossa, the shallow basin formed by the internal surface of the iliac portion of the innominate bone. It is the blind sacculated commencement of the large bowel, the terminal ileum opening into it at one side by an aperture guarded by the ileocolic valve. The appendix is attached a little lower down in the angle between the two. The caecum has no mesentery but is completely invested by peritoneum, and it has the considerable mobility of a flapping balloon tethered at one point.

The *appendix* is a worm-like tube of varying length, blind at its free end and with a tiny mesentery of its own. Its position is far from constant; it may lie tucked up behind the caecum, hang down over the pelvic brim, or turn up in front of or behind the terminal ileum, a range of variation responsible for the difficulties in diagnosing acute appendicitis.

Colon

The ascending, transverse and descending colon are arranged round the abdominal cavity like three sides of a square—⊓—but the two vertical limbs lie in the paravertebral gutter on each side and therefore behind the plane of the transverse limb, which sags down towards the pelvis in a redundant fashion. The junctions of transverse colon with the vertical limbs at either end are known as flexures, the *hepatic flexure* on the right under cover of the right lobe of the liver and the *splenic flexure* on the left in relation to the spleen. And, because of the greater bulk of the liver on the right side of the abdomen, the hepatic flexure is pushed down several centimetres lower than the splenic. The colon in general is distinguished by:

(1) three superficial bands of longitudinal muscle standing out under the serous coat and traversing the bowel from end to end—the *taenia coli*, which are spaced equally round its circumference;

(2) the scattered fatty tags or polyps which project as the stalked *epiploic appendages* on the surface;

(3) a segmentation into coarse *sacculations* vaguely resembling those of an earthworm.

The internal structure is like that of the bowel in general, except that the serous coat is incomplete wherever the colon is only partly peritonized, and that the mucosa is smooth and pale.

The *ascending colon* lies on the posterior abdominal wall, mainly on the quadratus lumborum muscle, and, since it is bound there by a peritoneal covering on the front and sides only, is relatively immobile.

The *hepatic flexure* lies in front of the lower pole of the right kidney and is itself overhung by the right lobe of the liver.

The *transverse colon* arches across the abdomen, lying immediately below the great curvature of the stomach and obscured by the great omentum of the latter. It is very

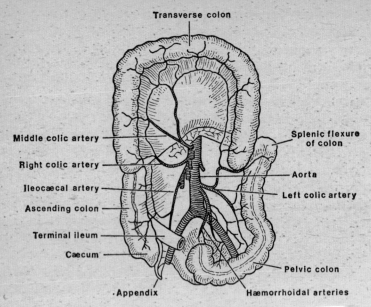

Fig. 110. The large bowel and its blood vessels. The transverse loop of colon, which is normally dependent, has been turned up for clarity.

mobile since it has a mesentery, the *transverse mesocolon*, derived, as we have seen, from the doubling back of the layers of the great omentum, which diverge to enclose the bowel and then rejoin to form its mesentery as they are reflected on the posterior abdominal wall (Fig. 107, page 190).

Plate 6 shows the transverse colon to lie in front of most of the structures of the posterior abdominal wall at this level—duodenum, pancreas, great vessels and portions of the kidneys on either side—and the prominence of the lumbar spine behind all these arches the colon forwards.

The *splenic flexure* lies in front of the left kidney, immediately below the lower pole of the spleen, with the diaphragm supporting it behind.

The *descending colon*, like the ascending, has no mesentery and is plastered onto the posterior abdominal wall behind the peritoneum. It runs down from the splenic flexure as far as the pelvic brim on the left side, where the shallow left iliac fossa gives way sharply to the deep true pelvis. And to reach this position it crosses in front of the left psoas muscle, and the left common iliac artery and vein.

The *pelvic colon*, the continuation of the descending colon, extends from the pelvic brim to the beginning of the rectum, opposite the middle of the sacrum in the depths of the true pelvis. It has a long mesentery, the pelvic mesocolon, and this part of the bowel is consequently loose and mobile; it may hang down in the pelvis or lie turned up in the main abdominal cavity.

The *rectum and anus* will be described in connection with the pelvis (page 231).

Blood vessels and lymphatics of the large bowel

These are indicated diagrammatically in Fig. 119. Each limb of the colon has a main artery of its own; the ascending and transverse colon have the *right* and *middle colic* vessels —branches of the superior mesenteric artery, which has already supplied the whole of the small bowel—while the descending limb receives the *left colic* division of another great branch of the aorta, the *inferior haemorrhoidal* artery, which passes down to the rectum after giving twigs to the loop of the pelvic colon. Note that, as the right, middle and left colic arteries approach the bowel (the middle between the layers of the transverse mesocolon and the other two behind the peritoneum of the posterior abdominal wall), they each fork into two main branches which run parallel with the bowel to anastomose with the adjacent vessels. They thus form a continuous arterial channel from one end of the large bowel to the other, a channel whose branches form a network of arteries whose twigs ultimately reach the intestine. The general peripheral pattern of the corresponding *veins* is similar; but, whereas the arteries have sprung from the great artery of the posterior abdominal wall, the *aorta,* the veins do not return to the companion inferior vena cava but enter an entirely separate venous trunk be-

hind the pancreas, the *portal vein*, which gathers up the
splenic vein on its way to run up in the lesser omentum and
enter the liver. This so-called 'portal circulation' ensures
that the venous blood draining from the bowel, and containing the products of digestion, passes through the liver
before entering the general circulation. This organ is thus
enabled to store certain of these products, such as glycogen,
and to effect certain biochemical changes in others, such as
the splitting-up of proteins and the formation of urea.

Liver

This is an appropriate place to consider the liver, a bulky
solid organ—the largest gland in the body—occupying the
upper right portion of the abdominal cavity. It has a
brownish friable substance, with a smooth peritoneal coat,

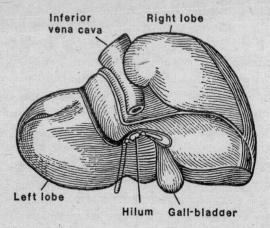

Fig. 120. The liver, seen from behind. (After *Gray*.)

and is slung from the undersurface of the diaphragm by
suspensory ligaments, which are the peritoneal reflexions
onto the undersurface of that partition. The large *right lobe*
and the much smaller *left lobe* lie directly beneath the right
and left domes of the diaphragm, and most of the organ is
under cover of the ribs. It reaches almost to nipple level on
each side, and only projects a little beyond the costal margin

in the epigastric and right hypochondriac regions, where it lies in contact with the back of the anterior abdominal wall.

Note that, because of the upward bulge it produces in the floor of the thoracic cavity, the lower borders of the lung and pleural cavity actually surround the upper part of the liver in a sort of rim, so that a penetrating wound in the lower part of the chest may traverse the lung and pleura and then enter the liver. The organ is $\mathcal{V}$-shaped in sagittal section, the sloping limb facing backwards, with superior, anterior and posterior surfaces, and bluntly rounded borders, of which the inferior is the sharpest.

The *anterior surface* is opposed to the backs of the lower ribs and to the back of the anterior abdominal wall in the epigastric region. On it the demarcation of right and left lobes is made by the *falciform ligament*, which attaches the liver to the back of the anterior abdominal wall in the midline between xiphisternum and umbilicus. The *superior surface* rests in contact with the diaphragm. The *posterior surface* is more complex. On the right it overlies the right kidney and the hepatic flexure of the colon; on the left it covers the front of the stomach, the upper pole of the left kidney with its suprarenal gland and the spleen.

The *gall bladder* is attached to the back of the right lobe and can just be seen from in front projecting below the inferior border. The *inferior vena cava* is embedded in the substance of the upper part of the back of this lobe, on its way up to pierce the diaphragm and enter the thorax.

At the very centre of the posterior surface of the liver, between the lobes, is the root of the organ, the *hilum*, where the main vessels and ducts enter and leave the organ. These structures are:

(1) the *hepatic artery* from the aorta, dividing into right and left branches to the two lobes;

(2) the *portal vein*, carrying nutriment from the bowel, again dividing into two branches;

(3) the right and left *hepatic ducts*, conveying the bile secreted by each liver lobe and joining to form the *common*

hepatic duct. The *cystic duct* from the gall bladder joins the common duct a little below its formation, and the main channel thus formed to pass the bile down to the duodenum is the *common bile duct*.

These structures—bile duct, hepatic artery and portal vein—have to run from the liver to the posterior abdominal

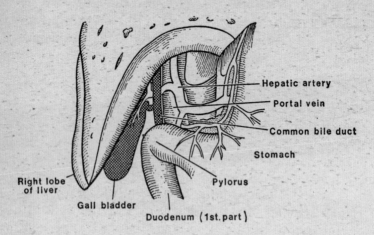

Fig. 121. The liver, seen from in front, with part of the right lobe cut away to show the structures entering and leaving the hilum. (After *Gray*.)

wall, and they do so between the layers of the lesser omentum, which is attached to the hilum of the liver at one end and to the lesser curve of the stomach at the other.

The *functions* of the liver include the preparation of carbohydrates and proteins for utilization by the body after their absorption from the bowel, the storage of excess carbohydrate as glycogen and the secretion of the bile. The latter is stored and concentrated in the gall bladder, a purely mechanical reservoir with no secretory function of its own. It has a muscular coat which contracts in reflex response to the entry of food into the duodenum from the stomach; the bile is then emptied through the common duct into the duodenum, where it emulsifies the fat of the food into tiny globules for easier digestion. (See pages 314 and 342 for further details.)

Spleen

This organ occupies a position immediately below the left dome of the diaphragm, corresponding to that of the right lobe of the liver on the opposite side; but it is a much smaller organ, entirely under cover of the ribs, of a soft pulpy consistency with a fibrous capsule. It is pyramidal or tetra-

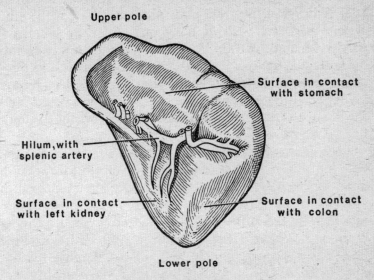

Fig. 122. The spleen.

hedral in shape, with a large, smooth, convex surface applied to the underside of the diaphragm, and three smaller surfaces facing inwards and converging on the *hilum*, where the splenic artery and vein are attached. One surface forms part of the stomach bed (Fig. 117, page 205); the others are in contact with the left kidney and the splenic flexure of the colon; and the tail of the pancreas reaches to the hilum.

The spleen is a reservoir of red blood corpuscles; since

there are muscle fibres in its capsule and substance, it is able
to squeeze more cells out into the circulation if effort or
lack of oxygen make it necessary, and then the organ shrinks
considerably. It is also a site for the formation of the lymphocytes of the blood and is concerned with the elaboration of
immune bodies in response to infection.

Posterior abdominal wall and related structures

Looking into the main abdominal cavity from the front
after removal of the great mass of small bowel, we see the
posterior abdominal wall, a bony and muscular boundary
partly obscured by certain organs and structures lying immediately in front of it. These are the pancreas and duodenum, stretching transversely across from side to side; the
kidneys and ureters, one in each paravertebral recess; and
the great vessels running vertically downwards in the midline (Plate 6).

The *posterior wall* itself is composed in the midline of the
bodies of the lumbar vertebrae, which project forward so
much as to render the abdominal cavity kidney-shaped in
cross-section, leaving deep *paravertebral recesses* on each side
in which lie the kidneys and the ascending or descending
limbs of the colon (Fig. 108, page 192). On each side of the
bodies is the psoas muscle and, more laterally, the flat
quadratus lumborum muscle, extending vertically between
twelfth rib above and iliac crest below. The plexus of
lumbar nerve roots, which emerge between the vertebrae
to form a network whose ultimate branches supply the leg,
lies embedded in the psoas muscle. The upper limit of the
posterior wall is the lowest attachment of the diaphragm
behind; the lower limit is the pelvic brim, and laterally it
is continuous with the abdominal muscles of the flank.

Great vessels: aorta and inferior vena cava
The *abdominal aorta* is the continuation through the aortic
aperture of the diaphragm, at the level of the twelfth

thoracic vertebra, of the thoracic portion of this great artery which has arisen from the left ventricle of the heart. It runs down the midline, immediately in front of the lumbar bodies, as far as the fourth lumbar vertebra, where it divides into the right and left common iliac arteries; a bifurcation marked on the surface by a point just below and to the left of the umbilicus. It therefore lies very deeply, and is crossed transversely by the body of the pancreas and the third part of the duodenum. There is a mass of sympathetic nervous tissue on each side of its commencement, the *coeliac ganglia* or *solar plexus*, just below the diaphragm, and branches of this form a network around the length of the vessel as the *aortic plexus*. The main sympathetic chain lies on the vertebral bodies to either side of the aorta.

The *inferior vena cava* is a great vein lying immediately to the right of the aorta, draining the venous blood from the lower abdomen and legs. It is formed by the union of the two common iliac veins in front of the body of the fifth lumbar vertebra, the junction lying behind the right common iliac artery. And it runs up the posterior abdominal to pierce the central tendon of the diaphragm and enter the right atrium of the heart.

The branches of the aorta include:

(1) Single unpaired branches from the front of the vessel in the midline: the *coeliac axis* to liver, spleen and stomach; the *superior mesenteric* to small bowel, ascending and transverse colon; and the *inferior mesenteric* to the remainder of the large bowel.

(2) Symmetrical paired branches on each side to the diaphragm, suprarenals, kidneys, testicles or ovaries. Of these, the largest are the *renal arteries*, and the right renal has to pass behind the vena cava to reach its kidney. Since the ovaries lie in the pelvis, and the testicles even lower, their vessels have to run longitudinally downward on the posterior abdominal wall to reach them, in company with the ureters. The paired branches of the vena cava correspond to those of the aorta, with certain adjustments for the slight

difference in position—e.g. the left renal vein is much longer than the right and has to cross in front of the aorta. But there is a considerable difference in the fate of the venous blood from the bowel, for this has to pass through the liver before entering the general circulation so that the products of digestion may be dealt with. This is achieved by means of a small secondary *portal circulation*. The superior and inferior mesenteric veins join behind the pancreas to form a main *portal vein*, which also drains the stomach and spleen. This portal trunk then ascends between the layers of the lesser omentum to the liver, there giving a branch to each lobe; in the omentum, it forms a common bundle with the bile duct and hepatic artery. After the portal blood has passed through the liver substance, it re-enters the general venous stream by a number of little *hepatic veins* opening directly into the vena cava as it lies in contact with the back of the liver, just before its passage through the diaphragm.

The main *lymphatic* drainage from the abdomen, including the lacteal vessels carrying digested fat from the bowel, is into a small vessel, the *cisterna chyli*, lying between the upper parts of aorta and vena cava. It enters the thorax with the aorta to become the *thoracic duct*, which is destined to travel right up to the root of the neck on the left side before discharging its contents into the venous circulation.

Pancreas

This, the sweetbread, is a soft, solid, lobulated organ of ⊂-shape, embraced in the concavity of the duodenum and stretching transversely across the posterior abdominal wall and great vessels. It is composed of a *head*, the large, rounded right-hand extremity, closely packed into the duodenal curve; a *neck*, connecting the head to the *body*, which stretches right across the midline to the left side, in front of the great vessels and the upper part of the left kidney; and a final *tail*, turning up to reach the hilum of the spleen.

The pancreas is an important organ, both of external and

of internal secretion. The *external secretion*, the pancreatic juice, aids in digestion of carbohydrates, fats and proteins; it is discharged through the main pancreatic duct into the second part of the duodenum. This duct forms a common channel with the lower part of the common bile duct so that bile and pancreatic juice are discharged simultaneously when required, the opening being kept closed between meals by a sphincter.

The *internal secretion*, from certain island-groups of cells, is *insulin*, which passes directly into the bloodstream and is essential for the proper utilization of sugar in the body. Deficiency of this secretion is the cause of diabetes (see also page 340).

Plate 6 shows how the superior mesenteric vessels emerge from behind the body of the pancreas and then pass in front of its head to enter the mesentery of the small bowel. The *splenic artery* runs in wavy fashion along the upper border of the organ from the coeliac branch of the aorta to the spleen. And the lower parts of portal vein, hepatic artery and common bile duct are all situated in or behind the substance of the head and neck.

Kidneys and ureters

The kidneys are responsible for the removal of waste products and excess water from the blood brought to them by the great renal vessels, and the urine they excrete is passed into ducts, the *ureters*, which carry it to the bladder. The kidneys lie one in each paravertebral recess, closely applied to the posterior abdominal wall. They are rounded, bean-shaped organs, convex at their outer borders and concave where they face the midline, with a main body and upper and lower poles. Perched on the upper pole like a little helmet is the *adrenal gland*, an organ of internal secretion.

The kidneys are rather flattened organs, lying obliquely with the upper pole rather nearer the midline. The general level of the right kidney is 2·5 cm or more lower than the left, owing to the great bulk of the overhanging right lobe of the liver, which, as we have seen, produces a similar difference between the two flexures of the colon. Thus, while the left kidney rests on the eleventh and twelfth ribs behind, the right is only in contact with the twelfth. At the

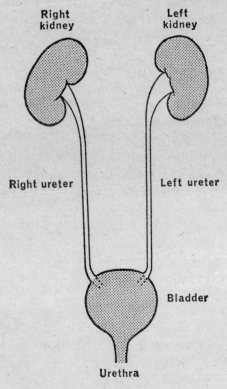

Fig. 123. The urinary system.

middle of the concave medial border is a puckered depression, the root of the organ or *hilum*, where the artery and vein are attached and the ureter makes its exit.

Longitudinal section (Fig. 124) shows the organ to have an outer solid substance enclosing an inner cavity, the *renal pelvis*, which collects the urine as it is formed. The renal substance has an outer rind or *cortex* and a deeper *medulla*, and the latter is arranged in pyramidal masses whose apices project into little bays of the pelvis called the *calyces*. The microscopic renal tubules, in which the urine is formed, discharge their content at the tips or papillae of the pyramids into the renal pelvis. The pelvis itself is partly enclosed within the kidney—the intrarenal portion—and partly

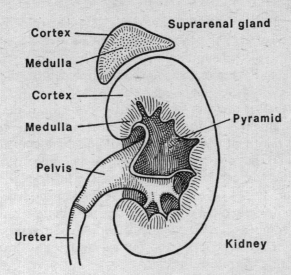

Fig. 124. Kidney and suprarenal gland, longitudinal section.

emerges at the hilum as the extrarenal portion to become continuous with the ureter at the pelvi-ureteric junction.

The kidney has a tough true capsule which forms the glistening outer surface, and lies embedded in a much looser and more voluminous false capsule of perinephric fat and fascia, in which it glides up and down with respiration. It is liberally supplied with nerves and consequently is very sensitive to overdistension if the free exit of urine is obstructed.

Since the kidneys lie on the lower ribs, they are in relation to the lowest part of the pleural cavity of the chest, which overhangs the back of the upper pole, separated only by the lowest fibres of the diaphragm; but most of the organ lies on the psoas and quadratus muscles behind. The other

relations have already been touched on, and differ somewhat on the two sides.

Right: The right lobe of the liver is in front of most of the
anterior surface; the second part of the duodenum lies before the hilum; the hepatic flexure of the colon is in front of
the lower pole.

Left: The body of the pancreas crosses the middle of the
anterior surface; the latter also forms part of the stomach
bed. The spleen is applied to the convex outer border, and
the splenic flexure of the colon is in contact with the lower
pole.

The *ureters*, the ducts of the kidneys, begin at the pelvi-
ureteric junction in the hilum of the kidney, emerge from
behind the renal vessels, and run obliquely downward and
inward on the psoas muscle behind the parietal peritoneum.
Reaching the pelvic brim, they cross in front of the common iliac vessels, run down the side wall of the true pelvis
and finally enter the bladder. They are hollow muscular
tubes, some 25 cm long and 4 mm in diameter, down
which urine is propelled in gushes by regular waves of
contraction, or peristalsis.

The diaphragm

This is the chief muscle of respiration and the most important muscle of the body after the heart. We have seen that
it forms a partition, part muscular, part tendinous, between
the thoracic and abdominal cavities. It has a right and left
muscular *dome* rising high into the thorax and separating
the base of each lung from the abdominal viscera; and there
is an intermediate flatter *central tendon*, on which rests the
heart in the pericardium.

The muscular portion must obviously have a number of
bony origins at different points around the periphery of its
lower attachments. Thus, in *front*, it arises from the back of
the xiphisternum; at the *sides*, from the deep aspect of the
lower ribs; and in the midline *posteriorly* it enters into the
formation of the upper part of the posterior abdominal wall

by arising in two muscular pillars from the sides of the upper three lumbar vertebrae, forming the *crura* of the diaphragm (Fig. 125).

There are several apertures for the structures passing between thorax and abdomen. The *aortic aperture* is where that vessel is embraced by the two crura as they cross in front of the body of the twelfth thoracic vertebra. The *oesophageal*

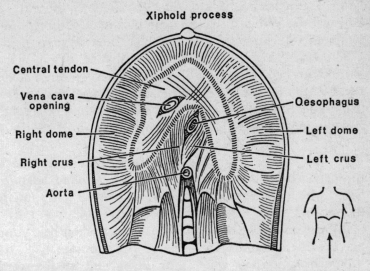

Fig. 125. The diaphragm, seen from below. (After *Gray*.)

aperture is in the muscular substance of the left dome at the level of the tenth thoracic body. And the opening for the *inferior vena cava* lies in the central tendon just to the right of the midline at the level of the ninth thoracic body (Plate 6).

It may be mentioned here that occasionally there are congenital deficiencies in the muscular substance of the partition, as clefts between its different heads of origin. Thus there may be a gap between sternal and costal attachments. And when this is so some of the abdominal viscera— the stomach, liver or small bowel—may herniate through and lie in the thoracic cavity in contact with the lung. This is a diaphragmatic hernia, and it is commoner on the left side as the bulk of the right lobe of the liver usually prevents such displacement.

The pelvis

We have already seen that the lower portion of the main abdominal cavity is known as the pelvic cavity and that the latter is set at such a backward inclination from the vertical that it forms rather a separate compartment, although the two are freely continuous at the pelvic inlet (Fig. 107).

The cavity is formed by the basin-shaped arrangement of the bony pelvis, the ring formed by the two innominate bones at the sides and in front, and the sacrum and coccyx behind—a ring that is interposed between the lower spine and the legs (Figs 21, 22). The bony pelvis falls naturally in two parts (Figs 127, 131):

(1) the *false pelvis*, the shallow edges of the basin above the pelvic brim on each side, i.e. the right and left *iliac fossae*, which are clothed by the iliacus muscles and support the caecum and pelvic colon respectively;

(2) the *true pelvis*, the deeply enclosed portion below the pelvic brim containing the essential pelvic organs.

The true pelvis is what is usually meant by the word 'pelvis' in general use and our further remarks will apply to it alone, as the false pelvis merely forms on each side the floor of the abdominal cavity proper.

It has an *inlet*, an *outlet*, a *cavity* and a *floor*. Its boundaries are only partially bony and are completed by soft tissues— muscles, ligaments and membranes—and the floor is composed entirely of soft tissues.

The *inlet* faces mainly forwards and only slightly upwards, owing to the marked backward inclination of the sacrum. Its boundary is the pelvic brim, which is seen from above to be heart-shaped, with the projecting upper part of the sacrum, the promontory, encroaching considerably behind and the obtuse angle of the pubic arch meeting at the symphysis in front (Figs 21, 22).

The *cavity* is a conical canal, much deeper posteriorly. Its walls are: *anteriorly*, the back of the pubic symphysis and rami; *laterally*, the pelvic aspects of the innominate bones, with the obturator foramina bridged across by the obturator membranes; and *posteriorly*, the anterior surface of sacrum and coccyx. These bony walls are very incomplete and are lined by various muscles, by pelvic fascia and, to some extent, by the pelvic peritoneum. The principal contents of the cavity are the pelvic colon and the rectum on the posterior wall, in the hollow of the sacrum, and the bladder in front, immediately behind the symphysis. In women the uterus and vagina are interposed between rectum and bladder; in men the bladder rests on the prostate gland and seminal vesicles.

The *outlet* is a diamond-shaped space when seen from below. The four bony points of the ◇ are: anteriorly, the symphysis; at each side, the ischial tuberosity; and posteriorly, the tip of the coccyx, the lowest part of the spinal column. The two anterior limbs of the ◇ are bony, the pubic arch formed by the inferior pubic rami running up to the symphysis; the two posterior limbs are ligamentous, the sacrotuberous ligaments spanning from sacrum and coccyx to ischial tuberosities.

Sex difference in the bony pelvis

These are more obvious than in any other bones, as the female pelvis has to allow the passage of the baby's head in the process of birth or parturition. As a result, it is roomier than the male pelvis and shallower. The female pelvis has its side walls more vertical; the iliac fossae are shallower; the inlet is large and nearly circular; the sacrum is short and wide, and projects but little into the pelvic cavity; and the outlet is wide, the pubic arch forming an obtuse angle, while the coccyx is usually very mobile.

All these characters are reversed in the male pelvis, a narrow cavity with sloping walls, a small heart-shaped inlet with marked sacral encroachment, a tighter outlet with the pubic arch an acute angle, and a more rigid coccyx.

The female pelvis may be *contracted* or *distorted*, either congenitally or from disease such as rickets, and then labour may be more or less obstructed and normal childbirth an impossibility.

Surface anatomy
The only region where the pelvis comes near the surface is at the *perineum*, the space between the legs which contains the external orifices of genital, urinary and digestive tracts. Fig. 126 shows the female perineal structures as they overlie

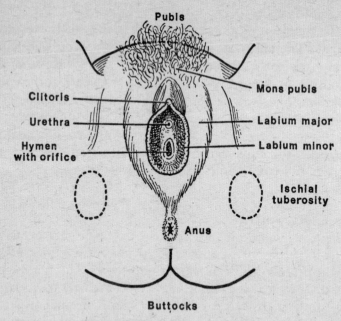

Fig. 126. The female perineum.

the framework of the pelvic outlet. There is a diamond-shaped area of perineal skin divided into two triangles by a line joining the ischial tuberosities. The posterior *anal triangle* contains the opening of the anus, continuous with

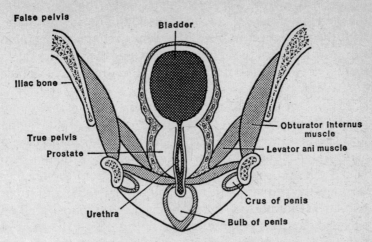

Fig. 127. Coronal section of male pelvis. Note the distinction between true and false pelvis; also the muscles of the pelvic floor. (After *Gray*.)

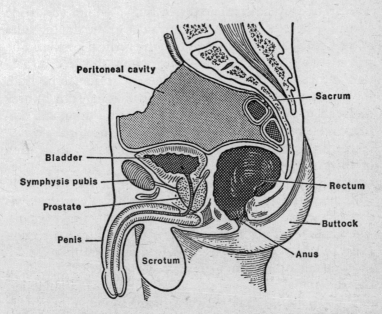

Fig. 128. Midline sagittal section of the male pelvis. (After *Gray*.)

the rectum above. The anterior *urogenital triangle* contains the opening of the *vagina*, the lower end of the genital tract; with the *urethra* just in front, the urinary channel which has run down from the bladder through the pelvic floor; further forward still is the *clitoris*, the diminutive female equivalent of the masculine penis.

In *men* the anal triangle is similar, but the urogenital region is very different, owing to the absence of the vagina and the presence of the *penis*. The latter has at its base a central *bulb*, attached to the midpoint of the perineum, into which the urethra passes after its passage from the bladder through the prostate and pelvic floor (Fig. 128); and it has two lateral supporting *crura*, attached to the pubic arch on each side and converging on the bulb to form the shaft of the organ, which is traversed throughout its length by the *urethral canal*, opening at the *external urinary meatus* at its tip.

The pelvic floor

The pelvis may be compared to a funnel with a very wide stem; the false pelvis will be equivalent to the sloping sides of the funnel and the true pelvis to the vertical portion. The pelvic organs proper are contained within the vertical portion and rest on its floor, a muscular partition slung from the sidewalls of the cavity and sagging obliquely downwards under their weight. Since it is a muscular partition and embraces the various canals—anal, vaginal, urethral—which pierce it to reach their perineal orifices, it surrounds each of these with an encircling sphincteric grip. The upper aspect of the pelvic floor is related more or less intimately with the parietal layer of peritoneum, which lines the pelvis as a closed sac and is reflected over the contained organs. The general arrangement is shown in longitudinal section (Figs 128, 129) and is considerably simpler in the male.

In the *male* the peritoneum is reflected from the back of the abdominal wall over the upper surface of the bladder, dips down in a recess between bladder and rectum (the rectovesical pouch) and then turns up again over the front of the rectum. Thus the greater part of the bladder is below peritoneal level and the prostate entirely so.

In the *female* the peritoneum, after covering the bladder, is thrust up by the projection of the uterus; it covers the

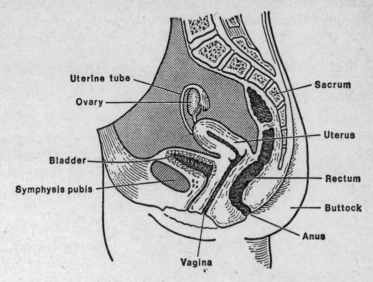

Fig. 129. Midline sagittal section of the female pelvis.

anterior and posterior surfaces of that organ, passing off the latter to form the recto-uterine pouch—the most dependent part of the abdominal cavity—before ascending again on the rectum.

In both sexes the lower limit of peritoneum is at the level of the third piece of the sacrum; the upper part of the rectum has a peritoneal covering in front, but no mesentery, and the lower part is entirely below peritoneal level.

Pelvic viscera

We can now describe the pelvic viscera in more detail; bladder and rectum are much the same in both sexes, but the genital organs need separate discussion and will be dealt with later (Chapter 22).

Bladder. This is a hollow, muscular organ occupying the anterior part of the pelvic cavity; it receives urine via the ureters from the kidneys and expels it into the urethra by the act of micturition. Its shape is inconstant, owing to the varying degree of urinary distension, but is roughly that of an inverted pyramid resting on its apex when, contracted and empty, it lies entirely within the pelvis. As it distends,

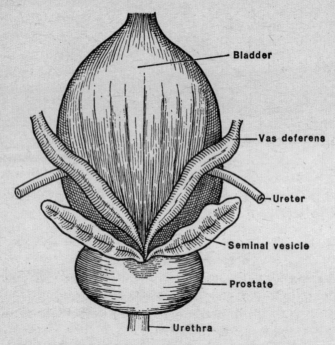

Fig. 130. Bladder and associated organs in the male.

its domed upper portion or *fundus* ascends into the abdomen, in contact with the back of the posterior abdominal wall, and may reach as far as the umbilicus in cases of chronic obstruction to the urinary outflow.

The upper surface is covered by peritoneum, and the ureters open into the organ at its upper lateral angles by valve-like oblique passages through the muscular wall, at the right and left *ureteric orifices*. The tapering dependent portion, the bladder neck, is continuous with the urethra at the *internal urinary meatus*. The two ureteric orifices and the internal meatus form the three points of the *trigone*, an important region of the bladder base and very sensitive to stimulation.

In *women* the organ is overhung from behind by the body of the uterus, and the urethra is a short canal of only 4 cm which pierces the pelvic floor to open immediately in front of the vagina. In *men* the neck of the bladder rests on the *prostate gland*, into which the male urethra passes; the *seminal vesicles* for the storage of sperm, and the ducts conveying sperm from the testicles, are also in close relation to the lower part of the male bladder.

The bladder wall is only partly peritoneal; there is a thick muscular coat and a mucosal lining, which is ridged and wrinkled when the organ is contracted, owing to the slack required to permit distension. The involuntary nervous system adjusts muscle tone to fluid content so that it relaxes as it fills, keeping the tension constant until a threshold level is reached, when the tension rises and is felt as a conscious urge to micturate (see also page 408).

Rectum and anus. The rectum is the lower part of the large bowel; faeces enter it from the pelvic colon and remain there until discharged by defaecation. It is some 13 cm long, begins as the continuation of the pelvic colon at the middle of the sacrum and conforms to the hollow of the sacral curve in its course, finally turning forward at the level of the coccyx to join the *anal canal*—a short wide passage 4 cm long that bends sharply backward to open externally at the anal orifice.

The rectum has a wavy course from side to side, and its lower portion, or *ampulla*, is capable of considerable distension; two or three valve-like folds of mucosa project into its lumen. Only the upper portion has a peritoneal coat, and that only on the front and back sides of the bowel; the main part of its wall is longitudinal and circular muscle, and the lowest part of the latter is thickened as a powerful ring gripping the ano-rectal junction, the *internal sphincter*.

The anal canal is also widely distensible for the passage of faeces; its upper part is lined with mucosa, but the lower 1·3 cm by inturned skin continuous with that of the perin-

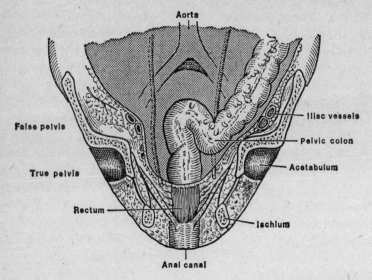

Fig. 131. Coronal section of pelvis to show the rectum and anus. Note the distinction between true and false pelvis; also the lower limit of the peritoneal cul-de-sac. (After *Gray*.)

eum. And immediately under the peri-anal skin, encircling the orifice, is another circular muscle, the *external sphincter*. Both sphincters are normally closed; in the act of defaeca-tion they relax, the abdominal wall tightens and voluntarily raises the abdominal pressure, and the rectum is lifted up over its contained faeces by the muscular floor of the pelvis.

Other pelvic structures

The *ureters* cross the pelvic brim by passing over the com-mon iliac vessels; travel down the sidewalls of the true pelvis under the peritoneum, 5 cm away from the rectum;

and then sweep inward and forward in a wide curve to reach the angle of the bladder.

Vessels. The *common iliac* vessels pass downwards and out from their origin, and divide at the level of the lumbo-sacral joint into external and internal iliac branches. The *external iliacs* continue the same course along the pelvic brim, together with the psoas tendon and the femoral nerve more laterally, and pass beneath the inguinal ligament to emerge as the femoral vessels in the front of the upper part of the thigh. The *internal iliac* vessels descend sharply down the sidewall of the true pelvis and break up into several branches supplying bladder, rectum and genital organs.

Nerves. We have already noted the *lumbar plexus* on the posterior abdominal wall, and this is continued in the pelvis as the *sacral plexus* of nerve roots, emerging from the foramina of the sacrum and lying on the posterior wall of the cavity. The main branch of the lumbar plexus is the *femoral nerve*, which accompanies the psoas tendon and external iliac vessels to the thigh; and the main branch of the sacral plexus is the *sciatic nerve*, which leaves the back of the pelvis to emerge in the depths of the buttock behind.

The *sympathetic chain* is continued down from the abdomen to run on the front of the sacrum and coccyx, ending as a single fused ganglion on the latter bone.

Uterus, uterine tubes, ovaries, vagina, prostate, seminal vesicles, testicles and their ducts are described in detail in Chapter 22, in connection with reproduction.

The thorax is a complete and intricate bony cage, containing the heart and lungs. Its apparent cylindrical shape is

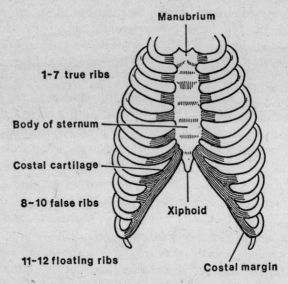

Fig. 132. The bony thoracic cage, seen from in front.

illusory and due to the upper projection of the shoulders on each side; when these are removed (Fig. 132), it is seen to be conical or barrel-shaped, with quite a narrow apex.

Bony wall

In the midline behind are the twelve thoracic vertebrae of the spinal column, and in the midline anteriorly the breast bone or sternum; between them are the encircling ribs. Note, in cross-section (Fig. 135), the paravertebral recess on each side of the spine due to the backward curve of the ribs before they turn forward.

Ribs

There are twelve pairs of ribs, separated by the *intercostal spaces* which contain the intercostal muscles, nerves and vessels. The first seven are *true ribs*, complete rings from spine to sternum; the eighth, ninth and tenth are *false ribs*, not reaching the sternum but turning up to join the ribs above; and the eleventh and twelfth are short *floating ribs*, ending in free blunt conical ends, embedded in the flank muscles of the abdomen. The first and twelfth ribs are very short, the intermediate ones reaching their greatest size at

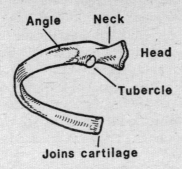

Fig. 133. A typical rib, seen from behind.

the seventh and eighth, so that the barrel contour of the chest slopes away in again above and below this level. Each rib also slopes somewhat downwards, its sternal end being lower than its vertebral end.

Each rib possesses a *head*, articulating with the side of the vertebral body; a short *neck*, lying on the transverse process of the vertebrae; a *body*, which runs back a little and then changes direction sharply at the *angle* to sweep forwards; and, finally, the *costal cartilage*, a gristly rod of 2·5 or 5 cm connecting bony rib to sternum. The framework is a little deficient below anteriorly, where the costal arch is formed by the right and left costal margins.

The *intercostal spaces* are filled by layers of muscle, between which the intercostal vessels and nerves encircle the chest, giving branches in their course. One nerve, artery and vein occupy each space, a good example of the pattern of segmentation (page 96).

The *sternum* or breast bone is dagger-shaped, with three main pieces. Uppermost is the broad, flat *manubrium*, to which are attached the inner ends of the clavicles and first ribs on each side. Below this is the main *body*, a flat bone of two compact layers enclosing spongy bone and red marrow, with the costal cartilages attached on either side. And inferiorly is the small pointed *xiphisternum*, already encountered in the upper part of the abdominal wall.

The thorax has an *inlet* above, through which the great vessels and nerves pass up into the neck or out over the upper surface of the first rib into the axilla and arm. The inlet is narrow and closely packed, only 6·5 cm separating the back of the manubrium from the front of the spine, with the first ribs on either side. The most important structures traversing the inlet are the *oesophagus* (gullet), *trachea* (windpipe), certain *nerves*, and the great arteries and veins. There is no inferior outlet as the thorax is sealed off below by the diaphragm.

Respiration

The purpose of respiration is to draw air into and push air out of the lungs. In inspiration the chest cavity is enlarged and air enters; in expiration the reverse occurs. Respiratory movements are both thoracic and abdominal, though either may occur separately. In *thoracic* inspiration the sternum is lifted up by the elevation of the ribs, which come to lie more horizontally, and both the transverse and antero-posterior diameters of the chest are thus increased. In *abdominal* inspiration the diaphragm contracts and its domes flatten out, pressing the abdominal organs down and bulging out the abdominal wall; in this way the vertical height of the thorax is increased. Thus inspiration is an active process, due to muscular exertion. Expiration, on the other hand, is passive; the chest wall subsides, the abdominal wall recoils, the diaphragm relaxes and air is driven out of the lungs.

The muscles of the chest wall

Anterior

In front the chest wall is covered by the great mass of the *pectoralis major*, to which we have already referred in connection with the arm. Arising from sternum, clavicle and ribs, its fibres converge laterally as they are inserted into the upper humerus; its function is adduction of the arm to the side. A smaller *pectoralis minor* lies under cover of the major, arising from three or four ribs and inserted into the coracoid process of the scapula. In the lower part of the chest wall, some of the abdominal muscles are attached to the ribs—the *rectus* near the midline and the *oblique* muscles more laterally.

Posterior

These are shown in Fig. 134. Two great superficial sheets of muscle are the trapezius above and latissmus below. The *trapezius* is a triangular muscle, with its base attached to the spinous processes of the thoracic vertebrae and extending up the back of the neck as far as the occipital region of the skull. The muscle narrows rapidly as it converges to its insertion on the spine of the scapula and the back of the clavicle. As we have seen, its function is to rotate the scapula on the chest wall (Fig. 54(b)) and thus aid the movement on abduction at the shoulder joint.

The *latissmus dorsi* is somewhat overlapped by the trapezius above. It originates from the lower thoracic vertebrae and from the lumbar spine via a dense sheet of lumbar fascia. And its fibres pass away and up to be inserted into the humerus at the same level as the pectoralis

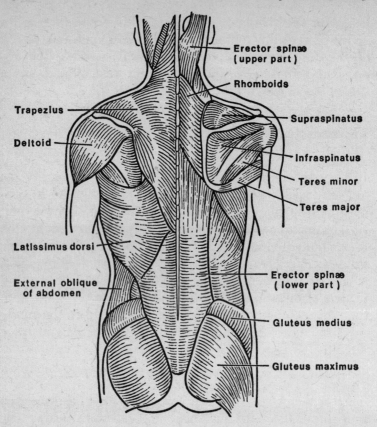

Fig. 134. Muscles of the back. On the right side the superficial muscle layer has been removed to expose the deeper structures.

major; like that muscle, it adducts the arm. Placing one's fingers in the armpit, a thick muscle boundary is felt both in front and behind; the one in front, the anterior wall of the axilla, is the pectoralis; the one behind, the posterior wall, is the latissimus.

Under cover of the superficial muscles behind are the *erector spinae* muscles, which, grouped together, form a thick, rounded belly stretching from skull to sacrum on each side of the vertebral column. It gains attachment to the ilium, vertebrae, ribs and skull at different levels, and by its tonus helps to maintain the normal curves of the spinal column in the erect posture. Its action is to extend, i.e. straighten out, the spine—including the head and neck.

Surface anatomy

Anterior
If the clavicles are traced to their medial ends, there can be felt between them, in the hollow of the base of the neck, the manubrium of the sternum. Five centimetres lower, in the midline, the junction of manubrium and body of sternum is marked by a prominent ridge, the sternal angle; it is easy to feel the costal cartilage of the second rib attached here on either side.

The main body of the sternum is subcutaneous, a broad plate of bone ending below as the little xiphoid process, which is rather depressed below the surface in the angle formed by the meeting of right and left costal margins. The ribs are attached on each side of the sternum and are best numbered by first locating the second rib, as indicated, and then counting downwards, for the inner end of the first rib lies so deeply beneath the clavicle that it cannot easily be felt. The ribs are palpable well out into the side and even round to the spine, depending on the amount of overlying fat and muscle, and the intercostal spaces are to be felt as depressions between.

The *nipple* lies in the fourth space in men, i.e. between fourth and fifth ribs, but is more variable in women owing to the size of the breast; the female breasts are described in Chapter 22. The beating of the apex of the heart is felt or seen just medial to the left nipple, i.e. in the fifth space on the left side, some 9 cm from the midline, but this again is inconstant.

The surface markings of the heart, lungs and pleura* are

*The pleura is the lining membrane of the chest cavity, *cf.* the peritoneum in the abdominal cavity.

shown in Plate 7(a). Note that the lungs do not extend quite
as far as the lower borders of the pleural cavity. This does
not imply an empty pleural space below; the space is potential only, comprising simply a narrow cleft here between diaphragm and chest wall where the two pleural
layers come in contact. The apex of the pleura just emerges
into the root of the neck on each side, and the two pleural
sacs touch each other behind the sternum.

Posterior

The back of the chest is obscured by the overhanging
shoulder blades above on each side and by the great muscle
masses of this region. The spinous processes of the thoracic
vertebrae are felt in the midline, but the ribs lie deeply and
can only be felt by heavy palpation. The eleventh and
twelfth ribs are quite short, extending outwards for only a
few inches on each side of the spine.

Thoracic cavity

The thoracic cavity is divided into right and left halves by
a massive partition, the *mediastinum*, which lies in the sagittal
midplane, stretching from the back of the sternum to the
vertebral column. Vertically, it extends from thoracic inlet
above to diaphragm below. The two halves of the cavity are
completely separate, and contain the right and left lungs.
Each lung is attached to the mediastinum at its root or
hilum by its bronchus (air-tube) and blood vessels.

The cavity is lined by a smooth serous membrane, the
pleura, which facilitates the gliding of the lung on the chest
wall. Each pleural space is thus a closed sac, for the pleura,
like the peritoneum, is divided into *parietal* and *visceral*
layers which are everywhere continuous. The *parietal pleura*
clothes the deep aspect of the ribs, the upper surface of the
diaphragm and the sides of the mediastinum; and the
visceral layer encloses the lung, the two layers being con-

tinuous along the root of that organ. The apex or dome of the pleural sac rises into the root of the neck, a little way above the clavicle on each side, and its base overhangs the liver in front and the upper poles of the kidneys behind. Although the lung closely follows the pleural distribution, it does not reach quite as far as its upper and lower limits. Normally there is no actual pleural cavity, for the two

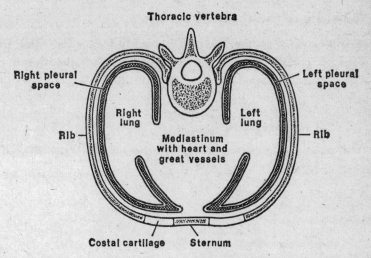

Fig. 135. Cross-section of the chest to show its division into right and left halves by the mediastinal partition.

layers are in contact, the lung entirely filling its side of the thorax. But the lung is a very elastic structure, constantly tending to shrink down and expel its contained air. This cannot occur normally, and a negative pressure is thus set up in the potential pleural space; the moment that air is admitted to the latter, by wound from without or perforation of the lung from within, the lung immediately shrinks into a small, solid, airless mass, collapsed against the mediastinum.

The *heart* is embedded in the mediastinum itself, occupying a central position in the chest between the two lungs, though extending more towards the left side. It is enclosed in a fibrous sac, the *pericardium,* which rests on the central tendon of the diaphragm below. Plates 8 and 9 show the right and left sides of the thoracic cavity after removal of the lungs; it can be seen how the intervening mediastinal partition is packed solidly with vital structures—heart, great vessels, trachea and oesophagus.

Contents of the thoracic cavity

The general arrangement of heart and lungs is indicated in Figs 39 (page 104) and 136. Fig. 39 shows the thoracic viscera after removal of the anterior chest wall, and it will

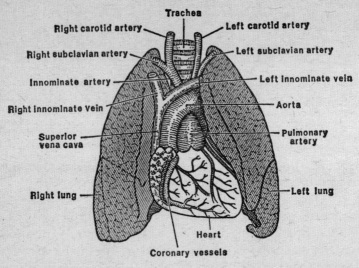

Fig. 136. Lungs, heart and great vessels, seen from in front after removal from the chest. The lungs normally overlap the greater part of the heart and have been retracted to expose that organ. (After *Gray.*)

be noted how the lungs overlap the heart so that only a small part is in contact with the back of the sternum. Fig. 136 shows the organs removed from the chest, with the lungs turned back to expose the great vessels and heart; the fibrous bag or pericardium enclosing the latter has been opened.

Trachea and bronchi

The *trachea* enters the chest from the neck at the thoracic inlet and runs down to the level of the sternal angle, where it divides into a *right* and *left* main *bronchus* for each lung. In its short course it has the oesophagus behind, separating it from the spine, and the great vessels—aorta and superior vena cava—in front, separating it from the back of the

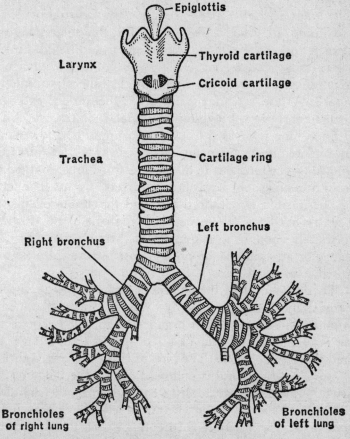

Fig. 137. The respiratory passages.

sternum. The bronchi enter the lungs at their roots, together with the pulmonary arteries and veins. All these structures are shown cut across in Plates 8 and 9 as they project laterally from the mediastinum to reach the hilum of the lungs. Trachea and bronchi have in their walls numerous cartilaginous rings, which keep the hollow tubes from collapsing and maintain a free air passage. In the trachea these rings are deficient behind, so that the windpipe is semicircular in section (see Fig. 148, page 267).

Lungs

The lungs are light, spongy organs designed to provide the maximum internal surface area for the interchange of oxygen and carbon dioxide between air and blood. Pink at birth, they become increasingly bluish-black with age in town-dwellers, owing to the carbon deposited from inhaled smoke. A fine reticular pattern on the surface indicates their underlying subdivision into small lobules.

Each lung is divided into a main upper and lower lobe by an oblique fissure running downwards and forwards; and the right lung has in addition a third lobe, marked off from the upper lobe by a short transverse fissure. The lungs are permeated by elastic tissue, and the action of this has been referred to. When removed from the chest, they immediately collapse to a quarter of their normal size. The whole or any part of the organ floats in water, unless rendered solid by disease.

The lungs lie free in the chest, enclosed in the visceral pleura, and are attached only by their roots to the mediastinum. The *apex* of each organ rises into the root of the neck, 4 cm above the first rib—i.e. 1·3 cm above the middle of the clavicle. The *base* rests on the upper surface of the diaphragm and is hollowed out by the bulge of that partition, which separates it on the right from the right lobe of the liver and on the left from the left lobe, stomach and spleen. The *outer surface* is in contact with the deep inner aspect of the chest wall and is therefore marked by the overlying ribs. The rounded *posterior border* fits into the recess at the side of the thoracic vertebrae, and the sharper *anterior border* overlaps the heart and pericardium on each side. Because of the preponderant bulk of the heart on the left side, there is a well-marked cardiac notch in the anterior border of the left lung.

The *medial surface* faces inwards towards the mediastinum and on it is situated the hilum. The relations of the medial surfaces differ on the two sides, for they lie opposed to the mediastinal structures, which are not arranged symmetrically. Plates 8 and 9 show more clearly than words the general pattern of these structures, which make deep indentations on the lungs. The main differences are:

(1) the curvature of the arch of the aorta over the left bronchus to the left side and the prominence of the descending aorta on the left side;

(2) the presence of the superior vena cava on the right;

(3) the right-sided position of the oesophagus below.

Internal structure of the lungs. Each main bronchus divides into a branch for each lobe, and these branches divide and subdivide within the lobes into a ramifying bronchial tree of fine *bronchioles*; all these subdivisions are accompanied by corresponding branches of the pulmonary artery and vein. Each bronchiole ultimately ends in a cluster of tiny air-filled sacs or *alveoli*, and the gaseous interchange occurs between their contained air and gases dissolved in the blood of the capillaries which encircle the alveoli. Each lung lobule is composed of a terminal bronchiole and its air cells, and the lung itself is composed of millions of lobules bound together by elastic connective tissue. Note that, whereas arteries usually convey fresh oxygenated blood to an organ and veins remove stale blood with excess carbon dioxide, these conditions are reversed in the lungs. For, since it is their function to renew the oxygenation of the blood of the whole body, the blood brought to them by the pulmonary artery is really venous, while that returned to the general circulation in the pulmonary veins is fresh and arterial.

Heart

The heart is a hollow, muscular organ lying between the lungs in the mediastinum; it is enclosed in the *pericardium*. This membrane has a tough outer *fibrous layer*, which is firmly blended below with the central tendon of the diaphragm, and a delicate inner *serous layer*. The serous pericardium lines the fibrous capsule and is reflected at the

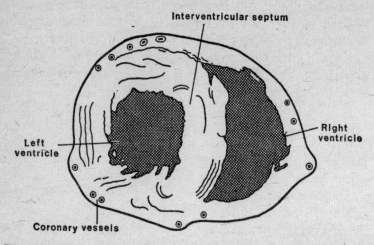

Fig. 138. Cross-section of heart in ventricular region. Note the greater thickness of the left ventricle, the massive septum and the coronary vessels cut across.

roots of the great vessels of the heart to cover the organ itself, i.e. there are parietal and visceral layers just as with the pleura and peritoneum, with a potential pericardial cavity.

The heart itself is composed of four hollow chambers: the two *atria* above, which receive blood from the great veins, and the two *ventricles* below, which expel it into the great arteries. Although the atrium and ventricle of the same side communicate freely, they are completely shut off from the opposite pair by a muscular *septum*, which is much thicker between the ventricles than between the atria. The latter are thin walled compared with the fleshy muscle of the ventricles, and the wall of the left ventricle is considerably thicker than that of the right. On the surface of the organ, grooves or sulci indicate the demarcation between the chambers, and in these grooves run the little, but all-important, *coronary vessels* which nourish the muscle of the heart itself (Fig. 136).

The shape of the heart is roughly conical, with the apex directed to the left and slightly downwards. Its surface projection has been indicated in Plate 9, and the anterior surface—behind the sternum and costal cartilages—is formed mainly by the right ventricle with a little of right atrium and left ventricle on either side. The right border is entirely right atrium, the left border is left ventricle and the apex is the tip of the latter. The left atrium lies entirely behind, on the posterior surface, which is directed to face the vertebral column, the oesophagus and descending aorta intervening.

The attachments of the great vessels. Two great veins enter the right atrium: the *superior vena cava* at its upper part and the *inferior vena cava* at its lower. This venous blood flows through the chamber to the right ventricle and is expelled into the *pulmonary artery* arising from its upper border, going to the lungs in the two branches of this vessel. At the back of the heart, four *pulmonary veins* return fresh blood from the lungs to the left atrium, which passes it on to the left ventricle; the latter expels it to the great *aorta*, the wide arterial trunk that is the commencement of the general body circulation. Although the aorta rises on the left and the pulmonary artery on the right, this relation appears reversed from in front because the two vessels are spirally intertwined.

The interior of the heart. The openings between the chambers, and between the ventricles and their great arteries, are guarded by a system of cardiac *valves*. Between each atrium and its corresponding ventricle is a parachute-like valve, the strands of which are attached to the ventricular wall, the flaps lying in the orifice between the chambers. On the right this is known as the *tricuspid* valve, with three flaps, and on the left as the *mitral*, with two flaps. When the atria contract, the bloodstream easily separates the valve flaps as

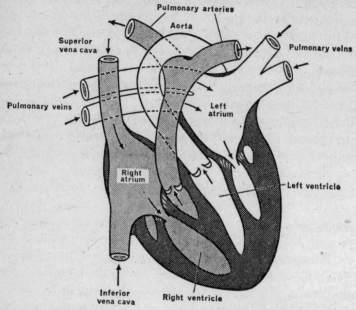

Fig. 139. Diagram showing the interior of the heart, entry and exit of the great vessels, and the direction of blood flow. The diagram is artificial in that no section normally exposes all four chambers in this way.

it passes into the ventricles; but when the latter undergo their contraction, or *systole*, the flaps are ballooned out, the guy-ropes of the valves are tautened and no backflow is possible, all the blood being directed into the arterial exits. These latter, the openings into the pulmonary artery and aorta from the corresponding right and left ventricles, are guarded by simpler valves composed of three semilunar pockets facing upwards. These are easily pushed apart by the blood surge, but fall together when the ventricles relax in *diastole*, the weight of the blood column in each pocket forcing them into a close contact and effectively blocking any reflux.

Great vessels of the thorax (Plates 8 and 9)

The *superior vena cava* is formed by the junction of the right and left *innominate veins*, each bringing venous blood from one arm and one side of the head and neck. The right innominate and the vena cava itself run vertically in a straight line continuous with the right atrium. But the left innominate crosses from one side to the other in front of the trachea and just above the arch of the aorta.

The *inferior vena cava* rises from the abdomen through the central tendon of the diaphragm and has only a very short intrathoracic course before it enters the lower part of the right atrium.

The *pulmonary artery* arises at the upper border of the heart from the summit of the left ventricle and twines round the aorta, so that the two vessels are related in the same way as are adjacent fingers when crossed over each other. Behind the aortic arch, it divides into the right and left pulmonary arteries, which pass out of the mediastinum into the roots of the lungs with the corresponding pulmonary veins and bronchi. The *pulmonary veins* are not seen from in front; two on each side, they drain into the back of the left atrium.

Aorta

This is the largest artery of the body, the beginning of the general or *systemic* circulation—as opposed to the local *pulmonary* circulation. It arises at the upper border of the heart from the summit of the left ventricle and its subsequent course is as follows:

(1) a short *ascending portion*, running up to the left of the superior vena cava;

(2) the *aortic arch*, curving horizontally backwards in

front of the lower part of the trachea and over the left
bronchus;

(3) the *descending aorta*, running downwards on the vertebral column at the back of the thoracic cavity to pass
through the diaphragm, where it becomes continuous with
the abdominal aorta.

Branches of the aorta. From the summit of the aortic arch
three great vessels arise. On the right is the *innominate*, which
divides into the right *subclavian* artery for the arm and the
right *common carotid* for the head and neck. On the left there
is no innominate artery, the left subclavian and carotid
springing directly from the arch. Only the lower parts of
these great vessels lie within the thorax, the level of the
innominate bifurcation being at the inner end of the
clavicle; they lie closely applied to the front and sides of the
trachea. The corresponding veins lie more superficially and
there is an innominate vein on both sides, not only on the
right; the vessel equivalent to the common carotid artery is
the internal jugular vein. The little *coronary arteries* arise
from the very beginning of the aorta, immediately above the
semilunar valves. There is a right and a left vessel, and any
disease of these seriously or fatally impairs the local circulation in the muscular wall of the heart itself; indeed, coronary disease after middle life is a common cause of heart
attacks and sudden death.

The *descending aorta* gives off on each side the pairs of
intercostal arteries that encircle the chest in the intercostal
spaces; note that the corresponding veins do not enter the
venae cavae directly but through separate intervening
channels, the *azygos* veins.

It can be seen that the aortic arch forms a great sweep
from the front of the chest to the back, with an upward
convexity from which arise the main arteries of the head,
neck and arms. Plate 9 shows this, and also how the concavity of the arch embraces the pulmonary arteries and
veins and the bronchi, i.e. the structures entering the roots

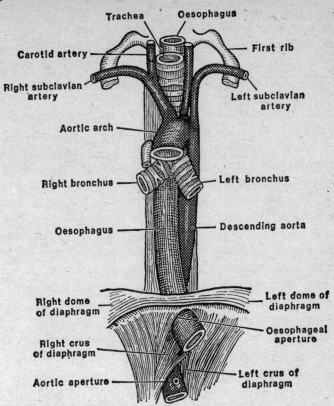

Fig. 140. Structures of the posterior thoracic wall: oesophagus and descending aorta. The heart and lungs have been removed. (After *Gray*.)

of the lungs. Note also in Plate 9 a little artery running the whole length of the back of the sternum from above downwards. This is the *internal mammary* artery, a branch of the subclavian in the neck, and we have seen that it ends below by entering the sheath of the rectus muscle as the superior epigastric artery to anastomose with the inferior epigastric coming up from below.

The arch of the aorta not only carries it backwards but also inclines it to the left, so that the descending portion is rather to the left of the midline and nearer the left lung than the right.

Remaining structures of the posterior thoracic wall

The *oesophagus*, or gullet, enters the chest from the neck at
the thoracic inlet, lying immediately in front of the vertebral
column; it travels down the middle of the back of the
cavity between the roots of the lungs to traverse the diaphragm below. It does not pursue a straight course, for it
inclines forward and to the left in its passage, allowing the
descending aorta to intervene between its lower portion and
the spine; and it is more closely related to the right than the
left lung. It is a hollow, muscular tube down which food is
propelled to the stomach by waves of contraction or
peristalsis; unlike many other parts of the bowel, it has no
serous coat as it lies entirely away from the pleura at the
back of the mediastinum. For this reason, any infection
arising from a perforation of this tube does not get sealed
off and usually spreads diffusely with fatal results.

The *thoracic duct* begins below, as we have seen, as the
upward continuation through the aortic aperture of the
diaphragm of the cisterna chyli of the abdomen, the channel
receiving digested fat from the bowel as well as the lymphatic drainage from the legs. The duct ascends on the
front of the vertebral column in the chest, lying just to the
right of the aorta, behind the oesophagus; it crosses to the
left side at the level of the aortic arch, enters the root of the
neck on that side, and discharges its contents into the junction of the left subclavian and internal jugular veins. Because this is the main channel for the absorption of fat, any
wound of the duct that disperses its contents by leakage
before they enter the bloodstream results in serious or fatal
wasting, even though the injury is not otherwise serious.

Nerves within the thorax (Plates 8 and 9)

The important nerve trunks within the thoracic cavity are:

(1) The *vagus* nerves, one on each side. These are the
tenth pair of cranial nerves which have travelled down the
neck and entered the thoracic inlet. The *right* vagus runs

alongside the trachea, passes behind the right lung root, where it breaks up into a pulmonary plexus, reforms again as a single trunk and accompanies the oesophagus through the diaphragm. The *left* vagus runs with the left carotid and subclavian arteries, crosses the arch of the aorta, passes behind the left lung root to form a similar plexus and finishes the rest of its thoracic course alongside the oesophagus, as does the right. The vagi are part of the great *parasympathetic* system described at page 444 concerned with automatic regulation of the viscera; in the chest they give branches to heart, lungs and oesophagus.

(2) The *sympathetic* trunks lie one on each side, 3–5 cm away from the vertebral bodies. They consist of a chain made up of a number of sympathetic *ganglia* lying on the necks of the ribs behind the parietal pleura and connected by the longitudinal sympathetic fibres. This chain is continuous with that in the neck above and leaves the thorax below with the psoas muscle to become the abdominal sympathetic. It also is part of the *autonomic* or *vegetative nervous system* and shares with the vagus the dual control of the viscera. Plates 8 and 9 show several branches running down from the main trunk; these are the *splanchnic nerves*, which pierce the diaphragm to joint the coeliac or solar plexus we have already noticed surrounding the beginning of the abdominal aorta.

(3) The *intercostal nerves*, derived from the spinal cord, run round the chest wall in the intercostal spaces with the corresponding vessels. At their origins they lie deeply behind the pleura and are connected to the sympathetic ganglia by little communicating twigs which link the voluntary and involuntary nervous systems into a whole.

(4) The *phrenic nerves* arise in the lower part of the neck for the supply of the muscle of the diaphragm. The right phrenic runs down alongside the superior vena cava and the right side of the heart; the left crosses the aortic arch and the left side of the heart. Both pierce the dome of the diaphragm and break up into twigs on its undersurface.

The head

The skull

The skull is an intricate framework of many bones, fitting into each other in jigsaw fashion at the immobile fibrous sutures. Although complicated at first sight, it is easily

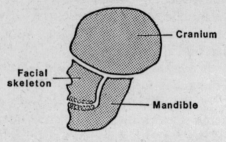

Fig. 141. The basic plan of the skull.

understood by reducing it to its three essential parts: cranium, facial skeleton and mandible.

The *cranium* is the bony box enclosing the brain and its membranes; it has a domed vault and a base set deep beneath the soft structures.

The *facial skeleton* is attached to the underside of the base anteriorly, and includes the nasal bone, maxilla (upper jaw) and others.

The *mandible* or lower jaw is entirely separate and is slung to the underside of the base at the back by a freely mobile joint.

The skull bones are mostly flat; in several places, particularly the vault, they have an outer and inner table of compact bone sandwiching a loose red-marrow-filled layer—the *diploë*. And at certain sites they are expanded by internal air spaces or *sinuses*.

The vault is made up of four main bones: frontal, two parietals and occipital. The *frontal* bone is the substance of the forehead and overhangs the orbital cavities in front. The *parietals* lie on each side above the temples and the *occipital* at the back of the head. The *coronal suture* separates frontal

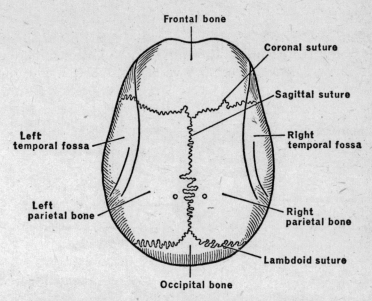

Fig. 142. The vault of the skull, or vertex, from above.

and parietals; the *sagittal suture* divides the two parietals in the midline; the *lambdoid suture* posteriorly is the meeting-place of parietals and occipital. At birth, the angle of junction at either end of the sagittal suture is a small space unclosed by bone and filled in with membrane, the anterior and posterior *fontanelles*; the posterior is closed by the end of six months, the anterior not till the second year. The vault is very resilient and often yields to violence without fracturing; but it may transmit a compression wave round to the rigid base so that the latter undergoes an indirect fracture.

Figs 143 and 144 are anterior and lateral views of the whole skull. Note, in the anterior view, the frontal bone coming down as a roof over the *orbital cavities*, the bony eyesockets; the openings of the *nasal* cavities, with the nasal bones above and the bony nasal septum separating the two

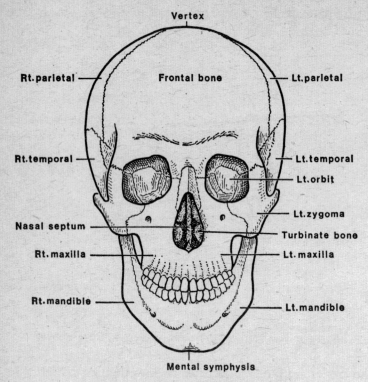

Fig. 143. The skull, anterior view.

sides; and the *maxilla*, or upper jaw, carrying the upper teeth. Note, in the *lateral* view, the side aspect of the vault and how the temporal bone helps to form the lower part of the sidewall in the hollow above the cheek; the cheek bone, or *zygoma*, arching across from maxilla to temporal; the *mastoid process* behind the bony external opening of the ear canal; and the *styloid process* jutting out from the base below.

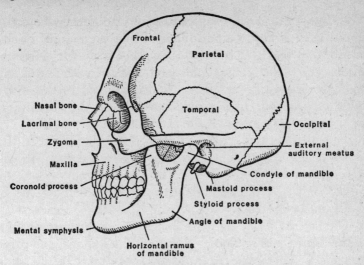

Fig. 144. Lateral view of skull.

The skull base. If we remove the skull cap and look in, our view of the upper surface of the base is shown in Fig. 146.

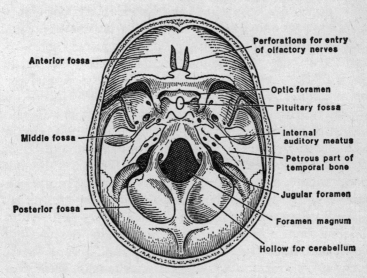

Fig. 145. The skull base, seen from within, after removal of the skull cap.

Note the great *foramen magnum* nearer the back, where the spinal cord becomes continuous with the medulla of the hindbrain; and the numerous scattered smaller *foramina*, where veins and cranial nerves make their exit and the cerebral arteries their entry. At various places the inner table is grooved by narrow channels for the meningeal arteries supplying the brain membranes, or marked by wider paths for the venous sinuses which flow between the layers of the dura mater, the outermost of these coverings. The base is divided on each side into three pockets or *fossae*, which become successively deeper and more extensive from before backwards.

The *anterior fossa*, right in front, houses the frontal lobe of the brain; and its floor is the roof of the underlying orbit and nose. Where the two anterior fossae meet in the midline, tiny perforations in their floor exist for the upward passage of the olfactory nerves from nose to brain.

The *middle fossa* houses the temporal lobe of the brain; through a little opening just behind the anterior fossa, the optic nerve connects eyeball and brain. In its floor can be seen large foramina for the exit of the fifth or trigeminal cranial nerve—concerned with sensation in face and jaws, and the movement of the masticatory muscles—and for the entry of the internal carotid artery into the skull. Note, between the two middle fossae, a little pocket embedded in a small raised platform; this is the *pituitary fossa*, in which lies the pituitary gland, slung from the undersurface of the brain.

The *posterior fossa* is separated from the middle by an obliquely-lying bony ridge, the petrous portion of the temporal bone, a mass enclosing the inner-ear cavity and containing the opening of the internal auditory meatus for the passage of the eighth (acoustic) nerve from ear to brain. The posterior fossa is mainly composed of the occipital bone; it contains the foramen magnum and a large jugular foramen for the exit of the internal jugular vein, accompanied by various nerves.

On the undersurface of the base, on each side of the

foramen magnum, are two articular processes or condyles, forming a joint with the first cervical vertebra, the atlas.

The *mandible* is formed by two symmetrical halves joining in the midline anteriorly at the chin, the *mental symphysis*. Each half consists of a horizontal *body*, carrying the lower teeth on its alveolar (gum) margin, and a more or less vertical *ramus*. Body and ramus are connected at the angle of the mandible, an angle that is rather obtuse in infancy, approaches 90° in the adult and recedes again in edentulous old age. At the upper end of the ramus are the pointed *coronoid process* and the rounded *condyle*, which articulates with the undersurface of the temporal bone at the *temporomandibular joint* in front of the ear (Fig. 144).

The *air sinuses* of the skull are mostly accessory to the nose, i.e. they are extensions of the nasal cavities, and their lining mucous membrane is similarly continuous, even though the connecting passages may be intricate and narrow. The main ones are the *frontal* and *maxillary*; there are also certain groups deep in the skull—*ethmoids* and *sphenoids*. The maxilla is entirely hollowed out by a great cavity, the *maxillary antrum*. All these are susceptible to infection derived from the nose, and because their drainage is so imperfect they may perpetuate the infection in chronic suppurating pockets. The *mastoid* air cells, which honeycomb the mastoid process, are independent structures but are often infected from the middle ear.

The *hyoid* is a little bone which can be felt in the neck in front, between chin and larynx, giving attachments to the muscles of the floor of the mouth and the tongue. It has a tiny body with spreading lateral wings.

Surface anatomy of the head

Most of the external surfaces of the cranium, facial skeleton and mandible are easily palpable through the skin, though the base is, of course, quite inaccessible. In the *cranium* the vault can easily be felt through the overlying scalp, which is freely mobile over the bone. In front the overhanging ridges of the frontal bone, which are the upper margins of the orbital cavities and lie immediately beneath the eye-brows, meet in the midline at the depression of the root of the nose. There are four bulges on the vault: the *frontal eminences* of the forehead and the *parietal eminences* a couple of inches above the ears. And the lowest accessible part of the skull in the midline posteriorly is the external protuberance of the occipital bone, which is also arrived at by tracing upwards the median furrow between the muscle masses at the back of the neck.

In the *face* the *nasal* bones are the firm skeleton of the upper part of the nose, but the softer bulbous portion at the tip is supported by cartilage only. A finger in the nostril touches the *septum* between the two nasal cavities, and this again is cartilage in front, the bony septum being further back out of reach. The *zygomatic arch*, the cheek bone, is felt running between the orbit and ear; above it is the de-pressed temporal fossa, in which lies the temporalis muscle which shuts the mouth and can be felt to contract on clenching the teeth.

The *auricle* is the external visible part of the ear, the actual aperture being the *external auditory meatus*; note that this channel is again cartilaginous to begin with, its bony por-tion being further in. Immediately behind the ear is the mastoid process, with its rounded tip below; and just in front of the ear, below the back of the zygoma, the con-dyle of the mandible can be felt moving at the *temporo-mandibular joint* when the mouth is opened and closed.

Body, ramus and angle of the mandible are all accessible, though the coronoid process is under cover of the zygoma and not palpable. The outer surface of the ramus is covered by the *masseter* muscle, which closes the mouth and stands out on firm biting. If this muscle is so contracted by clench-ing the teeth, a small, tubular structure can be rolled under the skin running parallel to the zygoma and a finger's breadth below; this is the duct of the parotid salivary gland running forward to open into the mouth.

The *scalp* includes all the superficial structures overlying the vault. Its skin is profusely supplied with hair follicles, and the subcutaneous fat is a thin, dense layer; beneath this is a muscle sheet, which is most developed in the frontal and occipital regions as the frontalis and occipitalis bellies, connected by an aponeurotic sheet—the *galea*—stretching over the vault. The galea is separated from the bone by loose areolar tissue, and all the layers of the scalp are welded into one which moves freely over the skull when its muscles contract, as in raising the eyebrows. Retraction of these muscles causes incised wounds of the scalp to gape widely, and wounds bleed freely as the arterial supply is abundant. The scalp veins communicate freely with those of the diploë, the venous sinuses of the dura and of the brain, so that an external infection may spread inwards to the brain.

The face

The *facial skin* is thin, mobile and vascular, and the facial muscles, or muscles of expression, lie immediately beneath its dermal layer.

These muscles are shown in Fig. 146. Note that both eye and mouth are surrounded by circular muscles, the *orbicularis oculi* and *orbicularis oris*, which shut the eyelids and lips. Small muscles attached to the nasal cartilages wrinkle the nose and dilate the nostrils. Others running up and down from the corners of the mouth lift or depress these angles; and the *platysma* muscle of the neck is a broad sheet which spreads up over the body of the mandible to take a share in the control of expression.

All these are supplied by the seventh (*facial*) cranial nerve, which leaves the base of the skull in front of the mastoid, runs forward in the parotid gland, and breaks up into diverging branches to face and scalp at its anterior border.

The *muscles of mastication* are a separate and deeper group, supplied by the fifth (*trigeminal*) cranial nerve. They include the *masseter* and *temporalis*—already referred to as

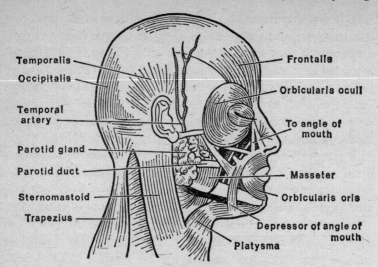

Fig. 146. Facial muscles and associated structures.

responsible for biting—and two others, the *pterygoids*, lying very deeply on the inner side of the ramus, responsible for the side-to-side and rotary movements of chewing. A very deep muscle, the *buccinator*, runs transversely between mouth and masseter in the substance of the cheek.

The *parotid salivary gland* lies under the skin in front of the ear, covering the back of the masseter, and reaches up to the zygoma and down to the mandibular angle. The facial nerve is embedded in it, and it overlies the internal carotid artery and internal jugular vein as they travel deeply between the neck and the skull base. The *parotid duct* leaves its anterior border just below the zygoma and runs forward on the masseter to open into the mouth by piercing the buccinator muscle opposite the second molar tooth.

The other external salivary gland is the submaxillary, which lies below, and partly under cover of, the angle of the mandible; its duct pierces the floor of the mouth.

Facial vessels and nerves. The *superficial temporal artery* runs up in front of the ear, where it can be felt pulsating against the bone, to supply the scalp. The *facial artery* ascends from the neck over the body of the mandible, and runs obliquely upwards and forwards to the angle of eye and nose, giving branches to the lips on its way. Both vessels are branches of the external carotid, and the veins correspond in distribution.

The seventh (facial) nerve has already been mentioned and is mainly motor to the muscles of expression.

The sensation of the face and scalp is supplied by the fifth (trigeminal) nerve, which also innervates the masticatory muscles.

Most of the lymph drainage of the face is to groups of parotid and submaxillary lymph glands—except from the lower lip, which goes to some small glands under the chin.

The mouth

This is the first part of the digestive tract. It is bounded by the cheeks and lips at the sides and in front, and roofed by the palate; and its floor is the muscular oral diaphragm, separating it from the neck, in which is embedded the root of the tongue. The oral cavity is everywhere lined by a glandular mucous membrane and is continuous behind with the cavity of the pharynx. A minor part of the main cavity is the *vestibule*, the space between lips and teeth.

The red surface of the *lips* is the mucous membrane, which is sharply demarcated from the skin at the mucocutaneous junction; each lip has a branch of the facial artery running in it parallel to its free margin, and this is felt pulsating when the lip is gripped between finger and thumb. Each is attached at its midpoint to the underlying gum by a fold of mucous membrane.

The teeth. These are set in the opposed (alveolar) margins of the upper and lower jaws, where mucosa and underlying periosteum are firmly blended to form the gum. The first, temporary, set of milk teeth erupt at intervals throughout the first two years; the second, permanent, set begin to re-place them at six years and are complete by twenty-five, except for the last molar or wisdom tooth which is often de-layed, or may never appear.

The teeth are named from before backwards as *incisors*, *canine* (tearing), *premolars* and *molars* (grinding), and are symmetrical as between either side and as between upper and lower jaws, so that each quarter of the whole set is the same. This may be expressed in the formulae:

	Mol.	Premol.	Can.	Inc.	Inc.	Can.	Premol.	Mol.	
Upper	3	2	1	2	2	1	2	3	
Lower	3	2	1	2	2	1	2	3	= 32,

for the permanent dentition, and:

Upper	2	0	1	2	2	1	0	2	
Lower	2	0	1	2	2	1	0	2	= 20,

for the milk dentition. Note the smaller number in the lat-ter owing to the absence of premolars.

Each tooth has a visible *crown*, projecting beyond the gum; a *root*, embedded in the alveolus, and an intermediate constricted *neck*. Longitudinal section shows the central *pulp cavity*, containing the soft vascular pulp and nerves; outside this the *dentine* or ivory, the main bulk of the tooth, and an outermost coating of thin hard *enamel*. The root is fixed in its socket by a layer of dental cement, and the vessels and nerves enter the pulp through a small foramen at the tip of the root (Fig. 147).

The *palate* is the roof of the mouth and the floor of the nasal cavity. It consists of the bony *hard palate* in front and the mobile, muscular *soft palate* behind, projecting back into the pharynx. The soft palate is an arched structure, with its supporting pillars attached on either side of the back of the tongue; each of these pillars splits to enclose the *tonsil*, a mass of lymphoid tissue. From the free posterior edge of the soft palate there hangs down, midway between the tonsils, a small conical process, the uvula.

In the *floor of the mouth*, if the tip of the tongue is elevated, a fold or frenulum is seen tethering the underside of the tongue to the mucous membrane; and, where the frenulum

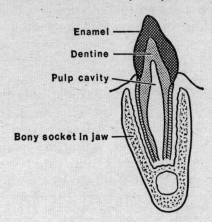

Fig. 147. Longitudinal section of tooth.

is attached to the floor, the duct of the submaxillary salivary gland opens on each side on a little papilla. Between the frenulum and the jaw on each side, a loose mucosal flap can be seen, or rolled by the tongue; this is the sublingual fold overlying the *sublingual salivary gland*, which has many small ducts opening directly into the mouth cavity.

Parotid, submaxillary and sublingual glands make up the three main salivary glands; they discharge saliva reflexly into the mouth with the stimulus of eating.

The *tongue* is a muscular organ concerned with swallowing, taste and speech. Its main anterior portion lies horizontally in the floor of the mouth; but the back of the tongue is curved round to lie more vertically and forms the anterior wall of part of the pharynx. The *root* of the tongue is embedded in the oral diaphragm; it is attached by various muscles to the hyoid bone and mandible, and indirectly to the soft palate by the arches of the latter structure.

The *dorsum* is the upper surface, and on it are a number of *papillae*, hair-like processes, and *taste buds*. The tip rests normally against the back of the upper teeth and the under-surface is attached to the floor of the mouth by its frenulum. In structure the organ is mainly muscular, with right and left halves separated by a fibrous partition; it has a mucous covering which is specialized for the sensation of taste.

The neck

The neck connects the head and trunk; its bony framework is made up by the seven cervical vertebrae behind, and it is traversed by the food and air passages on their way to the chest, and by the great vessels and nerves running between the thoracic inlet and the base of the skull.

In addition, in the root of the neck on each side there is an outflow of nerves and vessels over the upper surface of the first ribs into the arms.

Cross-section of the neck shows the relative disposition of these structures. But it must be remembered that, whereas a section above the level of the sixth cervical vertebra shows the air and food passages as larynx and pharynx, at a lower level these have become the trachea and oesophagus respectively.

The main points to be noted are (Fig. 148):

(1) the supporting *cervical vertebrae*, rather nearer the back of the neck;

(2) the thick *muscle masses* applied to the back of the spine, the upper part of the great sacrospinalis muscle;

(3) the *oesophagus*, immediately in front of the spine;

(4) the *trachea* between skin and oesophagus;

(5) the separate compartments of deep fascia on each side, the *carotid sheaths* in which run the *carotid artery*, *internal jugular vein* and the *vagus nerve*.

Longitudinal section indicates the vertical arrangement, and the level of transition from larynx to trachea and from pharynx to oesophagus (Fig. 150, page 271).

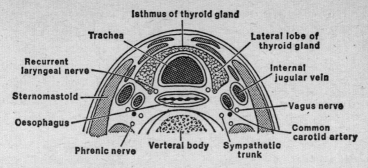

Fig. 148. Cross-section of the neck below the level of the sixth cervical vertebra. The section does not include the back half of the vertebra and associated spinal muscles. (After *Gray*.)

Triangles

The neck is conventionally divided into several triangles, which are marked off by the various muscles. The key to this pattern is the *sternomastoid*, a long, flat rectangular muscle on each side, whose origin is by two heads from the inner end of the clavicle and upper border of the manubrium; it passes up and out to its insertion on the mastoid process. The area in front of the muscle is the anterior triangle and that behind it the posterior triangle (Fig. 150).

The *anterior triangle* is bounded by the anterior midline of the neck and the anterior border of the sternomastoid, meeting at the apex below; its base, above, is the mandible. It is further subdivided as shown into four smaller triangles. The *muscular triangle* contains several small, flat muscles which depress the larynx in swallowing; the *carotid triangle* contains the upper part of the common carotid artery as it divides into internal and external carotids, the internal jugular vein and important nerves; the *digastric triangle* contains the submaxillary salivary gland; and the *submental triangle* contains only a few small lymph glands.

The *posterior triangle* is bounded by the back of the sternomastoid and the front of the trapezius, meeting at the apex at the occipital bone; its base is the clavicle. It is divided by the tiny omohyoid muscle into an upper *occipital* and a lower

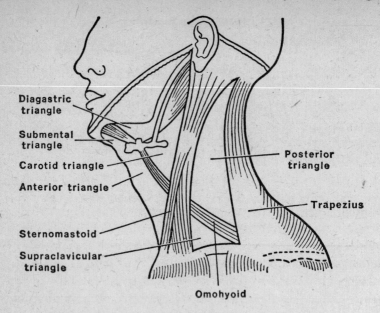

Fig. 149. Muscular triangles of the neck. (After *Gray*.)

supraclavicular triangle. The former is mainly muscular in content, but the latter contains the subclavian vessels and the brachial plexus of nerves, passing out on their way to the arm.

Surface anatomy of the neck

Anterior. The *sternomastoid* is very prominent, especially when rendered taut by turning the head to the opposite side, then its sternal tendon can easily be picked up between finger and thumb. Between the sternal heads of the two muscles is the *suprasternal notch*, a skin-hollow resting on the upper border of the manubrium. The *external jugular vein* can usually be seen running under the skin obliquely down and back, superficial to the sternomastoid.

The main structures in the midline of the front of the neck, from above down, are as follows.

Feeling backwards in the hollow beneath the chin one

reaches the small *hyoid bone*, with its wings on each side, felt when the throat at this level is grasped between finger and thumb. Immediately below the hyoid is the *laryngeal prominence*, or Adam's apple, composed of the two halves of the thyroid cartilage meeting at a V-angle in front, and with a notch palpable at their upper border. Below this is the *cricoid cartilage* of the larynx, at the level of the sixth cervical vertebra behind; and lower still are the cartilaginous rings of the upper part of the *trachea*.

The pulsation of the *carotid arteries* is felt by pressure along the anterior border of the sternomastoids; behind this muscle, the posterior triangle is marked by a broad depression of the overlying skin, and at the base of this, behind the upper border of the clavicle, the *subclavian artery* can be felt beating on deep pressure (Plate 10).

Posterior. At the back of the neck, the median longitudinal furrow between the muscle bellies overlies the indistinctly-felt spinous processes of the cervical vertebrae.

Fasciae and muscles of the neck

Just as elsewhere in the body, there is an envelope of deep fascia between the subcutaneous fat and the underlying muscles; in the neck this forms a continuous encircling sheet. But, in addition, there are deeper concentric layers encircling the trachea and, deepest of all, the vertebral bodies. The muscles are partitioned off between layers of fascia, and a local condensation of this membrane forms on each side of the cylindrical carotid sheath containing the common carotid artery, internal jugular vein and vagus nerve.

The cervical muscles are grouped as follows:

(1) *Superficial: platysma, trapezius, sternomastoid*. Trapezius and sternomastoid have already been met with. Both are supplied by the eleventh (*accessory*) cranial nerve. Each sternomastoid acts by bringing the ear down to the shoulder

on its own side and turning the chin away to the other side; both acting together flex the cervical spine, bringing the chin down to the sternum.

The *platysma* is a muscle of facial expression, a broad sheet replacing the subcutaneous fat, without bony attachment. It arises from below the clavicle, and spreads up the side of the neck and inward over the body of the mandible to blend with the facial muscles at the angle of the mouth.

(2) The *suprahyoid* muscles are a small group between hyoid and mandible, forming the floor of the mouth, or oral diaphragm, and giving attachment to the base of the tongue.

(3) The *infrahyoid group* are several slender strap-like muscles, descending from the hyoid bone and thyroid cartilage on each side of the midline to be attached to the back of the manubrium below. They depress the larynx in swallowing.

(4) The *prevertebral* muscles are the deepest group, running up the front of the vertebral bodies behind all the other structures of the neck; they help to flex the cervical spine.

(5) The *scalene* muscles are found in the posterior triangle on each side, for they run downwards and out from the transverse processes of the cervical vertebrae to the upper borders of the first and second ribs. They elevate these ribs in inspiration, and assume close relationship with the sub-clavian vessels and brachial plexus as these run out over the first rib to the arm.

Pharynx

In the lower part of the neck, and in the chest, the air passage (larynx and trachea) and the food passage (oeso-phagus) are quite separate. But in the upper part of the neck, above the entrances to larynx and oesophagus, there is a common cavity for food and air extending up as far as the base of the skull. This is the *pharynx*, situated immedi-ately in front of the vertebral column.

In longitudinal section of the head and neck (Fig. 151) it can be seen that the pharynx, from above downwards, is placed successively behind the nasal cavity, the back of the mouth and the opening of the larynx, and it is therefore

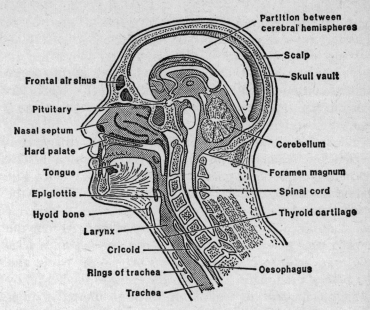

Fig. 150. Head and neck in longitudinal (sagittal) section.

divided into *nasopharynx*, *oral pharynx* and *laryngopharynx*. Its blind upper end reaches the undersurface of the base of the skull, and it is continuous below with the oesophagus at the level of the sixth cervical vertebra.

The pharyngeal wall itself is a muscular cylinder designed to propel food between mouth and oesophagus, and is composed of three *constrictor muscles*—*superior*, *middle* and *inferior*—roughly corresponding to the subdivisions of the cavity. In swallowing, it is essential to prevent food passing upwards into the nose round the free back edge of the soft palate, or down into the larynx. Therefore, the soft palate is opposed to the posterior pharyngeal wall so as to shut off the nasopharynx, and the *epiglottis*, a flap of elastic fibro-

cartilage situated at the summit of the larynx behind the
root of the tongue, is bent over backwards to shut off the
entry into the larynx. During the act, the larynx is pulled
bodily upwards and then recedes again.

Larynx (Plates 10 and 11)

The larynx is the organ of voice and one part of the air
passages of the respiratory system. It opens above into the
laryngeal portion of the pharynx at the back of the tongue
and is continuous with the trachea below. It projects forward in the upper part of the neck with little to separate it
from the overlying skin.

The skeleton of the organ is a framework of cartilages connected by ligaments and membranes to form a hollow
resonating chamber containing the vocal cords. The
thyroid cartilage is the largest; its two leaves meet in the
midline as the projecting Adam's apple, with a superior
notch opposed to the back of the hyoid bone. From the
back of this notch arises the stalk of the *epiglottis*, a leaf-
like fibrocartilage plate sticking up behind the hyoid and
the base of the tongue. It guards the entry to the larynx by
falling backwards to seal it off when swallowed food is
passing through the pharynx. The *cricoid cartilage*, the
lowest piece of the framework, is a signet-ring-shaped
structure, and immediately below is the first ring of the
trachea.

The *vocal cords* are a pair of mucosal folds, strengthened
by an underlying ligament, one on the deep surface of each
thyroid leaf, stretching from front to back. The muscles
associated with the larynx are either *extrinsic*, e.g. the infrahyoid group, designed to move the organ as a whole as in
swallowing, or *intrinsic*, concerned with the regulation of
tension in the vocal cords. The latter may lie relaxed at the
sides of the cavity, meet in the midline and shut off the airway completely, or occupy an intermediate position, giving
varying gradations of pitch. The interval between them is
the *glottis*; this is the narrowest point of the respiratory

tract and therefore the site most likely to be blocked by inflammatory swelling or the impaction of foreign bodies.

The *trachea* or windpipe is a hollow cartilaginous and membranous tube, some 10 cm long, extending from the larynx above through the thoracic inlet to its bifurcation into the two bronchi. It is not a cylindrical but a ◠ -shaped tube, with the convexity under the skin of the front of the neck and the base of the ◠ resting on the oesophagus behind, which separates it from the vertebral column. Its framework consists of a number of cartilaginous rings, incomplete behind, with connecting membranes. In the neck, the second, third and fourth rings are crossed by the narrow isthmus connecting the two lobes of the thyroid gland, one of which lies on each side. And the common carotid arteries lie beside the trachea, the thoracic relations of which have already been discussed (page 243).

The trachea has a lining mucous membrane which, like that of the other parts of the respiratory tract, is *ciliated*— i.e. each cell bears a hair-like process that wafts such foreign irritants as particles of dust upwards toward the base of the tongue, where they are all swallowed.

The *oesophagus* has already been discussed, in part, in Chapter 10. About 25 cm long, it is a muscular channel for the passage of food between pharynx and stomach. In the neck it begins at the level of the cricoid cartilage as the continuation of the lowermost portion of the pharynx. It descends on the front of the vertebral column behind the trachea and enters the thoracic inlet.

The *thyroid gland* is one of the group of endocrine or ductless glands that includes the pituitary, suprarenals, etc. It lies in the lower part of the neck as a right and left lobe on each side of the trachea, with a connecting isthmus crossing that tube (Plate 11). Each lobe is pyramidal in shape, apex upwards, and is applied to the sides of the cricoid and thyroid cartilages; laterally, it overlaps the carotid sheath. The gland is exceedingly vascular and is composed of numerous tiny vesicles containing *thyroxin*, the secretion essential for normal growth, bodily and mental activity; deficiency of this leads to cretinism in infants, and to slowness or idiocy in adults. Any enlargement of the organ is called a *goitre*, which may be associated with overproduction or underproduction of the internal secretion.

The *parathyroid glands* are four pea-like bodies embedded in the lobes of the thyroid, an upper and a lower on each side. They are ductless glands controlling the utilization of calcium, and their overaction causes rarefaction of the skeleton and spontaneous fractures.

Great vessels of the neck

We have seen that, on the right side, the innominate branch of the aorta divides at the level of the sternoclavicular joint into subclavian and common carotid arteries; also that, on the left side, there is no innominate vessel, the subclavian and carotid arising directly from the arch of the aorta.

In the neck the *common carotid* ascends on either side of the trachea and thyroid gland to the level of the upper border of the thyroid cartilage, where it divides into *internal* and *external* branches (Plates 10 and 11). It runs in the carotid sheath, with the internal jugular vein lateral to it and overlapping it in front, and with the tenth cranial (*vagus*) nerve. Just before the bifurcation there is a dilatation, the *carotid sinus*, where the walls of the vessel are liberally innervated and very sensitive to changes in the blood pressure; this carotid sinus automatically maintains a reflex control of the circulation to the brain. Apart from its terminal division, the common carotid has no branches.

The *internal carotid* continues its course to the base of the skull, enters the cranium and supplies the brain as well as the contents of the orbit.

The *external carotid* is mainly responsible for supplying the neck, and the outside of face and head. It gives large branches to thyroid gland, larynx and tongue; gives off the facial artery, as we have seen; and ends by running up behind the ramus of the mandible in the substance of the parotid gland to finish as the superficial temporal artery to the scalp, which can be felt pulsating in front of the ear. It also supplies the main artery to the upper and lower jaws.

The *internal jugular vein* corresponds to the common and internal portions of the carotid artery (Plate 11). It leaves the base of the skull as a continuation of the venous sinuses of the brain through a special foramen—in company with the ninth, tenth, eleventh and twelfth cranial nerves—and runs down the neck, lateral to the carotid artery and more under cover of the sternomastoid. It ends by joining with the subclavian vein behind the sternoclavicular joint to form the innominate vein, and this pattern is the same on both sides. There is no vein corresponding to the external carotid artery, but some veins from the face and scalp join to form the *external jugular vein*, which we have met running superficially over the sternomastoid and which ends by piercing the deep fascia above the clavicle to enter the subclavian vein. The latter arches in over the first rib on each side from the arm.

The nerves of the neck

The main nerves of the neck fall into the following groups:

(1) *Cranial nerves*. The ninth (*glossopharyngeal*), tenth (*vagus*), eleventh (*accessory*) and twelfth (*hypoglossal*) pairs of cranial nerves enter the neck from within the cranium by traversing foramina in the base of the skull. The first three of these travel in company with the internal jugular vein.

The *glossopharyngeal* is concerned with sensation and taste in the pharynx and back of the tongue, and pierces the side-wall of the pharynx in its upper part to supply its mucous membrane.

The *vagus* runs down in the carotid sheath, between and behind the vein and artery, giving branches to pharynx and larynx; we have noted it within the chest and abdomen supplying the viscera, and its place in the autonomic system is referred to later on page 446. An important branch to the intrinsic muscles of the larynx is known as the *recurrent laryngeal nerve*, for it hooks round the underside of the sub-clavian artery on the right to turn back and reach the larynx; the right recurrent laryngeal nerve is therefore entirely within the neck, but the left descends within the thorax to hook round the aortic arch before ascending again. Each recurrent nerve is intimately related to the back of the thyroid gland, and disease or injury here may result in permanent loss of voice.

The *accessory nerve* supplies the sternomastoid muscle and the trapezius; after leaving the first, it runs down and back in the upper part of the posterior triangle to reach the anterior border of the trapezius.

The *hypoglossal nerve* supplies the muscles of the tongue and the floor of the mouth.

(2) *Cervical spinal nerves.* The roots of the spinal nerves emerge in pairs between the vertebrae on each side. The upper roots form a small *cervical plexus* supplying various muscles. And the lower roots, C. 5, 6, 7, 8, together with the first thoracic root, form an important network, the *brachial plexus*, which lies in the lower part of the neck in the posterior triangle and gives rise to the main nerves of the arm. This plexus forms a bundle with the subclavian artery and vein, the whole collection passing out over the upper surface of the first rib to reach the arm. Thus the plexus is partly in the neck above the clavicle and partly in the axilla below the clavicle.

(3) *Sympathetic chain.* In the neck, the sympathetic chain consists of three *ganglia*—the *superior, middle* and *inferior*—with a connecting *sympathetic trunk*; these lie on the pre-vertebral muscles on each side, very deeply behind the carotid sheath. And the trunk is continuous with the thoracic portion of the sympathetic chain through the thoracic inlet.

Lymph glands of head and neck

The main collecting glands for the head and face are those associated with the ear, the pre- and postauricular; those on the surface of the parotid and submaxillary salivary glands; the submental glands under the chin; the occipital glands at the skull base behind; and deep retropharyngeal glands lying between pharynx and vertebral column.

In the neck itself there are superficial and deep chains of glands and lymphatic vessels. Roughly, the superficial glands are grouped around the external jugular vein and accessory nerve, and occupy the posterior triangle; the deep glands surround the internal jugular vein and carotid vessels.

At the base of the neck, on the right side, a major lymphatic trunk is formed from all the vessels draining the right arm and the right side of head and face; this discharges into the angle of union of subclavian and internal jugular veins. On the left side, it will be recalled that the thoracic duct, which has collected all the lymph from both legs, and from the abdominal and thoracic viscera, arrives in the posterior triangle. Here it adds to itself the vessels of the left arm, head and neck, and empties itself likewise into the corresponding venous angle. The thoracic duct is thus much larger and more important than the right lymph trunk, but even so it is often only just visible and even when distended is rarely thicker than a matchstick.

12 Regional anatomy: the spinal column

The spine is the central bony axis of the body and, to allow flexibility, is composed of numerous individual vertebrae in a segmental pattern. These articulate with each other, and the sum of the limited motion between the individual pairs is a considerable range. The column is traversed throughout by the central canal, which encloses the spinal cord; and it supports the weight of the trunk and transmits it to the legs.

The vertebrae are grouped regionally as:

> 7 *cervical* (neck)
> 12 *thoracic* (chest)
> 5 *lumbar* (abdominal)
> 5 *sacral* (hip)
> 4 *coccygeal* (tail).

Although modified considerably in the different regions, the essential features of a vertebra are the same everywhere (Fig. 153). In front a massive, rounded *body* projects forward, with upper and lower surfaces facing those of the vertebrae above and below. Behind is the *neural arch*, which, with the back of the body, forms a complete bony ring—the *neural canal* for the spinal cord. The arch is attached to the body by two pillars or pedicles; projecting at either side are the two *transverse processes*; and the *spinous process* sticks out backwards and downwards from the back of the ring to be felt under the skin of the back. In addition, there is a pair of small *articular facets* on the upper surface of the arch and a pair below for articulation with adjacent vertebrae. The vertebral body consists almost entirely of spongy bone, with a thin, compact shell.

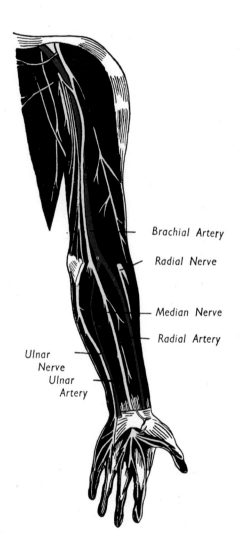

Brachial Artery

Radial Nerve

Median Nerve

Radial Artery

Ulnar
Nerve
Ulnar
Artery

Plate 1. Plan of the main nerves and vessels of the upper limb.

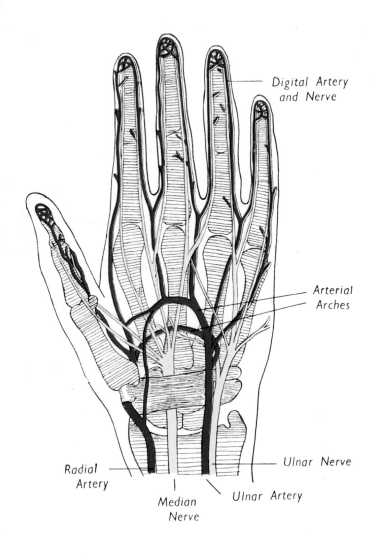

Digital Artery
and Nerve

Arterial
Arches

Ulnar Nerve

Radial
Artery

Median
Nerve

Ulnar Artery

Plate 2. Nerves and vessels in the hand. (After *Bruce and Walmsley*.)

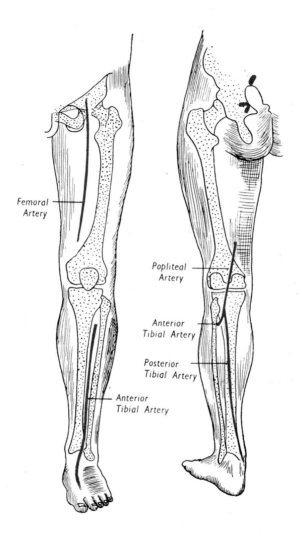

Femoral
Artery

Popliteal
Artery

Anterior
Tibial Artery

Posterior
Tibial Artery

Anterior
Tibial Artery

Plate 3. Surface marking of the arteries of the lower limb.

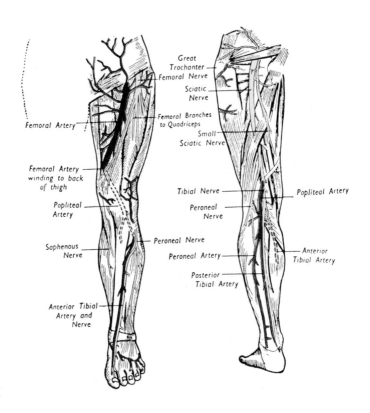

Great
Trochanter
Femoral Nerve

Sciatic
Nerve

Femoral Branches
to Quadriceps

Small
Sciatic Nerve

Femoral Artery

Femoral Artery
winding to back
of thigh

Popliteal
Artery

Saphenous
Nerve

Anterior Tibial
Artery and
Nerve

Tibial Nerve

Peroneal
Nerve

Peroneal Nerve

Peroneal Artery

Posterior
Tibial Artery

Popliteal Artery

Anterior
Tibial Artery

Plate 4. Plan of the main nerves and vessels of the lower limb.

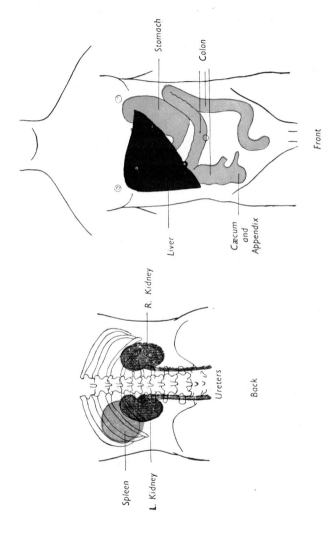

Plate 5. Surface markings of certain abdominal organs.

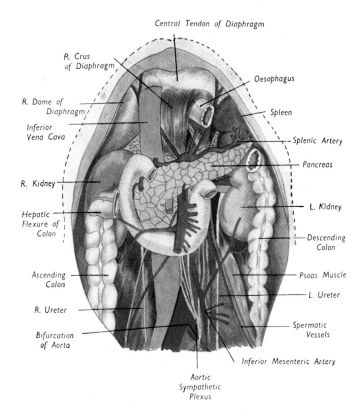

Central Tendon of Diaphragm

R. Crus
of Diaphragm

Oesophagus

R. Dome of
Diaphragm

Spleen

Inferior
Vena Cava

Splenic Artery

Pancreas

R. Kidney

L. Kidney

Hepatic
Flexure of
Colon

Descending
Colon

Ascending
Colon

Psoas Muscle

L. Ureter

R. Ureter

Spermatic
Vessels

Bifurcation
of Aorta

Inferior Mesenteric Artery

Aortic
Sympathetic
Plexus

Plate 6. Structures of the posterior abdominal wall after removal
of the liver, stomach, small intestine and transverse colon. The
duodenum, which is unlabelled, is the C-shaped loop of bowel in
the centre. The transverse colon normally arches across it to connect the ascending and descending limbs of the large bowel, here
shown cut across.

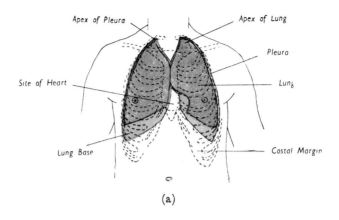

Apex of Pleura

Apex of Lung

Pleura

Site of Heart

Lung

Lung Base

Costal Margin

(a)

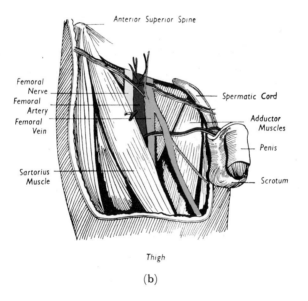

Anterior Superior Spine

Femoral
Nerve
Femoral
Artery
Femoral
Vein

Spermatic Cord

Adductor
Muscles

Penis

Sartorius
Muscle

Scrotum

Thigh

(b)

Plate 7. (a) Surface markings of lungs and pleural cavities in the chest. (b) Contents of the right groin. Note the femoral nerve and vessels, the associated muscles and the spermatic cord connecting the testicle in the scrotum with the abdomen. (After *Gray*.)

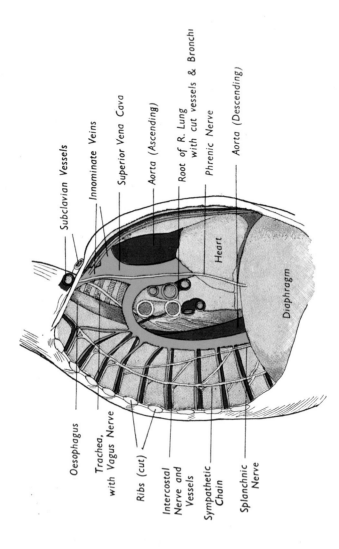

Subclavian Vessels

Innominate Veins

Superior Vena Cava

Aorta (Ascending)

Root of R. Lung
with cut vessels & Bronchi

Phrenic Nerve

Aorta (Descending)

Heart

Diaphragm

Oesophagus

Trachea,
with Vagus Nerve

Ribs (cut)

Intercostal
Nerve and
Vessels

Sympathetic
Chain

Splanchnic
Nerve

Plate 8. Contents of the right side of the chest. The lung and pleura have been removed to exhibit the mediastinal structures.

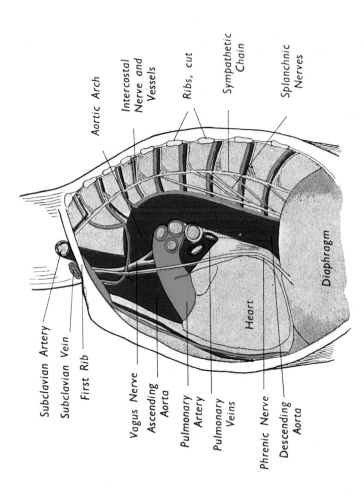

Subclavian Artery

Subclavian Vein

First Rib

Vagus Nerve

Ascending Aorta

Pulmonary Artery

Pulmonary Veins

Phrenic Nerve

Descending Aorta

Aortic Arch

Intercostal Nerve and Vessels

Ribs, cut

Sympathetic Chain

Splanchnic Nerves

Heart

Diaphragm

Plate 9. Contents of the left side of the chest. The lung and pleura have been removed to exhibit the mediastinal structures.

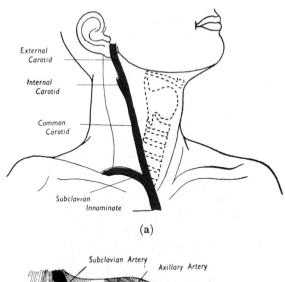

(a)

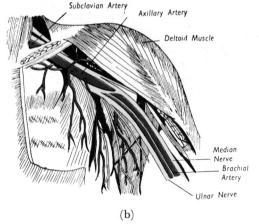

(b)

Plate 10. (a) Surface markings of the great arteries of the neck on the right side. (b) Great vessels and nerves in their course from trunk to arm. Both pectoral muscles have been removed. (After *Gray*.)

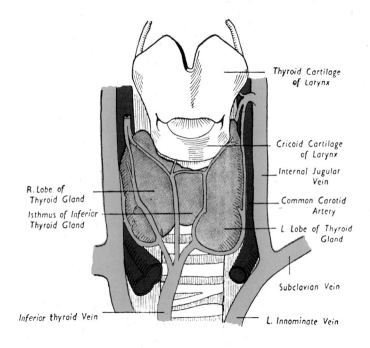

Thyroid Cartilage of Larynx

Cricoid Cartilage of Larynx

Internal Jugular Vein

R. Lobe of Thyroid Gland

Common Carotid Artery

Isthmus of Inferior Thyroid Gland

L. Lobe of Thyroid Gland

Subclavian Vein

Inferior thyroid Vein

L. Innominate Vein

Plate 11. Larynx, thyroid gland, great vessels of the neck. (After *Bruce and Walmsley*.)

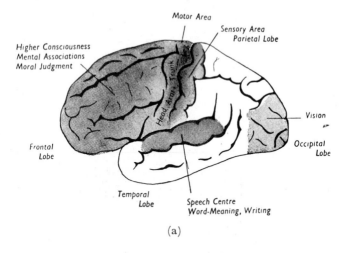

(a)

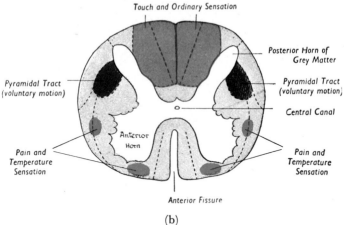

(b)

Plate 12. (a) Functional areas in the left cerebral hemisphere. The right side is identical, save for the absence of a speech centre. (b) Cross-section of spinal cord, showing the important nerve tracts.

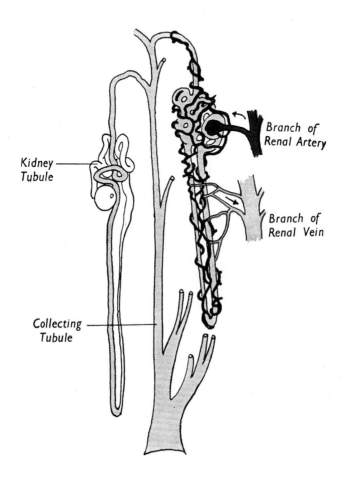

Kidney
Tubule

Branch of
Renal Artery

Branch of
Renal Vein

Collecting
Tubule

Plate 13. Two kidney nephrons—the microscopic coiled tubes in which the urine is formed—draining into a common collecting tubule. The blood vessels which carry blood to and from each tubule for purification are shown on one side.

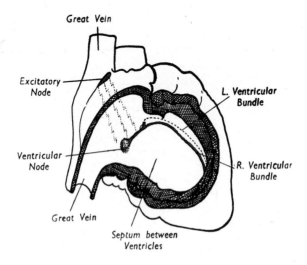

Great Vein

Excitatory Node

L. Ventricular Bundle

Ventricular Node

R. Ventricular Bundle

Great Vein

Septum between Ventricles

Plate 14. The right side of the heart, opened to show the mechanism which excites regular contraction of the muscular chambers. The stimulus begins at the excitatory node and spreads to the ventricular node, whence it is conveyed by a special bundle of fibres to each ventricle. (Adapted from Starling's *Principles of Human Physiology*.)

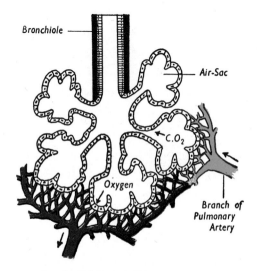

Plate 15. The interchange of respiratory gases in a cluster of air sacs of the lung.

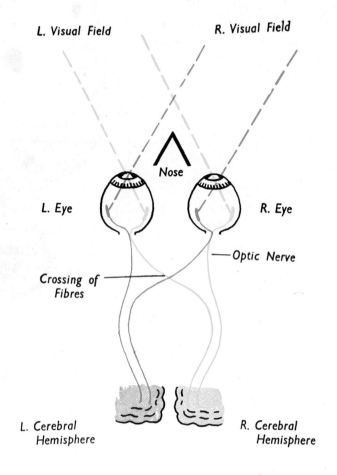

Plate 16. Diagram showing the sorting out of the nerve pathways from the retina, which secures that the visual centre in one cerebral hemisphere receives all the impulses arising from the opposite field of vision.

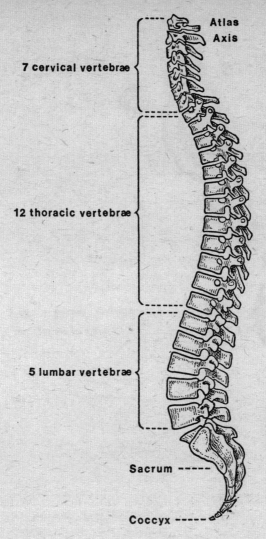

Atlas
Axis

7 cervical vertebræ

12 thoracic vertebræ

5 lumbar vertebræ

Sacrum -----

Coccyx -----

Fig. 151. The spinal column, lateral view.

A pair of vertebrae so fit together as to leave on each side an *intervertebral foramen* for the exit of a spinal nerve, so that there are thirty-four pairs of spinal nerves all told, since the first cervical nerve emerges between the base of the skull

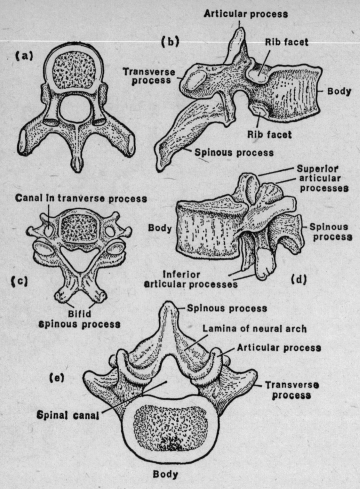

Fig. 152. Some typical vertebrae. (a) Thoracic vertebra. (b) Thoracic vertebra, side view. (c) Cervical vertebra. (d) Lumbar vertebra, side view. (e) Lumbar vertebra, from above.

and the uppermost cervical vertebra, while there is only one coccygeal nerve below.

The *cervical* vertebrae are delicate, with small bodies, as there is little weight-bearing for them to do, and a very large neural canal, as the cord is thickest at its upper end before it has given off most of its roots. The transverse processes are pierced by a foramen transmitting a vessel, the *vertebral artery*, which runs up the neck from the subclavian and enters the foramen magnum to supply the hindbrain. The spinous process is short and bifid. The first cervical vertebra is the *atlas*, supporting the head. It is a simple bony ring without a body, and its upper surface articulates with the condyles of the occipital bone to form the joint at which the movement of nodding occurs. The second cervical vertebra is the *axis*; from its upper surface a peg-like process sticks up within the axis ring. Skull and atlas together rotate on this peg in sideways turning of the head.

The *thoracic* vertebrae increase in size down the column. The bodies are heart-shaped seen from above, with facets at the sides for articulation with the heads of the ribs; a similar facet at each end of the transverse process articulates with the neck of the rib. The neural canal here is relatively small; the spinous process is long and projects well down.

The *lumbar* vertebrae are under considerable strain and are consequently of great size, with massive bodies. The neural canal is triangular, intermediate between the cervical and thoracic in size, and the blunt, square spinous processes project almost horizontally backwards.

The *sacral* vertebrae have become fused into one bone, the sacrum, intervening between the two innominate bones as the posterior segment of the pelvic ring; this fusion is essential since stability is so important here. It is a large, flattened, triangular bone with anterior (pelvic) and posterior (subcutaneous) surfaces. The base above articulates with the fifth lumbar vertebra at the lumbosacral joint, the apex below with the coccyx. It is placed obliquely, so that its long axis is directed backwards as much as down, and the anterior surface is concave—the sacral hollow in which lies the rectum.

Although a single fused mass, the general features of the component vertebrae are still obvious, the bodies being marked off by transverse ridges. Four pairs of foramina transmit the sacral nerve roots, and the body of the bone is traversed by the sacral canal, which, at this level, contains not the spinal cord (this has come to an end in the upper lumbar region) but a mass of nerve roots.

Above, the promontory of the sacrum is a prominent ridge of bone at the upper end of the anterior surface which

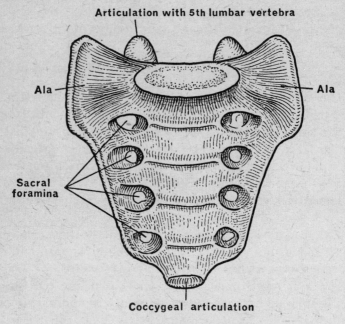

Articulation with 5th lumbar vertebra

Ala — — Ala

Sacral
foramina

Coccygeal articulation

Fig. 153. The sacrum, pelvic aspect.

forms the back of the pelvic brim. And on each side of this a spreading wing reaches out to form with the innominate bone the *sacroiliac joint*, a large, irregular interlocking pair of surfaces at which little motion occurs. The back of the sacrum is the lower origin of the sacrospinalis muscle, which extends up beside the spine as far as the skull.

The *coccyx* is the rudimentary tail appendage of man, a small irregular mass of four fused vertebrae. It is triangular, angled forward at the sacrococcygeal joint to parallel the anal canal, and is quite solid with no central channel.

The spinal column as a whole

The vertebral bodies are not in contact but separated by thick cushions of fibrocartilage, the *intervertebral discs*, which act as shock-absorbers. They increase in size down the column until in the lumbar region they are 13 mm thick. Together, the discs make up a fourth or fifth of the length

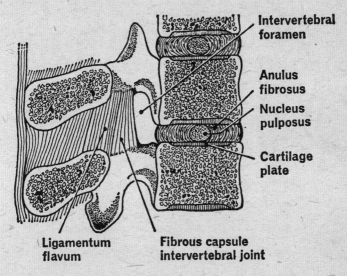

Fig. 154. Median section of articulating thoracic vertebrae. (From *A Companion to Medical Studies, Vol. 1.*)

of the column. Each consists of an outer fibrous ring enclosing a marble-like gelatinous core under considerable tension; this is the pulpy *nucleus*, which acts as a ball-bearing on which pivot a pair of adjacent bodies. The intervertebral joint thus consists of the disc between the bodies, plus the simple plane synovial joints between the pairs of articular processes on the arches. Long strap-like *ligaments* bind the

bodies together, the anterior and posterior ligaments which
traverse the length of the column on the front and back of
the bodies; and shorter bands connect the spinous and
transverse processes. The whole makes up a flexible rod
which is one single entity.

Although the spine should be quite straight looked at
from in front or behind, there are in side view a series of
normal curvatures alternating in direction from above down.

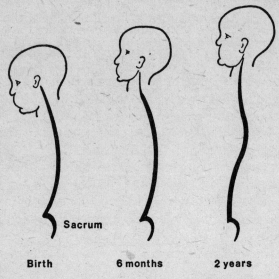

Sacrum

Birth   **6 months**   **2 years**

Fig. 155. Spinal curves at different ages.

The cervical and lumbar curves are forward convexities or
lordoses; the thoracic and sacral curves are forward con-
cavities or *kyphoses*. And this arrangement of successive arcs,
spanned by the chords of the spinal muscles, is obviously
sounder than a more rigid, straight column would be. The
child's spine at birth does not possess the cervical and lumbar
curves; there is only the anterior concavity of the whole
spine (the crouched foetal position) and that of the sacrum,
the two primary curves. As the head is held up in the first
six months, the cervical lordosis develops, and the lumbar

curve results from sitting up and standing in the second year of life.

In the adult the vertebral bodies are somewhat wedge-shaped to conform to the shape of the curves—except in the lumbar region, where the intervertebral discs are considerably thicker in front than behind.

Spinal movements

These are:

(1) *flexion*, forward bending, maximal in the cervical region, though also occurring to a considerable extent in the lumbar spine;

(2) *extension*, backward bending, maximal in the lumbar region but also taking place in the neck;

(3) *rotation*, a twisting of the column on the longitudinal axis traversing the nuclei of the discs, freest in the upper thoracic region and negligible elsewhere;

(4) *sideways bending*, maximal in lumbar and cervical regions;

(5) *circumduction*, a swaying movement combining all the above, in which the trunk describes the surface of a cone, apex down.

13 Foodstuffs and vitamins

Food and energy

The energy of *light*, absorbed by plants for photosynthesis, initiates every biological process; the *heat* given off by animals and plants is the final waste. The conversion of one to the other makes useful work possible. In performing work the muscles are chemical as well as physical engines, their bound chemical energy being transformed partly into mechanical work and partly into heat. The basic physical unit of heat is the calorie, the amount of heat required to raise the temperature of 1 gramme of water from 14·5°C to 15·5°C, equivalent to $4·2 \times 10^7$ ergs. In physiological calculations we normally use the kilocalorie (1000 calories), stated as kcal.*

The fuels of the body are protein, fat and carbohydrate (and alcohol) taken in the diet. Most of the energy bound in their carbon and hydrogen linkages can be made available to the body by mainly oxidative metabolic processes. The heats of combustion of these foods can be measured in the laboratory by burning them in a bomb calorimeter; the values so obtained are slightly larger than the energy actually available in the body, since digestion and absorption are never complete and oxidation in the tissues is only partial. Nevertheless, the alimentary tract is remarkably efficient. 99% of the carbohydrate, 95% of the fat and 92% of the protein taken are absorbed; only 5% of the calorie intake is lost in the faeces. Since, normally, a considerable proportion of the nitrogen intake is excreted in the urine as

* Under the metric (SI) system, the joule replaces the calorie as the basic unit of heat, and thus the megajoule is coming into use instead of the kilocalorie in physiological calculations. 1 kcal (1000 cal) = 4·186 MI.

organic compounds—urea, uric acid and other substances—
about a quarter of the calorie value of ingested protein is so
lost to the body. The actual amounts of heat actually available
to the body when it burns 1 gramme of a basic foodstuff are:

1 g of protein	4 kcal
1 g of fat	9 kcal
1 g of carbohydrate	4 kcal

It is possible to calculate the energy value of a particular
diet by analysing its various components into the three
main groups and applying these factors, as in Table 1.
Applying these values, it is possible to prescribe diets of
known calorie content for particular purposes, e.g. for the
rationing of armies and for the clinical treatment of obesity
or diabetes. This is the function of the dietitian.

	Water %	Protein %	Fat %	Carbo-hydrate %	Energy kcal
Wheat flour	15·0	13·6	2·5	69·1	3·4
White bread	38·3	7·8	1·4	52·7	2·4
Rice	11·7	6·2	1·0	86·8	3·6
Milk	87·0	3·4	3·7	4·8	0·7
Butter	13·9	0·4	85·1	trace	7·9
Cheese	37·0	25·4	34·5	trace	4·3
Steak	57·0	20·4	20·4	nil	2·7
Haddock	65·1	20·4	8·3	3·6	1·8
Potatoes, raw	80·0	2·5	trace	15·9	0·7
Peas	72·7	5·9	trace	16·5	0·9
Cabbage, boiled	96·0	1·3	trace	1·1	0·1
Orange	64·8	0·6	trace	6·4	0·3
Apple	84·1	0·3	trace	12·2	0·5
Sugar	trace	trace	trace	100·0	3·9
Beer	96·7	0·2	trace	2·2	0·3
Spirits	63·5	trace	nil	trace	2·2

Table 1. Energy value and nutrient content of common
foods (per gramme).

These foods not only supply energy but also replace wear and tear in the tissues, and here the proteins occupy a special position. Whereas any of the foodstuffs will be used for the production of energy, only proteins can be used for cell building. As far as energy production is concerned, fat is more economical than either of the others, for not only does it provide twice as much heat but it is also commonly used in much purer form, as in butter or lard.

In practice, nitrogenous foods such as protein are not an economical way of obtaining energy; in a normal mixed diet, the fats and carbohydrates are used to meet the ordinary energy requirements of the body and proteins serve only to repair tissue breakdown. Man cannot obtain his necessary energy from protein alone; he would have to eat some 2 kg of lean meat daily to obtain his basic minimum of 2000 calories and 3 kg for his normal 3000-odd calories, an amount impossible for most people to digest. A diet of lean meat and green vegetables is certainly a cure for obesity, but only by producing a modified form of starvation. Nor is protein stored as such to any extent in the body, except during growth, when some is laid down in the muscles. So we see that protein is not only an essential of the diet, but it must also be a regular daily constituent. However, it has no advantages as a source of energy over fats and carbo-hydrates, which are capable of considerable variation and mutual interchangeability.

Nevertheless, since most of us eat more than we strictly need, it is an advantage to include as much protein in the diet as possible and the surplus calories obtained from fat and carbohydrate are sometimes referred to in this context as 'empty' calories.

The diet

The average daily energy requirement of a working man is 3000 calories. An adult woman needs 2500 calories, simply by virtue of her smaller size and surface; and sedentary workers need correspondingly less. In fact, because of an inevitable loss of calories in preparing and cooking food, and in digestion, an extra allowance is necessary and the calorific value of the food for an adult man, *as purchased*, is some 3500. To maintain health a diet should possess this calorie value; it must also contain protein, fat and carbohydrate in the proportions indicated below; it must include enough fresh food to supply the necessary vitamins; it should have enough mineral content; and, finally, *it must be palatable*.

Protein should supply one-sixth of the calorie needs, i.e. there should be at least 100 g a day, at least half in the form of first-class protein of animal origin. If this minimum is provided, an excess serves no particular purpose. It is conventional to assume that extra meat is necessary for heavy workers, but this is not, in fact, the case. What protein does do, more than other foods, is to give a fillip to metabolism generally, a sort of cocktail which increases the rate of heat production and helps to keep one warm in cold weather.

Carbohydrates form the largest part of most diets, for they are cheap and easily obtainable in the form of bread and other filler foods such as potatoes. The optimum daily need is some 500 g. Although it is unlikely that anyone would eat less than the minimum amount, there *is* an absolute minimum; below this absolute minimum the carbohydrate: fat ratio is disturbed, so that fats are not completely burnt in the body and acid products accumulate in the tissues.

Fat is valuable because of its high energy value. A normal diet contains only one-seventh of its weight of fat (some 100 g), but this supplies a quarter of the total calories; and this concentration of bulk is exaggerated still more by the small amount of water it contains compared with other foods. Although interchangeable to a considerable extent with carbohydrate as an energy source, fat has certain desirable qualities of its own. For instance, its very slow absorption after meals prevents the recurrent hunger which would otherwise be felt with a largely carbohydrate diet;

and for really heavy work in cold or Arctic conditions a large amount of fat is the only practical way of supplying calorie needs, which may rise as high as 8000 a day.

Respiratory quotient

We can obtain a clue to the particular substance that is being oxidized in the body to provide energy by comparing the oxygen used up in the process (and contained in the inspired air) with the carbon dioxide evolved (and got rid of in the expired air). Thus the burning of carbohydrates may be represented by the simple chemical equation:

$$C_6H_{12}O_6 + 6O_2 = 6CO_2 + 6H_2O,$$

and here the respiratory quotient $\dfrac{CO_2 \text{ expired}}{O_2 \text{ inspired}}$ would be

exactly 1. Since different materials are being burnt at the same time, the actual respiratory quotient estimated by analysis of the inspired and expired air depends on the relative proportions of the foodstuffs, whose respective quotients are:

carbohydrates = 1·0
protein   = 0·8
fat   = 0·7

Thus the actual quotient is always less than 1, since a purely carbohydrate diet is never met with in practice; the quotient represents the gross result of oxidation of the average mixture of foodstuffs consumed in the body.

From the energy point of view, any of the foodstuffs may be utilized, and carbohydrates and proteins are spoken of as *isodynamic* because an equivalent weight of each gives rise to the same amount of heat. Carbohydrates and fats in the proportion of rather more than 2:1 also provide the same number of calories, and so they can replace each other in the diet in these proportions without altering the available energy. We have seen that, although protein is theoretically interchangeable in the same way, an upper limit to its use is set by the capacity for its digestion and assimilation. Since protein is not stored in the adult body, any taken in the food is rapidly broken down and excreted as nitrogenous compounds in the urine and faeces, i.e. there is a normal state of nitrogenous equilibrium and, with an already adequate mixed diet, any increase in the amount of, say, meat protein simply results in a corresponding increase of nitrogenous wastes. This is quite different from the situation with fats and carbohydrates, for an excess of either of these is stored in the body as *fat*.

Protein is broken down by digestion into its constituent amino-acids; these enter the bloodstream, where they partake in continuous equilibrium and interchange with the protein of the body cells, the latter taking up such amino-acids as they require for repair. Residual amino-acids are disintegrated in the liver; their nitrogenous part is lost as urea through the kidneys, the rest being burnt down to carbon dioxide and water. We have already seen that all food proteins are not of the same biological value, for only some contain those essential amino-acids which cannot be synthesized in the body; these are the *first-class proteins* of meat, wheat, etc. The other *second-class proteins* lack some of the essential amino-acids or are made up largely of inessential ones; gelatin is an example of this class. A deficiency of the essential acids results in a loss of appetite and failure of growth; to make this vital provision animal protein is more economical than vegetable, though the latter may be adequate if the diet is sufficiently varied.

Foodstuffs in detail

Foods may be comprehensively grouped as follows: cereals; starchy roots; pulses and legumes; vegetables and fruits; sugars and derivatives; meat, fish and eggs; fats and oils; and beverages. No one of these is essential, but a good diet usually contains foods from all these classes.

Cereals

These form the chief food of the majority of human beings,
contributing four-fifths or more of the total calories in the
diet of peasants; even in rich countries, they provide a
quarter of the total calories and are the largest single item
of diet. All cereal seeds have much the same structure and
composition (see Fig. 156). The main bulk is the starchy
endosperm, with the germ or embryo attached at one end.
Surrounding the grain are the outer endosperm, aleurone
layer and pericarp, the whole enclosed in a woody husk.
The grain contains 80% of starch, a little fat and as much
as 10% of protein; but the protein of any one cereal is not
particularly suited to human requirements, so that a mixture of plant sources is required. Cereals contain quite a
large quantity of calcium and iron, but these are poorly absorbed because the phytic acid also present forms insoluble
phytates with these cations. *Whole* grains are a valuable
source of the B group of vitamins. They contain no vitamin
C, and neither vitamin A nor vitamin D; but oily grains
like maize do contain vitamin A precursor and much
vitamin E.

Wheat and rice are usually eaten after removal of the
outer husk by milling to yield white flour or rice, which
makes them more attractive and palatable. However, the
material removed contains most of the B vitamins and
nearly all the thiamine, and this results in an important
deficiency in poor rice-eaters, who may suffer from beriberi (page 301). The losses are made good to some extent
in certain countries (such as the U.K.) by enriching white
flour with thiamine, nicotinic acid and iron; it is very
difficult so to enrich polished rice.

The poor man's cereal is *millet*, which, though husked, is
not highly extracted. *Maize* is an excellent food, although if
it forms too high a proportion of the diet it may be associated with pellagra (page 302). *Wheat* is the only cereal
(apart from rye) that can be made into bread, thanks to the
stickiness of its protein content. Most bread contains a third

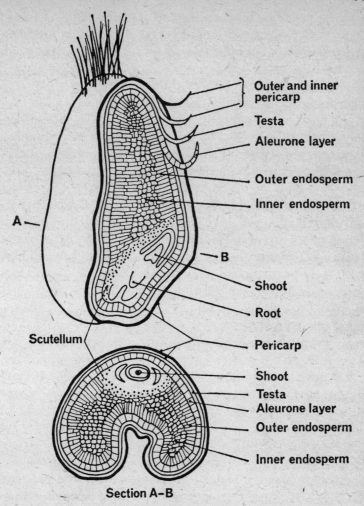

Fig. 156. The structure of the wheat grain. (From McCance, *Lancet* i (1946), 77.)

of its weight in water and much of this is expelled by toasting, so that, weight for weight, toast is considerably more fattening than bread.

The nutritive value of cereals is not to be underestimated. Bread and rice are excellent foods, palatable, providing calories, proteins, B vitamins, and some calcium and iron. Only when a diet consists of an excess of cereals, or an excess of any one cereal in particular, does it become unbalanced.

Starchy roots

The potato is an excellent and energy-rich food, and has, in some places, completely replaced cereals in agriculture. On no other single food can an adult man survive indefinitely, and this is because, unlike cereals, potatoes contain significant quantities of vitamin C and carotene, and adequate amounts of the B vitamins. They have rather less protein than cereals and some 70% of water. Tropical roots are the cassava and the yam, easily cultivated and a rich source of energy, though their protein content is not high. Because of this they may be adequate for adults, but children fed largely on them are liable to develop kwashiorkor and other protein-deficiency diseases.

Pulses and legumes

Pulses, lentils and dhals have a protein content double that of cereals, and are a valuable protein supplement in the diet in tropical countries. They are also rich in the B vitamins, in which cereals are often deficient. The *soya bean* contains up to 20% of fat.

Vegetables and fruits

These include green leaves, flowers, seeds, stems and roots. They contain much water, few calories and little protein; but they are filling, increase the bulk of the faeces and prevent constipation. They contain some vitamin B and also C, but the latter is often lost in storage and cooking. The chief vitamin content is carotene, the precursor of vitamin A, carrots being a particularly rich source. Vegetables are valuable in increasing the nutritive value of poor cereal diets in the tropics but do not have much to contribute to the average mixed Western diet.

The main nutritional value of *fruits* lies in their vitamin C content; they are also a useful source of potassium. The many organic acids present are not of great metabolic significance, but the banana provides carbohydrate in abundance and in easily digestible form, although, for a tropical main diet, it is very short of protein.

Sugar

There has been a great increase in our sugar consumption over the last century, and some authorities believe this to be an important factor in relation to the increased incidence of cardiovascular disease. Without other nutrients, sugar is a typical provider of 'empty' calories, and it contributes to the development of dental caries, obesity and diabetes. Syrups, preserves and honey are little more than concentrated sugar; honey has no particular medicinal value and contains no vitamins.

Meat, fish and eggs

Meat, as eaten with fat, contains about one-fifth protein, one-fifth fat and three-fifths water, and the protein is of high biological value. Meat is also a useful source of iron and the B group of vitamins, but not of vitamins A and C. Meat from all sources has approximately the same nutritive value; 'toughness' is due to contained sinews, which increase with the age of the animal. *Offal*, such as liver and kidneys, generally resembles meat in biological usefulness, but liver has a greater content of vitamin A and iron. Meat extracts are mostly water and possess few nutrients, but their meaty flavour promotes appetite. *Fish* is a rich source of protein and oil, and also of vitamins A and D; the bones of small fish may be a useful source of calcium in tropical areas. *Hens' eggs* contain about 10% of first-class protein, 10% of fat, much vitamin A, and some calcium and iron. They are a most useful item of the diet, though it must be mentioned that their high cholesterol content has been regarded as contributing to the development of degenerative arterial disease.

Milk

Human milk is a complete food for the small infant. Cow's milk, which contains more protein and less carbohydrate (Table 2), can be made suitable by diluting and adding sugar.

	Human milk	Cow's milk
	values per 100 *grammes*	
Calories	70	66
Carbohydrate	7 g	5 g
Protein	2 g	3·5 g
Fat	4 g	3·5 g
Calcium	25 mg	120 mg
Phosphorus	16 mg	95 mg
Iron	0·1 mg	0·1 mg
Vitamin A	170 units	150 units
Vitamin C	3·5 mg	2·0 mg
Vitamin D	1·0 units	1·5 units
Thiamine	17 μg	40 μg
Riboflavine	30 μg	150 μg
Nicotinic acid	170 μg	80 μg

Table 2.

Milk, though rich in calcium, is poor in iron, and infants may develop iron-deficiency anaemia after the first six months if mixed foods rich in iron are not added to the diet. If all milk is boiled, so destroying the vitamin C, the baby may develop scurvy. Mothers usually produce adequate milk of good quality, even though they may themselves be receiving an inadequate diet; this may be at their own expense, e.g. as when the calcium content of the milk is maintained at the expense of that of the maternal skeleton. Milk is probably essential for growing children and adolescents unless there are abundant alternative sources of protein, but it is not essential for adults. Indeed, it is quite widely believed to be unsuitable, since its high content of saturated fatty acids may be related to the development of arterial disease. *Sour milk* is a product of high nutritive value obtained by fermenting the lactose content using the *lacto-bacillus acidophilus*; the protein, calcium and vitamins are unaffected. *Skimmed milk*, from which the butter fat has been removed, can be made as a stable powder and has proved invaluable as a supplementary, or even a chief, protein source for growing children in substarvation areas. *Cheese* is made by clotting milk using the enzyme rennet; the separated clot, dried and pressed, contains most of the original protein and some of the fat, and is highly nutritious, free of 'empty' calories and so suitable for reducing diets.

Fats and oils

These may be of animal or vegetable origin. The former are mainly dairy fats—butter and ghee—beef suet and pork lard; the latter include the oils of the olive, groundnut, cottonseed and red palm, and corn oil from maize. Margarine is an artificial fat made from mixing vegetable oil with whale oil. Animal fats (except those of fish) are solid and contain a high proportion of saturated fatty acids. Vegetable oils are liquid and contain a high proportion of unsaturated fatty acids, this also being true of fish oil. Fats are a concentrated source of proteins, keep the diet from being too bulky where calories matter and are vehicles for the fat-soluble vitamins. They are valuable in cooking, and the result is that in rich countries far more fat is taken than is required to supply the essential fatty acids needed. Butter contains vitamins A and D, in amounts depending on the time of year and the quality of pastures. These vitamins have to be added to margarine; but once this has been done there is little or nothing to choose between the two as regards nutritive value. Vegetable oils contain no vitamins, except for the carotene content of red palm oil.

Beverages

These are drunk for their flavour or their alcohol content, or both. *Ethyl alcohol* (ethanol) is produced by fermenting grain or fruit with wild or cultivated yeasts. It is a nutrient, readily absorbed, and an immediate source of energy. Its surplus calories may, however, cause obesity. It is a sedative, and promotes social intercourse and well-being in moderation, but it may cause addiction and disintegration of the personality, and have toxic effects on the nervous system, when taken in excess. It is also a disinfectant and a preservative.

Beer is made in two stages: grain starch is first converted to maltose by maltase and this is then fermented by yeast. Beer contains up to 6% alcohol, yielding a calorific value of 25–70 kcal/100 ml; but it contains no vitamins, with the exception of riboflavine. *Wine* is made by fermenting grapes or other fruits; it contains 8%–10% of alcohol and yields 70–80 kcal/100 ml. Wines may be fortified with added alcohol as port or sherry to reach a concentration of up to 15%. They are devoid of vitamins but may contain quite large amounts of iron, and this may cause chronic liver disease in wine bibbers. *Spirits* are made by distilling fermented liquor; they may contain some 30% of alcohol and provide 220 kcal/100 ml but are devoid of other nutrients. Illicit or imperfect distillation may yield spirits containing methyl alcohol (wood alcohol, methanol) which is very poisonous and can cause blindness or death.

Coffee and *tea* contain caffeine, theophylline and theobromine. Caffeine is the most important, stimulates the nervous system and increases the output of urine. Both contain tannins, which are protein precipitants, and these may be responsible for gastritis in habitual drinkers. Although tea and coffee are of no nutritive value whatsoever, they are usually taken with milk and sugar, and this often contributes to obesity in the middle-aged woman who declares that she 'eats nothing'.

Soft drinks may contain much vitamin C, and their calorie value depends on their added sugar. *Fruit juices* are rich in potassium and so useful to convalescents. *Mineral waters* have a varied mineral content but in such small amounts as to be of no significance. However, as the water is natural and has not been through a sewage system, it forms a pleasant addition to the normal water intake.

The vitamins

If an animal is fed on a diet that contains adequate amounts of fat, carbohydrate, protein and salts, *but in which these constituents have been purified*, it becomes ill, stunted and eventually dies. Yet these changes may be entirely prevented by the daily addition of a teaspoonful of milk to this artificial diet. This is because natural foods contain minute quantities of organic substances known as vitamins or accessory food factors—compounds that the body is unable to synthesize (except, to some extent, in the case of vitamin D) and that must be supplied in the diet. Many of them are components of essential enzyme systems. They are essential for the proper utilization of food, to allow normal growth and development in the child, and to maintain health in the adult. Yet, in themselves, they contribute nothing towards the energy requirements or tissue repair of the body. There are five major diseases that arise from lack of one or other vitamin: scurvy, beriberi, pellagra, keratomalacia and rickets. Rare in properous communities, these diseases remain major scourges in less developed parts of the world.

The vitamins are divided into two main classes: the fat-soluble and the water-soluble. The water-soluble vitamins cannot be stored in surplus for they are readily excreted by the kidneys, and deficiency disease will soon arise if the intake is inadequate. In contrast, the fat-soluble vitamins are stored in adipose tissue, particularly in the liver, and a previously well-fed individual will remain free of deficiency symptoms for many weeks or months, even if his external supply is completely cut off. A surplus of fat-soluble vitamins *is* toxic, and overdosage may be a real problem in the treatment or prevention of certain disorders.

Fat-soluble vitamins

Vitamin A (retinol). This is present only in certain foods of animal origin—dairy produce such as butter, cheese and milk, liver, and particularly in fish-liver oils; but it is interesting to note that it has originally been formed in green plants, either in pasture or in the minute diatoms of the sea. In addition, however, the small intestine can manufacture the vitamin from a precursor, the carotene pigment found in some plant sources. In the U.K. we obtain about half our vitamin A from animal sources and about half from precursors in fruit and vegetables, and deficiency disease is rare. But it is common in the tropics, where its effects include stunting of growth, susceptibility to infective disease, defective repair of epidermal tissues and certain progressive eye changes—horny thickening of the cornea, night-blindness due to absence of the visual purple (see page 454), collapse and disorganization of the eye causing permanent blindness—all from lack of, say, 1 mg of vitamin A a day. Pure vitamin A is a pale yellow substance, quite insoluble in water, a highly unsaturated aromatic hydrocarbon stable to ordinary cooking processes but destroyed in shallow frying.

Vitamin D (calciferol). This, too, is found in dairy produce, and also in eggs and liver; it is added to margarine and often to infant foods. It is also formed in the skin from the action of the ultraviolet rays of sunlight on a sterol compound present there, so that the body is not entirely dependent on dietetic sources. However, heavily pigmented children from Latin countries and the tropics sometimes develop rickets when moved to a smoky temperate zone where the sun's rays are much less direct.

Vitamin D promotes the absorption of calcium from the bowel and the deposition of bone by the formation of calcium salts. It therefore exerts a major influence on the proper growth and hardening of bones and teeth, and any deficiency in childhood leads to rickets and dental caries. Rickets is a disease in which imperfect, soft osteoid tissue, poorly calcified, is laid down instead of normal bone so that deformities develop—spinal curvature, bow-legs or knock-knees. It is readily cured by giving the vitamin, but if this is given in excess its calcifying properties may produce dangerous deposits in the internal organs, particularly the kidneys. There is an adult form of rickets called *osteomalacia*—usually occurring in elderly women living alone on tea and bread and butter whose vitamin D intake is low—and, here again, deformity and fractures of the bones are both common and readily curable. The vitamin seems to consist of two similar sterol compounds and can be made artificially by irradiating ergosterol. As stated, it is toxic in large doses.

Vitamin E. This is essential to normal fertility. There is no definite evidence that vitamin E deficiency ever occurs in humans, but in rats fed on artificial diets the females may abort and the males become sterile, and embryos fail to develop and die in the womb. The vitamin is found in quantity in oil of wheat germ and is a yellow oily liquid, a mixture of tocopherols.

Vitamin K. This is a complex of fat-soluble naphthoquinone derivatives which are essential to the complex mechanism of blood clotting. It can be synthesized. Dietary deficiency does not exist, but the vitamin may not be properly absorbed from the gut and the resultant haemorrhagic tendency may require administration of the pure vitamin.

Water-soluble vitamins
Vitamin B. This is a complex of three factors:
(1) Vitamin B_1 (thiamine) is abundant in the husks of cereal grains, pulses, yeast, milk and meat; it is lost when wheat and rice are subjected to machine milling. Thiamine catalyzes the oxidation of carbohydrates in cells, particularly in nerve cells, and a deficiency causes the disease of beriberi. In *dry beriberi* there is peripheral neuritis, causing

muscle wasting and paralysis; in *wet beriberi* there is heart-failure and severe waterlogging of the tissues. Wherever rice is a staple diet and is finely milled, there is a risk of the disease, unless yeast or pulse supplements are taken, and even in Western countries synthetic thiamine is often added to white flour as the bran has been removed. The required daily intake for the average adult is of the order of 1·5 mg.

(2) *Nicotinic acid* is another member of the B complex of vitamins, deficiency of which causes the disease of *pellagra*, a scaly eczema accompanied by mental changes and diarrhoea—dermatitis, dementia and diarrhoea—that is associated with the consumption of maize in the Deep South of the U.S.A., in some parts of southern Europe, and in Asia and North Africa. This is because maize, as a staple, is a poor source of nicotinic acid. But as it happens, the body has some power to synthesize this compound from the amino-acid tryptophane present in rice, so that rice-eaters are protected even though rice itself does not, if milled, contain much of the natural factor. Pellagra may also result from interference with intestinal absorption, as may happen in chronic alcoholism; and some of these cases are initially diagnosed as suffering from mental disorder until the true state of affairs is recognized. Nicotinic acid is present in whole cereals and rice, in meat, offal and fish; it is very abundant in yeast and yeast extracts but not at all in dairy products, fruit or vegetables. In this country and the U.S.A., it is legally required to add the vitamin to white flour. The mean necessary daily intake is of the order of 10–12 mg.

(3) *Riboflavine* is a yellow pigment that is an essential co-enzyme for tissue respiratory processes. It is present in most foods and is the only vitamin present in significant quantity in beer. Lack of riboflavine may cause *angular stomatitis*, a fissured ulceration of the mucous membrane at the corners of the mouth, but not any specific major disorder. The daily dietary requirement is 1–2 mg.

The anti-anaemic B vitamins. Red blood cells only live for 100 days or so, are continually destroyed and must be continually reproduced. Certain vitamins are essential to this replacement process, and if they are inadequate there will be anaemia due to a reduced number of cells.

Vitamin B_{12} (cyanocobalamin) occurs in animal foods but is absent from all plant products. It is a complex porphyrin derivative containing cobalt. It is also produced by certain moulds and fungi, and is therefore obtained as a by-product in the manufacture of certain fungal antibiotics; this also implies that strict vegetarians obtain their B_{12} only because of contaminating moulds. The daily requirement is less than 1 μg. For the proper absorption of the vitamin by the intestine, an *intrinsic factor* secreted by certain stomach glands is essential; when this is absent, the individual will develop a severe, sometimes fatal, form of anaemia—pernicious anaemia. This may be cured by eating large amounts of animal liver, in which the vitamin is normally stored, or, more normally nowadays, by giving regular injections of the vitamin, on which the patient is as dependent for his ability to lead a normal live as a diabetic is on his insulin injections. Severe B_{12} deficiency causes degeneration in the nerves and spinal cord.

The other anti-anaemic B vitamin is *folic acid*. Folic acid is widely found in plants and animals, but the main source is from the leaves of green vegetables. The average daily requirement is some 50 μg. Anaemia due to deficiency is common in tropical countries, particularly in pregnancy.

Other water-soluble vitamins exist, such as pyridoxine and pantothenic acid, which are components of essential, tissue-enzyme systems. They are widely distributed in food, deficiency is rare and no specific deficiency diseases exist. But the most important remaining water-soluble factor is *ascorbic acid—Vitamin C.*

Vitamin C. Ascorbic acid is a sugar, a crystalline solid, readily synthesized, very soluble and easily oxidized. The main source is from fruit, vegetables and milk. Meat and eggs contain only a trace; potatoes in quantity are a useful source, though this declines during storage. The vitamin is easily destroyed in the preparation and cooking of vegetables, and when diets contain little or no fruit or vegetables, or when vegetables are routinely boiled, the deficiency disorder of scurvy may result, characterized by tissue haemorrhages manifest externally as swollen, bleeding gums, and also by the indolent healing of wounds and fractures, owing to defective formation of new fibrous tissue. This disease can be readily prevented—as Captain Cook discovered—with citrus fruits. Because milk, in any case, contains little ascorbic acid and what exists is destroyed by boiling, infants are particularly liable to the disease; hence the advisability of utilizing welfare orange juice. The minimum daily intake is of the order of 25 mg daily, and this is provided by an ordinary helping of (not overcooked) cabbage or some other green vegetable, a good helping of potatoes twice daily, or by 42·6 g of fresh oranges or orange juice.

To sum up, the majority of the vitamins are of known chemical composition and can be synthesized artificially or otherwise obtained in pure crystalline or oily form. The only evidence we have of their effects is from what happens when the body is deprived of them; then, just as the smooth running of a machine is interrupted for lack of lubricating oil, certain deficiency disorders make their appearance. It must be emphasized that, although the vitamins are essential to health and the normal utilization of food, they are not themselves foodstuffs and will not compensate for an inadequate diet. Further, in the ordinary adult living on a good mixed diet, there is little or no need for the addition of vitamin preparations. This is not the case with the relatively greater needs of growing children, and for them special supplements of fish oil and orange juice may be desirable. The chronic alcoholic, too, because his drug interferes with the proper absorption of vitamins and their storage in the liver, may well need large doses of multivitamin preparations.

14 Digestion

The alimentary tract consists of a long, muscular tube, extending from the mouth to the anus and lined with an epithelium, which is specialized in some places for secreting digestive juices and in others for absorbing the products of digestion. This internal lining is the mucous membrane. But, in addition, there are major extrinsic glands lying outside the bowel altogether—the salivary glands, liver and pancreas—whose secretions are discharged via their ducts into the intestine. These secretions amount to many litres in twenty-four hours and most of this fluid is reabsorbed; so it is obvious that bowel disorders, accompanied by vomiting or diarrhoea, may cause profound dehydration and electrolyte disturbance. For the most part, secretion takes place in the upper or proximal part of the tract, absorption in the middle zone and excretion in the distal or lower segment, but there is some overlap of function.

While we are aware of the desire to eat, drink and defaecate, digestion and absorption are unconscious processes. Over most of the bowel, movement is also involuntary and effected by smooth muscle; but chewing, the initiation of swallowing and defaecation are conscious processes.

The *object of digestion* is to convert the complex and largely insoluble constituents of the food into simpler substances that can dissolve and then diffuse through the lining of the intestine to enter the blood or lymph, and that the tissues can deal with. Thus all carbohydrates must first be reduced to the simplest monosaccharide form or the cells will not be able to assimilate them. The fats must be hydrolysed to fatty acids and glycerol—though some fat enters the lymph unhydrolysed as finely emulsified globules —and the proteins are split into their constituent amino-acids. In dealing with the latter, the body displays considerable versatility and ingenuity; to convert the protein of cheese into one of the human forms of protein is equivalent to taking a meccano model to pieces and picking out the right bits to build a new model of one's own liking. Digestion is accomplished by the enzymes contained in the digestive juices formed in the various glands stationed along the length of the alimentary tract. Each specific foodstuff needs a corresponding enzyme to break it down, just as every lock can be opened only by its own key.

The mouth

Entry of food into the mouth promotes a flow of saliva. This is a reflex or automatic action; the stimulation of the taste nerve endings is transmitted to the brain and relayed back along the nerves controlling salivary secretion. But there may also be what is termed a 'psychic' secretion on the mere sight or anticipation of food, and this is an example of a conditioned reflex in which an association of ideas has the same effect as the original physical stimulus.

The saliva that lubricates the oral cavity is a mixture of the secretions of the three pairs of salivary glands—the *parotid* in the cheek, the *submandibular* below the angle of the jaw and the sublingual beneath the tongue in the floor of the mouth—as well as of many minute glands in the mucous membrane of the mouth and palate. Saliva contains *mucin*, which makes the food a slippery, easily-swallowed mass and also intensifies taste by dissolving solids; *ptyalin*, an enzyme that digests, or initiates the digestion, of starch to maltose; and minerals, particularly *calcium phosphate*, which is deposited on the back of the teeth as tartar. It is a curious fact that much of salivary digestion actually occurs in the stomach. Food does not remain long in the mouth, but the swallowed mass still contains ptyalin, which continues to exert its digestive action in the stomach until complete penetration of the bolus by the acid gastric juice brings the process to a halt after half an hour. There is normally a continuous slight secretion of saliva between meals which is suppressed by fear and by certain drugs, such as atropine, that dry up the mouth.

No absorption of nutrients occurs in the mouth, but certain drugs may be absorbed rapidly and directly across the thin mucous membrane into the bloodstream.

Chewing breaks up the food and mixes it with saliva, and small amounts are passed back at intervals into the pharynx to be swallowed. Chewing is initially a voluntary, but eventually an involuntary, activity; it is doubtful whether thorough mastication makes digestion any more efficient.

Swallowing

The oesophagus, in which no digestion takes place, is a long, muscular tube, beginning in the neck behind the lower part of the larynx and travelling down along the back of the thoracic cavity to pierce the diaphragm and join the cardiac end of the stomach. The object of chewing is the production of a round mass, or *bolus*, of food which is passed down the oesophagus to the stomach in the act of swallowing. Now, although in the lower part of the neck, and in the chest, the air and food passages are quite separate, there is in the upper part of the neck a common cavity for food and air, the pharynx (page 270), extending up as far as the base of the skull immediately in front of the vertebral column. In swallowing it is essential to prevent food passing upwards into the nose round the free back edge of the soft palate, or downwards through the airway into the lungs. Therefore, in swallowing the soft palate is apposed to the back wall of the pharynx so as to seal off the nasal cavities, and the epiglottis—the elastic flap of fibrocartilage situated behind the root of the tongue—bends over backwards to close the upper entry into the larynx. At the same time, the larynx as a whole is pulled bodily upwards and then descends again (Fig. 157). The initiation

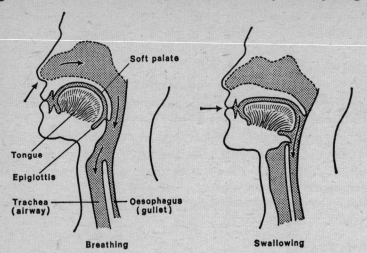

Breathing   **Swallowing**

Fig. 157. The mechanism of swallowing. Note how the palate moves upward to block the back of the nose and the epiglottis down to shut off the windpipe.

of swallowing is largely voluntary, as the chewed mass is passed backwards through the mouth; but once the food touches the pharynx, soft palate and epiglottis, the process continues as a reflex activity. Once in the oesophagus, the food is propelled by *peristalsis*, a wave of contraction in the muscular wall preceded by a wave of relaxation controlled by the nerve plexuses in the oesophageal wall. At the lower end of the tube, the cardiac sphincter relaxes reflexly and food enters the stomach. This sphincter is normally closed to prevent any reflux of acid gastric juice into the oeso-phagus. The mean transit time along the oesophagus is some two seconds.

The stomach

The stomach is a hollow, muscular organ lined with a glandular mucous membrane which secretes the gastric juice. The organ is not essential to life and is often removed surgically without causing too much disturbance. **It acts as** a reservoir in which a semi-solid mass known as the *chyme* is formed from the ingested food; digestion continues until the chyme is transferred to the duodenum, but there is little absorption, except for water, alcohol and simple crystalloids.

Small amounts of gastric juice are present in the resting state. The mere sight or anticipation of food, or its entry into the mouth, excites a reflex or conditioned secretion, which is highly susceptible to emotional factors. More is formed when the food actually enters the stomach, and as much as a pint of highly acid juice may be secreted after a heavy meal, the process continuing for up to four hours. The stimulus to this secretion is not mere contact with food but a hormone, *gastrin*, which is released into the general circulation from the pyloric region of the stomach, and thus reaches and stimulates the main body of the organ. Protein, alcohol and coffee are much more potent excitants of this phase of gastric secretion than fat or carbohydrate. The secretion is strongly acid due to the presence of up to 0·5% of hydrochloric acid; but some of this is neutralized by combination with mucus, so that only some 0·3% is present in the free state. There is a good deal of mucus, which protects the stomach lining, and also certain digestive enzymes. A few normal individuals are completely devoid of acid in their secretion; although in itself of little importance, this may be associated with the absence of certain other essential substances—the cause of pernicious anaemia. The enzymes present are *pepsin*, which converts proteins into polypeptides, and *rennin*, which curdles milk into a casein clot, which is then digested by pepsin in the usual manner. Gastric juice acts on the food in two ways: by virtue of its acidity and through its enzyme action. The acid alone is responsible for minor breakdown of certain foods— e.g. the more complex sugars—but most of the work is done by the pepsin, which also helps to liberate the fat of the food in liquid food by dissolving the fibrous framework round its globules, thus preparing it for digestion in the

duodenum. The response of gastric secretion to food may be
studied experimentally by giving a standard meal of gruel
and then withdrawing the stomach contents by tube at
intervals for analysis. The findings may then be plotted as a
curve. Fig. 158 shows the results of such an experiment,
with the usual rise and fall of hydrochloric acid levels. It
also shows the findings in certain diseases, for such 'test

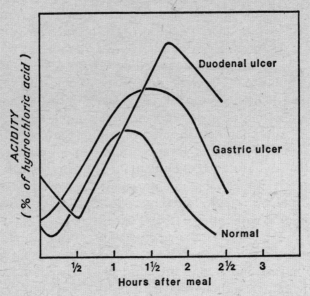

Fig. 158. Gastric secretion of acid after a meal in normal individuals and in ulcer cases.

meals' also give us a clue to the nature of stomach de-
rangements. Thus in gastric ulcer acid secretion reaches an
unduly high level, and in duodenal ulcer the resting juice
in the organ, before any food has been taken, contains a
considerable amount of acid, which aggravates the ulcer
by its action on the mucous membrane. In the latter case, the
first effect of a meal is to reduce the acidity, hence the need
for small frequent feeds in duodenal ulceration to protect
the stomach lining from overacidity.

Even though the empty stomach is small and relaxed, it

is not entirely still; small contraction waves travel from the cardiac end towards the pyloric exit. An ingested meal distends the organ and stretches the stomach wall, and this stimulus to the nerve endings therein excites a series of regular contraction waves which propel the food from the fundus of the organ on the left towards the pylorus on the right. For the first thirty to sixty minutes after a meal, the opening into the duodenum at the pyloric sphincter remains closed; thereafter, it opens at intervals and permits gastric contents to pass in intermittent spurts into the duodenum. By the end of gastric digestion, in three to four hours, all the food has passed through; the opening then relaxes, and some duodenal contents and bile regurgitate back into the stomach. Fluids leave the stomach long before solid food and, taken alone, may enter the duodenum in a few minutes; fats remain in the organ much longer than carbohydrates.

All in all, the stomach is not perhaps a very important organ in digestion and absorption in man. We know that digestion can take place efficiently even after removal of the organ, the chief role of which appears to be to receive and store food at irregular intervals, and to pass on at more regular intervals partly digested food in smaller quantities to suit the capacity of the small intestine, where the bulk of digestion takes place.

Although the standard anatomical position of the stomach is given as the upper part of the abdominal cavity, rather to the left, it is important to realize that, in fact, the organ undergoes considerable changes in position. These are due not only to changes in the position of the body as a whole but also to emotional influences; we speak of 'our heart being in our boots', but what actually happens is that the stomach drops considerably, perhaps even as low as the pelvis. The general mobility of the muscular wall and the rate of glandular secretion are so intimately affected by emotional states that it is not surprising that many cases of dyspepsia or actual ulceration are of nervous origin.

Small intestine

It is here that the greater part of the digestive process occurs, effected not so much by the small glands of the bowel itself, which produce mainly mucus, as by the secretions of two great extrinsic glands, the liver and pancreas, which discharge their secretions into the duodenum. The principal function of the bowel wall is absorption. The three digestive juices, then, are the *pancreatic juice*, the *bile* formed by the liver and the intestinal juice, or *succus entericus*, produced by the lining of the bowel. The bile and pancreatic juice are discharged simultaneously into the duodenum via ducts from the pancreas and liver, which open in common; the intestinal juice is formed along the entire length of the small intestine. Although it undoubtedly contains certain enzymes, the intestinal juice itself probably plays little intrinsic part in the digestive process, despite the fact that it contains the activator for pancreatic trypsin. But it does contribute a large amount of water to facilitate the solution and absorption of digested food. On the whole, secretion and digestion predominate in the upper coils of the small intestine and absorption of the digested products into the portal circulation in the lower part; but, to some extent, both sets of processes are everywhere going on simultaneously.

Pancreas

The bulk of this organ secretes the pancreatic juice into the duodenum; but there are small islets of contained tissue that produce the hormone insulin (see page 340).

The pancreatic juice is formed in two ways. There is a response to the ingestion of food that is due to a nervous reflex; but there is also a much more important response to the entry of acid gastric contents into the small intestine. This latter mechanism is a beautiful example of the work of chemical messengers, or *hormones*, in the body; and, indeed, the discovery of this mechanism was the foundation of modern endocrinology. When stomach contents enter the intestine, some of the acid passes through the bowel wall and extracts from it a potent stimulating agent called *secretin*, which enters the bloodstream and is carried to the pancreas, where it acts as a specific stimulus to the secretion of the pancreatic juice. Alkaline pancreatic juice then enters the duodenum and neutralizes the acid influx, thus automatically bringing to a stop the mechanism for evoking pancreatic secretion. The next spurt of gastric contents starts a fresh cycle, and so on. Thus by this chemical arrangement pancreatic secretion is exactly suited to the needs of digestion.

The pancreatic juice is about as alkaline as gastric secretion is acid. It contains three powerful enzymes: *trypsin*, which furthers the digestion of proteins to their component amino-acids; *amylase*, which hydrolyses starch and other carbohydrates to glucose or disaccharides; and *lipase*, which breaks up fats into fatty acids and glycerides. It is interesting to note that the pancreatic juice will not function properly without the presence of the other two secretions. Thus protein digestion by trypsin has to be initiated by the conversion of a precursor—trypsinogen—into trypsin proper by another enzyme—enterokinase—liberated by the duodenal mucosa, while fat digestion is enormously helped by the bile, which emulsifies the globules and so exposes a far greater area to the action of the lipase. The freed fatty acids will form soaps if the intestinal contents become sufficiently alkaline.

Liver and bile

The liver is composed microscopically of myriads of lobules, each of which receives a dual blood supply—of ordinary arterial blood from branches of the hepatic artery and of blood from the intestine, containing absorbed products of digestion via branches of the portal vein. Bile is continuously secreted by the cells of these lobules to a total of some 852 ml a day; but, as it is only required at intervals for digestion, it is stored in the gall bladder, where it is concentrated by reabsorption of water and released into the bile ducts when required. The most potent stimulus to bile secretion is the presence in the blood of the bile salts themselves.

Bile is not only a digestive juice but also a means of ridding the body of the breakdown products of the haemoglobin contained in the red cells of the blood. It is a slightly alkaline, viscid green fluid. The colour is due to contained *bile pigments*, breakdown products of the haemoglobin from effete red cells, which are disintegrated in the liver and spleen. It is these pigments, still further altered and mixed with undigested food, that give the faeces their characteristic brown colour. And in *jaundice*, due to obstruction of the bile duct which prevents the pigments entering the bowel, the faeces remain pale and clay-coloured, while the dammed-up bile overflows into the circulation and stains the skin its typical yellow colour. (Jaundice may also result from disease of the liver itself, as in infective hepatitis, or from over-production of haemoglobin breakdown products, as when red cells are abnormally fragile.)

The *bile salts* are the sodium compounds of certain complex acids and are of the greatest importance in fat digestion. For, although the bile contains no enzymes of its own, its salts finely emulsify the fat of the food to facilitate its digestion by pancreatic lipase. The fats, now split into glycerides and fatty acids (or soaps), traverse the bowel wall, together with the bile salts; and here, in the individual cells, globules of neutral fat are reformed and absorbed, while the salts are returned by the bloodstream to the liver to be used over and over again.

The muscular gall bladder is normally full between meals and is particularly distended in starvation. Entry of food into the duodenum relaxes the spasm of the sphincter muscle guarding the opening of the common bile duct into the bowel, and the gall bladder contracts and discharges bile.

Certain swallowed substances are particularly effective in exciting this response, e.g. Epsom salts and fats.

Further information on the functions of the liver is contained on page 342.

The large intestine

The *large intestine* is little concerned with the digestion or absorption of food, for most of this has been accomplished by the time food reaches the lower end of the small bowel. This is certainly true of an ordinary mixed diet containing plenty of meat. On a mainly vegetarian regime, however, a considerable part of the food reaches the large intestine; a little may be absorbed, but most simply goes to swell the bulk of the faeces. However, there is one absorptive function of importance—the taking-up of water from the faeces. These enter the colon in a fluid state; there, they lose four-fifths of their water content, becoming quite solid by the time the lower (left) half of the colon is reached. The further passage of the hard faeces is lubricated by the scanty mucous secretion of the glands in the wall of the large bowel. A considerable portion of the excreta is composed of the enormous numbers of bacteria found in the large intestine; this is in contrast to the situation in the stomach and the upper small intestine, where the contents are relatively sterile. As a result, an abdominal wound that penetrates the large intestine may be very dangerous, since it liberates bacteria into the peritoneal cavity, setting up a virulent *peritonitis*. The *faeces* themselves are derived for the most part from substances formed within the intestines. They contain 70% of water, some 15% of mineral salts, such as calcium phosphate and iron salts, and a considerable quantity of nitrogenous material, mainly from dead bacteria. This composition is altered if there is an increased cellulose (vegetable fibre) content in the food; this increases the bulk of the faeces, stimulates bowel passage and offers less opportunity for the extraction of water. The faeces also contain shreds of bowel mucous membrane and are passed together with gaseous *flatus*, much of which is ordinary methane.

Movements in the digestive tract

The general principle of movement everywhere in the bowel—stomach, small and large intestine—is that food is propelled in one direction, and that onwards and downwards, by *peristalsis*. This is a rhythmically recurring wave of contraction travelling down a segment or loop of bowel at the rate of about 2·5 cm a second, presenting a localized ring-like spasm at any given moment at a particular point, and preceded and succeeded by a wave of relaxation. Peristalsis is not dependent on any central control and continues in an experimentally isolated coil of bowel, for it is brought about by the regulating action of a nerve plexus lying within the muscle coats of the intestinal wall itself, forming a continuous network along the whole length of the intestine. It is, however, subject to modification by influences from the central nervous system, particularly those of emotional origin, which may affect both the rate of successive waves and their amplitude.

In the small intestine there also occurs another type of movement, *segmentation*, in which a length of bowel is demarcated into a number of sausage-shaped segments by localized areas of ring spasm, the segments changing and reforming from time to time. This movement does not propel the food along the digestive tract. Finally, there are *pendular movements*, waves of contraction passing backwards and forwards over the intestine and causing churning of the contained food rather than its propulsion.

Regular easy peristalsis does not occur in the colon except, to some extent, in its right half, for here the problem is primarily one of storage of the faeces until the ingestion of a new meal makes it necessary to move on the contents of this lower end of the intestine. The entry of food into the stomach excites what is known as the *gastrocolic reflex*, a mass contraction of the large bowel that transfers the faeces from left (descending) colon to rectum, preliminary to the act of *defaecation*. Defaecation itself consists of the expulsion of the faeces outside the body, and the rectum should normally remain empty until just before the act. Faeces enter the rectum following mass contraction of the large bowel after a meal, and the resulting distention is felt as a call to defaecate. Although the act itself is a reflex one, it is under conscious initiation and control, and the urge passes off if the call is not obeyed and the rectum relaxes again. As a result, in the chronically constipated the rectum always contains some faeces, and loses its power of responding regularly and effectively to distention. During the act itself, the levator ani muscle (Fig. 127) contracts and lifts the anal canal over the descending faecal mass, while at the same time evacuation is assisted by voluntary straining—contraction of the diaphragm and abdominal muscles to raise the intra-abdominal pressure. The external opening of the lower end of the anal canal is normally kept closed by the contraction of its two guarding circular muscles or sphincters.

It is a simple matter to determine how long it takes for the food to arrive at different stages in its passage through the bowel. The method is to give a 'meal' of barium sulphate, which is opaque to X-rays and which, in consequence, can be watched during its passage along the alimentary canal (Fig. 159). The bulk of the stomach contents have passed on after three hours, though a small residue may persist for a further hour or two. Food arrives at the lower end of the small bowel and enters the large intestine through the ileocaecal passage after four to four and a half hours; fills the right half of the colon by six hours; and occupies the left half, ready for expulsion into the rectum, after twelve to eighteen hours. These, however, are only average times and vary considerably; part of the food may be detained in the large bowel for several days.

Vomiting is a reflex action which may be initiated by many different stimuli. These may take the form of irritation of

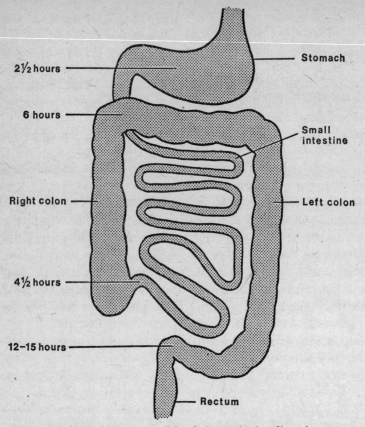

Fig. 159. Rate of passage of food through the digestive tract.

the stomach lining by food, alcohol or drugs; touching of the palate or pharynx; distension of the stomach or bowel; inflammation or irritation of the bowel; stimuli from the vestibular apparatus of the inner ear (page 461), as in sea-sickness; and severe pain or emotional disturbance.

The vomiting centre is in the medulla of the hindbrain (page 436). There is a violent contraction of the muscles of the abdominal wall which expel the gastric contents through a relaxed oesophagus. Vomiting is usually accompanied by nausea, sweating, salivation and an accelerated heart rate.

15 Absorption, utilization and storage of digested food

Absorption

Digestion achieves the transformation of complex in-soluble food substances of high molecular weight into simpler soluble particles. These are capable of entering the circulation, and their transference from the bowel to the blood or lymph vessels is known as *absorption*. Absorption is dependent on a number of physical factors, such as con-centrations and osmotic pressure, but for the most part nutrients are actively transported, and only water and alcohol cross the intestinal cells by simple diffusion.

Most of the nutrients are absorbed in the small intestine, and it will be recalled that the lining of this part of the bowel is not a smooth surface but a velvety pile made up of innumerable hair-like processes, or *villi*. These project in-wards and enormously increase the surface area available for absorption of bowel contents into the bloodstream. Each villus (Fig. 160) is richly supplied with blood capillaries, and each is also traversed by a central lymphatic channel, or *lacteal*, for carrying digested fat globules to a lymphatic plexus in the bowel wall and thence to the main lymphatic trunks of the body. In the fasting state the villi are at rest, but during absorption there is a constant pumping motion as they become alternately fully elongaged and contracted or club-like. This produces a rhythmic squeezing out of absorbed materials into the general circulation.

Water

A large quantity of water enters the digestive tract daily. In addition to the *external load* of water taken with the diet,

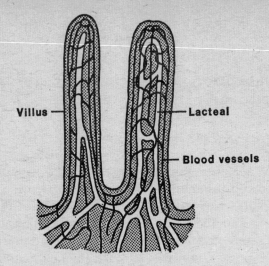

Fig. 160. Villi of the small intestine with their lacteals and blood vessels.

say 1500 ml, there is the *internal load* of water contained in the various digestive juices: saliva 1500 ml, gastric juice 3000 ml, bile 500 ml and pancreatic juice 200 ml—a grand total of 8500 ml. All this water is absorbed from the bowel, the majority from the small intestine and the remainder in the large intestine, where the fluid faeces are dried up and become more solid. Most of the absorbed water enters the portal venous blood and only a little enters the lymphatics. The transference of water (and dissolved electrolytes) across the bowel wall into the vessels of the villi is a complex procedure and is not fully explicable in terms of simple osmosis; it is greatly influenced by the selective activity, or ionic pumping action, of the intestinal cells themselves. Certain ions, such as magnesium and sulphate, are not absorbed easily and remain in the intestine holding water there, so exerting a purgative action. The absorbed water from the bowel constitutes a load on the circulation and kidneys; hence the importance in heart disease and kidney disorder of restricting water intake.

Vitamins and minerals

The water-soluble *vitamins* are relatively simple, easily soluble molecules and are readily absorbed, although it will be recalled that the absorption through the ileum of vitamin B_{12} is entirely dependent on a specific intrinsic factor secreted by the stomach. Most of the *calcium* ingested is lost in the faeces, but some is absorbed to supply the needs of the skeleton, and this proportion is greater in growing children and in poorly nourished races. This absorption is absolutely dependent on the presence of vitamin D; if this is defective or if the calcium intake is inadequate, rickets will develop in children and the equivalent bone softening —called osteomalacia—in adults. The absorption of *iron* is related to internal requirements; an excessive intake leads to damaging storage deposits in the liver and tissues.

Carbohydrates

Carbohydrates are broken down by digestion into a mixture of simple monosaccharides and disaccharides, and the final reduction of the latter into the monosaccharide form is effected by the cells of the bowel wall, followed by active transport across the cell into the portal blood. The reduction of carbohydrates is effected by various specific enzymes for the different sugars, and if these enzymes are congenitally absent disaccharides may pass on unchanged into the colon and cause diarrhoea; thus some infants possess no lactase enzyme to digest lactose and so cannot tolerate ordinary milk.

Fat

Very little fat normally escapes digestion and it is absorbed only in the small bowel. But there sometimes exist conditions when the bowel mucosa is defective or atrophic and then undigested fat will pass out with the faeces, rendering them pale and greasy, a condition known as *steatorrhoea*. Fat leaves the stomach as large globules and is finely emulsified in the duodenum by the action of the bile salts, and so rendered more easily subject to being broken down (hydro-lysed) by pancreatic lipase into fatty acids and glycerides. These products enter the cells of the bowel wall as a clear micro-emulsion and recombine within the cells of the villi into small globules, or *chylomicrons*, of neutral fat, which are transferred to the lacteal vessels and thence to the lymphatics of the bowel. After a fatty meal, these channels and the main thoracic duct are distended by a milky fluid, the *chyle*, rendered opalescent by the tiny fat particles. The thoracic duct empties this fat into the bloodstream, whence it joins the great veins of the neck, so that the blood serum may contain obvious fat a couple of hours after a meal. The whole process is accelerated by post-prandial exercise.

This is the fate of reconstituted neutral fat. In addition, a certain amount of split fat—glycerol and fatty acids—finds its way directly into the bloodstream via the blood capillaries of the villi. But the major part of absorption is by the lacteals and thoracic duct, and division or injury of the duct cuts off the flow of fat into the circulation, resulting in starvation, emaciation or even death.

Proteins

Under the influence of the pepsin of the gastric juice and the trypsin of the pancreatic secretion, proteins are broken down into a mixture of amino-acids and peptides. Some further digestion may take place in the bowel wall itself. The amino-acids are readily absorbed and enter the general circulation; absorption is very efficient and little of the dietary nitrogen is lost with the faeces.

Utilization (metabolism) of absorbed material

Protein metabolism

We have earlier (page 289) referred to the place of proteins in the diet; now we must consider their fate within the body in more detail. It will help if we first take a closer look at what proteins, chemically, actually are. They are large-molecule, nitrogen-containing compounds that subserve many differing physiological functions. Some, like the keratin of hair and skin, are insoluble and inert; others fulfil in their structure the mechanical requirements of certain tissues—muscle and connective; some are soluble and used for transporting oxygen, minerals and hormones about the body; and every enzyme is a protein catalyst. Because proteins are essential to life and contain nitrogen, and because man is incapable of directly utilizing inorganic nitrogen, he depends on ingesting the protein of other life forms, which he then degrades into their simple amino-acid constituents to be recombined in specifically human patterns.

The amino-acids may be regarded as the protein currency of the body, just as glucose is the carbohydrate currency—small, universally recognized coins, freely interchangeable and capable of being built up at different sites into the larger, more complex, individual proteins of particular tissues. It would appear, therefore, that it might be possible to live on a diet containing not the natural proteins of milk and meat but their constituent amino-acids, formed by preliminary artificial digestion in the laboratory. And this is, in fact, the case; digests of casein and other proteins have been found of value in feeding patients after operation or in starvation whose digestive systems could not cope with the original complex materials.

All proteins can be hydrolysed to yield their constituent amino-acids, either by boiling for several hours or by treatment with a series of enzymes for a lesser period. *Simple proteins* so treated yield amino-acids only. *Conjugated proteins* yield amino-acids, plus a prosthetic group, linked to the main molecule, that confers a specific character on that molecule. The total of amino-acids obtainable from protein hydrolysis is around twenty, and the basic chemical formula of an amino-acid is:

$$R—CH—COOH$$
$$|$$
$$NH_2$$

where, in the simplest form, $R = H$ (glycine).

In more complex forms, R may be an aliphatic chain, may include a thiol grouping or may be a cyclic radical. All amino-acids are optically active, but only their laevorotatory forms are yielded by hydrolysis. They are also amphoteric, i.e. they can combine with both acids and bases, and this means that they can act as neutralizing buffers. Soluble proteins can be isolated and identified from biological fluids by chromatography, which depends on their differing rates of diffusion along moist filter-paper, yielding separate stainable protein 'spots', and also by electrophoresis, which depends on their differential migration in filter-paper to one or other of a pair of electrodes.

The amino-acids in a protein are linked by peptide bonds, —CO—NH—, so that the basic structure of a protein molecule may be represented as:

$$
\begin{array}{ccc}
R_1 & R_2 & R_3 \\
| & | & | \\
\end{array}
$$
$$
-NH-CH-CO-NH-CH-CO-NH-CH-CO
$$

where R represents specific amino-acid radicals. During hydrolysis polypeptides are formed, which are intermediate between proteins and amino-acids.

Proteins may be purified by dialysis, by gel filtration, by electrophoresis, by ion exchange, or by precipitation or centrifuge techniques; for details of these, a textbook of organic chemistry should be consulted. Their molecular weights vary from a few thousand to over a million, which indicates that the simplest protein molecule must contain over a hundred amino-acid residues and some several thousand polypeptides. Although there are only some twenty constituent amino-acids, there are many more different kinds of protein in the body; this is possible because proteins differ according to their proportions of constituent acids and how these are linked together. Proteins may be 'finger-printed' by hydrolysing them to mixtures of peptides and separating these by electrophoresis, followed by paper chromatography, which yields a specific visual pattern for each protein (Fig. 161).

The shape of the protein molecule may be compact and globular in the case of the soluble proteins or rigid and elongated in the case of the fibrous proteins, often insoluble.

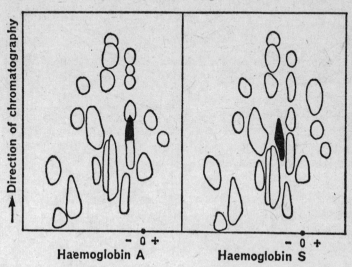

Fig. 161. Fingerprints of tryptic hydrolysates of haemoglobins A and S. O, point of application of hydrolysate; + —, direction of electric field. The peptides which differ in behaviour and composition are marked black, the one containing valine (in HbS) has moved slightly further towards the cathode. (From *A Companion to Medical Studies, Vol. 1.*)

The molecules are frequently folded or twisted in helical manner, so identifiable by X-ray crystallography (Fig. 162), and most proteins have some evidence of a helical structure, though not those of milk casein or the plasma gamma-globulins. The helix itself is subordinate to the final overall shape of the total molecule.

Proteins may be *denatured* by heat, acids or alkalis, which weaken the internal bonds of the molecule, so causing loss of their typical final structure, though not disrupting their basic primary structure.

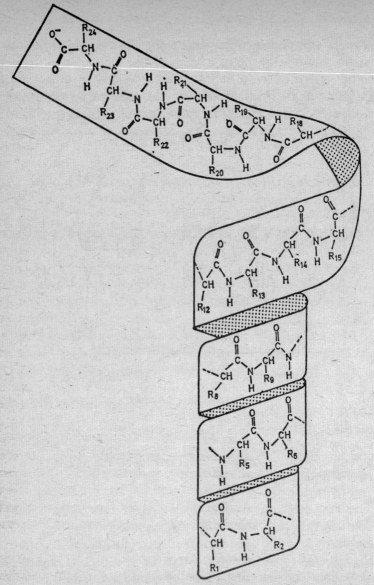

Fig. 162. The right-handed α-helix. The lower part of the ribbon, with the polypeptide structure drawn upon it, shows the helical structure. Note that the relative positions of —NH— and —CO— in successive coils are favourable for hydrogen bond formation. Further up the ribbon there is random folding, with full extension of the chain at the upper end. (From Haggis, G., *et al.*, *Introduction to Molecular Biology*, Longmans.)

Proteins in nutrition. Energy is obtainable from all three basic food components—fats, carbohydrates and proteins; but only proteins will serve for protein synthesis for the needs of growth, and tissue repair and maintenance, i.e. the ultimate fate of the constituent amino-acids of proteins is partly to enter into the composition of actual living protoplasm and partly to serve as a fuel source of energy. In this latter function, protein is obviously on a par with the fats and carbohydrates; in the former, it is irreplaceable.

There is great variation in the capacities of specific proteins to support life and growth. Thus young rats fed on maize protein (zein) in the laboratory lose weight because zein contains neither lysine nor tryptophane among its amino-acids. If these are given as a supplement, growth resumes normally, i.e. these two acids evidently cannot be synthesized in the body, however much nitrogen is available in the diet. Thus they are known as indispensable or *essential* amino-acids and must be supplied as such in the diet. For man, the essential amino-acids are: isoleucine, leucine, lysine, phenylalanine, tyrosine, histidine (infants only), methionine, cystine, threonine, tryptophan and valine. Hence the importance of a well-balanced diet with a wide variety of foods; and hence it is that peoples living on a single staple, such as yams, rice or cassava, run the risk of developing a state of protein deficiency. Note carefully that the description 'essential' refers only to whether amino-acids are indispensable in the diet because they cannot be synthesized in the body; the so-called 'non-essential' amino-acids which *are* capable of synthesis are nonetheless vital to the metabolism.

Milk and egg proteins are most effective in promoting growth in infants and young animals, and can be used as reference proteins in assessing the biological value of other proteins. The biological value, or net protein utilization, is:

$$\frac{\text{nitrogen retained in body}}{\text{effective dietary intake of nitrogen}}$$

a factor that can be calculated as:

$$\frac{\text{dietary } N_2 \text{ intake} - \text{faecal loss of } N_2 - \text{loss of } N_2 \text{ in urine}}{\text{dietary } N_2 \text{ intake} - \text{faecal loss of } N_2}$$

The biological value of the reference protein is standardized at 100%, and the value is less for less well-utilized proteins. Any constituent of the diet can be evaluated for its biological value by analysis, and this is best done in the form in which it is habitually taken, because preparation and cooking may partly destroy some amino-acids. Using this method, it is possible to calculate a *protein score* for different foodstuffs containing protein.

The nitrogen required in the daily intake is equal to the daily nitrogen loss, plus that required for growth or other demands, such as those of pregnancy or lactation. The nitrogen lost with the faeces and from the body surface is basic and largely independent of intake, but the nitrogen lost in the urine varies considerably with variation in intake. The *obligatory nitrogen loss* is that lost daily in the faeces and urine, and from the body surface while the subject is on a protein-free intake. This obligatory loss, which is a direct measure of protein breakdown, or catabolism, is increased by stresses such as infection, injury (including accidents and surgical operations), pregnancy, lactation, fever, convalescence and emotional strains. Hence the need for a protein-rich diet after injuries or surgical procedures. But added protein, contrary to some popular beliefs, is *not* required for heavy manual workers or athletes. Because of their growth requirements and their higher metabolic rate, infants require more protein than adults per kilogramme of body weight.

There is a relation between protein nutrition and calorie supply. Protein is always utilized most effectively when the bulk of the body's energy requirements is met by other foodstuffs, particularly carbohydrates. If the carbohydrate content of the diet is inadequate, some protein must be diverted to supply energy and there is a risk of protein deficiency developing. Paradoxically, one's appetite, and consequently one's food intake, is determined by calorie needs, and if these are met by a diet that happens to be poor in protein it is impossible to eat enough food to supply the intrinsic protein need, i.e. there is an optimum proportion of total calorie intake to be supplied by dietary protein. Thus it is quite possible to supply all the calorie needs of a growing child with deproteinized milk; but protein deficiency will develop unless a supplement is given in the form of dried, skimmed milk.

Protein transformations in metabolism. Proteins take part in a wide range of chemical processes; many different enzymes are involved and each protein is metabolized in a specific manner. The proteins of the tissues are not static once formed but constantly receive fresh amino-acid constituents from the blood and shed their old ones to be excreted. They are like a reservoir that remains at a constant level though continually filling and overflowing. In addition to this dynamic equilibrium, some amino-acids play a part as intermediate metabolites of fat or carbohydrate, and others are converted to yield nitrogenous non-protein compounds such as adrenaline and thyroxine. The ultimate end-products of amino-acid degradation are CO_2 and water, plus nitrogenous compounds such as urea and creatinine, which are lost in the urine. Amino-acids are present in the blood plasma and enter the interstitial (extracellular) fluid, from which they are taken up by tissue cells; the amino-acid concentration within the cells may be five to ten times as great as the concentration in the extracellular fluid. Little amino-acid is normally excreted *as such* in the urine because of selective reabsorption back into the blood in the renal tubules (page 406), but in some diseases of the kidneys (and liver) amino-acids do pass out in the urine in the entire state. Amino-acids are degraded by a process of *deamination*, an enzyme-catalyzed oxidative process in which the nitrogen is converted to ammonia. This does not remain in the free state but is converted in the liver to urea, which is then excreted by the kidneys. Most of the amino-acid carbon is eventually lost as CO_2; but, in addition, a significant amount of glucose is formed from the split protein. The sulphur of certain amino-acids (cysteine and methionine) is excreted in the urine as about 1 g of inorganic sulphate daily, but some sulphur becomes linked to polysaccharides and forms the ground substance of cartilage.

Protein degradation in the body is mainly a process of hydrolysis, liberating amino-acids, some of which are utilized to replace effete intracellular proteins. Some idea of the length of life of protein molecules is obtainable by tagging them with identifiable isotopes; human serum albumin, for instance, is replaced at the rate of 10% a day. But it is difficult to distinguish between protein turnover in ordinary cell metabolism and that required to replace dead cells. It has been stated that the rate of turnover of protein in the body as a whole is some 250 g a day, i.e. some $2\frac{1}{2}$% of the total body protein.

This leads to the concept of *nitrogen balance* as between daily dietary intake and loss by excretion, mainly in the urine; in the steady state these are in equilibrium. But on a protein-free diet the balance becomes negative and the urinary output falls over several days to the basal obligatory loss; the weight and protein content of the liver and other viscera fall, and cell division and replacement slow down. Return to a normal diet reverses these changes, i.e. there is a positive nitrogen balance until equilibrium is restored. The proportion of the body protein that varies with varying dietary intake is known as the *labile body protein*—only some 5% at most of a total body protein of around 600 g. Finally, there is no reason to believe that any advantage is to be gained by eating a diet rich in protein surplus to actual need.

Fat metabolism

Fats are compact, readily mobilized and rich stores of energy. They are soluble in certain solvents, such as ether, but not in water. The main groups are:

(1) fatty acids;
(2) triglycerides and waxes;
(3) phospholipids;
(4) cholesterol and its derivatives;
(5) vitamins A, D, E and K.

The *fatty acids* are long or short, unbranched aliphatic chains with one terminal carboxyl (COOH) group, of which the simplest form is acetic acid, CH_3COOH. The type formula of the complex varieties is

$$CH_3—(CH_2)_n—CH_2COOH$$

They form salts (soaps) with alkalis and combine with alcohols to form esters, of which ethyl acetate is a simple model:

$$CH_3\ COOH + C_2H_5\ OH = C_2H_5\ COOCH_3 + H_2O$$

Fatty acids may be saturated or unsaturated, the latter having an unsatisfied capacity to accept hydrogen; and, commercially, hydrogenation is used to convert unsaturated vegetable oils to more solid fats resembling animal fats. It is thought that unsaturated fatty acids may be valuable in preventing certain forms of arterial disease.

The *triglycerides* comprise the bulk of dietary fat, and the basic model is an ester of glycerol with three fatty acid chains. All fats are triglycerides, whether of animal or vegetable origin. They differ only in the nature and amount of the fatty acids esterified to the glycerol base. Referring to the number of carbon atoms in the fatty acid chain, human fat consists mainly of C_{16} and C_{18} acids, both saturated and unsaturated. Fish oils have a high proportion of poly-unsaturated fatty acids and are therefore liquid. More saturated lard and suet are solid at room temperature.

Incidentally, a fat such as butter is not a chemically pure substance but a mixture of different triglyceride molecules.

The *phospholipids* contain phosphorus and sometimes nitrogen. They are important components of cell membranes, plasma, egg-yolk and of the nervous system.

Cholesterol is a very important compound of a basic ring structure:

Some is supplied in the diet; more is synthesized in the body, notably in the liver. It is an example of a class of lipids, all with the same basic ring structure, known as *steroids*. These are most important compounds and include the hormones of the adrenal and sex glands, vitamin D and the bile acids.

The *fat-soluble vitamins* have been discussed at page 300.

The fate of absorbed fat. The daily absorption of fat is of the order of rather more than 100 g—most as triglycerides, some as cholesterol and phospholipids. These have entered the duodenum as a crude emulsion, rendered more finely divided by the bile salts, and then split by pancreatic lipase to fatty acids, monoglycerides and some glycerol. These products are not water-soluble and their transfer across the bowel lining is facilitated by bile salts, which form them into tiny aggregates, or *micellae*, which enter the villi. Within the villus cells they recombine, though the compounds they form are not identical to the original triglycerides. But some glycerol and free fatty acids enter the portal blood as such and reach the liver. The cholesterol esters are also hydrolysed, enter the bowel epithelium and are re-esterified.

The lipids in the villi enter the fine lymph vessels in small particles known as *chylomicrons*, consisting of triglycerides mainly, some cholesterol and phospholipids, and a little protein. These particles enter the central lacteal of each villus and travel in the lymph up the thoracic duct to enter the great veins at the root of the neck. Once in the bloodstream they mingle with serum lipids and lipoproteins formed in the liver.

Lipids are transported in the plasma as *lipoproteins*, which are large molecules composed of triglycerides embedded in cholesterol, phospholipids and protein. The lipoproteins convey the triglycerides to the tissues, where they are separated for local use, while the conveyor system shuttles back to the liver to pick up more triglycerides—partly those arriving from the bowel and partly those made endogenously. The *serum lipids* consist of:

(1) cholesterol, in amounts of 150–250 mg/100 ml in youth, rising to 300 mg or more in later life;

(2) phospholipids in amounts of 150–300 mg;

(3) triglycerides in amounts of 50–300 mg in the fasting state, but reaching much higher levels after a fatty meal, when the separated plasma may be milky with chylomicrons.

The *cholesterol* goes to the liver, where it is converted into bile salts and excreted into the bowel. However, there is a disease known as atherosclerosis, common in middle-aged and elderly men but less common in women, where cholesterol is deposited in the lining of the arteries and may obstruct them, as in coronary heart disease.

In the tissues—mainly fatty tissue and muscle—the arriving lipoproteins are cleared from the plasma and the triglycerides are split by a lipase enzyme. The fatty acids enter the fatty tissue and are reconverted to triglycerides; or they enter muscle and are used to supply energy by oxidation. The freed glycerol returns to the liver. Thus part of the transported serum fatty acids is stored in the body fat and part is burnt to supply the energy for muscular contraction. And during physical exertion the fatty acid content of the plasma rises and more is taken up by muscle. This mechanism is less important if there is ample free glucose available. In brief exertion the initial source of muscle energy is the glycogen actually stored in the muscle cells, which is split to yield glucose. But after some minutes the muscles begin to take up free fatty acids from the blood (see page 82).

The energy store represented by the triglyceride content of adipose tissue is available for use when required, the glycerides being split back to glycerol and fatty acids which enter the bloodstream, from which the acids are taken up by various internal organs and there oxidized to yield energy.

It is also possible for the carbohydrates of the diet to contribute to the fat store—as obese persons know only too well—partly because the capacity to store carbohydrates is very limited. The glucose molecule is broken down and the carbon atoms reassembled in long, fatty acid chains.

Cholesterol is synthesized in all the body tissues, and that present in the blood is partly the free compound and partly various esterified forms. The *phospholipids* are essential to the integrity of cell membranes, which separate the extra-cellular fluid from the cell cytoplasm and control the in and out transfer of various small molecules, nutrients, wastes and water.

Fatty acids and carbohydrates. Both fatty acids and carbohydrates can be, and are, normally broken down in the tissues to yield CO_2, H_2O and energy. In the resting state the major part of the energy requirement is obtained from the fatty acids, a much smaller part from glucose, and an even smaller quantity from amino-acids, ketones and glycerol. But this situation changes with eating or exercise. After a carbohydrate meal the respiratory quotient (page 290) moves nearer to 1, i.e. glucose rather than fat is being used to provide energy. During brief violent exertion, as we have seen, the main initial source of energy comes from the glycogen store of muscle, and only if the exertion continues do the muscles begin to utilize free fatty acids from the blood, at which point the respiratory quotient falls.

During fasting, when no glucose is being absorbed, the liver content of glycogen becomes depleted, the blood level of glucose falls, less glucose is used in the tissues and the metabolism turns over to fat, rather like a motor being switched over to a different fuel tank. But note that the brain always preferentially uses glucose for energy-production transformations.

Ketosis. If insufficient glucose is available, either because of fasting or because of prolonged exertion which has used up the glucose stores, energy is obtained predominantly by the mobilization of fat stores; this metabolism yields *ketone bodies*, such as acetoacetate, and the presence of these in the bloodstream lowers the pH of the blood—a condition of *acidosis*. Ketone bodies also appear in the urine. Similar changes may occur in uncontrolled diabetes; the mechanism here is referred to on page 340.

The liver plays a cardinal role in fat metabolism; it normally contains some 5% of fat, but more may appear in certain disease states—for instance, fatty degeneration—in starvation and certain forms of chemical poisoning; in many other conditions, such as the severe anaemias, the depot fat is transferred to the liver before utilization.

Fatty (adipose) tissue. The basic fat is widely distributed and makes up some 10% of the body weight. About half is in the thick layer under the skin, most of the remainder occurring in great sheets within the abdominal cavity, or as a packing round the internal organs, and a little in the intermuscular planes. There is no fat within the cranial cavity. A healthy woman normally has about twice as much fat as a healthy man, but the quantity actually present in any one person depends on nutritional and constitutional (genetic) factors. Its distribution is affected by the sex hormones; there is more around the hips and thighs in women, more in the upper half of the body in men.

From the physical aspect, the depot fat insulates the body

against temperature changes, especially cold, and buffers the internal organs against shock and injury; it is also an electrical insulator and serves that purpose as a special layer around the individual fibres of nerves. Microscopically, the tissue is seen to be divided into lobules by fibrous septa— connective tissue partitions. The individual fat cells are derived from ordinary connective tissue cells that have become deformed by a large central fat globule, which pushes the nucleus to one side.

Chemically, fat may be regarded as an organ for the storage of fatty acids in the form of triglycerides as a source of energy when required under nervous and hormonal control in response to changes in nutritional and environmental conditions. It consists of some 85% fat (mostly as neutral triglycerides, though there are also some free fatty acids, glycerol, cholesterol and phospholipids), 12% of water and 3% or so of protein. But adipose tissue should not be regarded as a largely inert or passive storehouse of triglycerides, called on only when required. Rather, it is an entrepôt where there is a continuous arrival of new material, which is broken down and resynthesized to pass out to the blood and tissues. There is a constant activity and turnover. Fat can be formed even on a fat-free diet from glucose in the blood, the synthesis occurring both in the liver and in other tissues, a process facilitated by insulin (page 340). Fat is also formed from the dietary lipids and from lipoproteins in the blood.

Fat is added to the adipose reserve when the food is more than the body needs for energy. Different species of animals have their own characteristic fats, e.g. human fat is liquid at body temperature, mutton fat solid. To a certain extent this depends on the capacity of the body to modify in its own way fat taken in the food, but the character of the depot fat may be altered experimentally by providing large amounts of an unusual fat in the diet; however, the process is a slow one and may take years.

Fear, cold and other stresses mobilize fatty acids, and this response is mediated partly by hormones secreted by the adrenal glands and partly by the nervous control of the autonomic system over both fat cells and their blood vessels.

There is also a tissue known as 'brown fat' found in hibernating animals and the human foetus. It is a rather dense, pigmented tissue, whose cells contain finely dispersed fat granules instead of a single large droplet. It seems that this tissue is more freely available as a source of energy than ordinary adipose tissue.

Carbohydrate metabolism

Basically, as we have seen, glucose is manufactured from CO_2 and water by photosynthesis in the green plant, using the energy of sunlight. With the addition of amino-acids, it constitutes a totally sufficient nutrient. In fact, glucose is rarely found as such in nature but occurs in polymer form; and starch is a polymer present in cereals that man can digest, whereas he is barely able to utilize the cellulose polymer of plants.

Chemically, the basic building block of carbohydrates is CH_2O, and the general formula of a carbohydrate is $(CH_2O)n$. The commonest carbohydrates are monosaccharides, where n = 5 or 6, pentoses and hexoses. These are usually soluble and sweet-tasting. Disaccharides, such as maltose, are made by the linkage of two monosaccharides. Saccharides can be artificially linked to other groupings to form sugar alcohols, such as mannitol and sorbitol; these are not metabolized in the body and can be used to replace sugar in diabetics. Ascorbic acid (vitamin C) is a hexose derivative.

One interesting fact is that saccharides exist as stereo-isomers, i.e. the same chemical compound is present in different asymmetrical spatial arrangements of atoms within the molecule. The number of possible isomers of any given molecule depends on the number of carbon atoms and may be large; there are sixteen possible isomers of a hexose. Now, although the chemical analysis of isomers yields identical results, they differ in their solubility and taste—and particularly in their physical and biological properties. All rotate the plane of previously polarized light, either to the right (dextrorotatory) or to the left (laevorotatory). It is a fascinating fact that the biological processes in the living body can make use of only one of the isomers of any given compound, and the other isomers, though chemically identical, exert little or no physiological action. Thus l-ascorbic acid is active against scurvy, whereas d-ascorbic acid is not; and l-thyroxine is a potent stimulator of general metabolism, whereas d-thyroxine is only a third as effective. When such a compound is manufactured chemically in the laboratory, it usually consists of a *racemic mixture* of equal proportions of both isomers which is, of course, optically inactive, since the rotations produced by the two constituents cancel out, and has only half the biological activity of the natural isomer.

In the metabolism of glucose, a basic chemical reaction is with adenosine triphosphate or ATP (adenosine is a nitrogen-containing purine body that yields glucose-6-phosphate and adenosine diphosphate (ADP)). This reaction is catalysed by an enzyme called hexokinase. Now glucose-6-phosphate is, as it were, the basic carbohydrate currency of the body, just as the amino-acids are the basic protein currency. It is the true starting-point for glucose/carbon metabolism, and there are many pathways and intermediate products of great complexity involved, in which organic phosphates and phosphatase enzymes are prominent. But the basic final path for the degradation of glucose to yield energy is by degradation to pyruvate with three carbon atoms (pyruvic acid is $CH_3(CO)COOH$, which is then split in the tissue cells to CO_2 and water).

Glycogen is found everywhere in the body but mostly in liver and muscle. It is a high polymer of glucose with a molecular weight of several million.

Lactose is the disaccharide of milk and is synthesized from the monosaccharides glucose and galactose in the breast tissue, and nowhere else. As already mentioned, lactose in the infant diet is hydrolysed in the bowel wall to glucose and galactose, and the latter is eventually converted to glucose. If the last sequence is impossible because of congenital lack of the appropriate enzyme, galactose accumulates in the tissues and may cause stunting or mental defect; the only cure is an artificial diet containing no lactose.

The liver is the only organ that can form free glucose; and this is vital as the blood sugar level cannot be permitted to fall below a certain threshold without grave consequences. This is because there is a continual demand for glucose from the blood by the various organs—particularly the brain, which cannot use other sources of energy. Therefore, even in starvation or on a carbohydrate-free diet, the liver must continue to manufacture glucose for as long as possible.

We have earlier, in discussing protein metabolism, used the simile of a reservoir that remains at the same level though it is being constantly emptied and as constantly refilled. This applies equally to the sugar of the blood. There is a constant intake from the food and also from the liver glycogen, which is mobilized to enter the circulation as glucose. And there is a constant output into the tissues, where some sugar is burnt for energy and some stored in the cells as a glycogen reserve. The blood sugar is thus in equilibrium with that of the tissues, and there is a very efficient regulating mechanism which maintains its concentration within fairly constant limits. The blood sugar level rises somewhat after a meal but is back to normal within two hours as the added sugar diffuses out into the tissues. This regulation is so effective that the blood sugar remains much the same in severe starvation as after a heavy meal, and it is only exceptionally that the level rises to such an extent that some sugar spills over and is excreted in the urine.

How does this regulating mechanism work? The controlling agent is *insulin*, an internal secretion or hormone of the pancreas, a gland we have already met with in digestion. Insulin is essential to the proper utilization of sugar by the tissues and in a normal individual is formed by the pancreas in amounts just sufficient to deal with the sugar added to the circulation from the food; its action is to keep the blood level down to normal by enabling the tissues to make use of this added sugar. Without insulin, the tissues cannot properly burn glucose, and sugar is dammed up in the blood and spills over into the urine. As a result of the loss of this source of energy there is relative starvation; also, as we have seen, fat metabolism is interfered with, since sugar is not being simultaneously oxidized, and acidosis results. This combination of circumstances is the condition we know as *diabetes*; it is due to some unexplained deficiency in the secretion of insulin. The proper treatment of diabetes depends, therefore, on giving insulin artificially by injection, and on working out such a balance between the insulin given and the sugar allowed in the food that the blood level is just kept normal. In minor cases of the disease, where the capacity to utilize sugar is only slightly impaired, it may, however, be possible to deal with the situation merely by reducing the sugar intake in the diet. Now, since diabetes is a problem of too much sugar in the blood, due to insulin deficiency, we might expect that an excessive dose of insulin would abnormally depress the level of blood sugar; this is, in fact, the case and the consequences may be serious or fatal—feelings of acute anxiety, great hunger, pallor and sweating, convulsions, even ultimate coma and death. The same results occur in the rare condition of a tumour of one of the cell-islands in the pancreas in which insulin is formed; so much insulin is produced by the tumour that the patient's life may depend on its location and removal. Furthermore, it is quite common for a slight overproduction of insulin in otherwise normal individuals to give rise to mild spontaneous depression of the blood sugar, marked by recurrent restlessness and hunger, or even mental disturbance; such a condition is rapidly alleviated by eating sugar or sweets.

The liver is the principal place of storage for glycogen, and it is constantly converting it into glucose and vice versa; it is also capable of forming glycogen from protein and fat. Just as insulin regulates the blood sugar on the outflow side by enabling the tissues to burn it up, so the liver regulates it on the inflow side by adding just enough from its own glycogen stores. This liver store is a reserve against starvation, and it can be quickly converted into soluble glucose and passed into the general circulation when necessary; of all the reserve materials of the body, this is the one most rapidly available and most easily made use of by the tissues. In its mobilization from the liver *adrenaline*, the internal secretion of the adrenal glands, plays some part; thus in fear and anxiety, in apprehended or actual struggle and in severe cold, adrenaline acts as a potent chemical messenger to the liver, and sugar is rapidly formed and poured into the blood to serve as an immediate source of energy.

The muscles of the body are the other main storage receptacles for glycogen; this they need as an essential agent to supply their energy demands in contraction. In this process they convert the glycogen not to glucose but into lactic acid, which enters the blood and is reconverted to glycogen either in the liver or in the muscles themselves again. Although, in fact, more glycogen is stored in the muscles than in the liver, it is kept for the particular needs of muscular exertion and is not so readily available for the general use of the body.

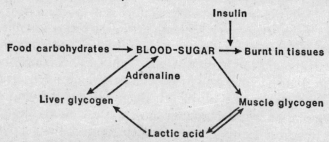

The liver in metabolism

A good deal has already been said about the metabolic functions of the liver, and, indeed, the liver is of cardinal importance for metabolism generally. It plays an important role—shared, it is true, with other tissues—in the synthesis, storage, degradation and mutual conversion of protein, fat and carbohydrate. Bile formation, the synthesis of urea from ammonia radicles split off in protein breakdown, and the metabolism of certain drugs, hormones and toxic agents, all these are the exclusive province of the organ. So, if the liver ceases to function normally, serious metabolic disorder and illness—even death from liver failure—may follow. Yet, as with other organs like the kidney, the liver has a huge functional reserve and there may be no obvious disturbance, even when the organ is quite extensively diseased or damaged.

We have seen that the liver is important in the synthesis of proteins, in the conversion of amino-acids one to another and in the production of urea as the chief nitrogenous end-product of protein metabolism. In liver failure the amino-acid content of the blood rises and there is a fall in urine urea and blood urea. Protein is formed in the liver partly from dietary amino-acids and partly from amino-acids derived from the general metabolic pool fed by tissue breakdown. Protein breakdown occurs by the deamination of amino-acids, and the ammonia so yielded is a potentially toxic product of nitrogen metabolism which the liver converts into the non-toxic product urea. Other tissues, particularly the kidney, also form some ammonia, and this is brought to the liver for conversion.

The liver also plays a special part in the formation of the *plasma proteins*. These include: albumin, the most abundant protein in the blood plasma; alphaglobulins and betaglobulins in much smaller amounts; certain proteins concerned with the mechanics of blood coagulation; and certain enzymes. The plasma proteins have quite a short life and have to be constantly renewed. Their quantitative analysis may be effected by electrophoresis, which yields a definite pattern (Fig. 163), and this pattern is altered in particular ways in particular diseases. If the diet is poor in protein, the plasma proteins will fall, and their osmotic action of holding water in the blood and body fluids will be reduced; in famine conditions, fluid will seep out of the blood to cause starvation oedema of the subcutaneous tissues.

The liver also has an essential function in maintaining the level of blood glucose, as already stated. We have noted the part played by hormones—insulin and adrenaline—in

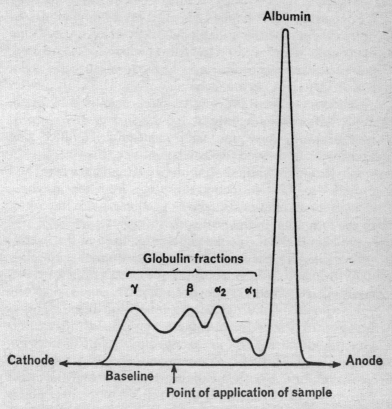

Fig. 163. Densitometer tracing of an electrophoretic separation of normal serum proteins. (From *A Companion to Medical Studies, Vol. 1.*)

this mechanism, both at rest and during exercise. The liver normally stores about 100 g of glycogen, which is immediately available for the maintenance of blood glucose levels. But the organ also actively manufactures glucose, and this from several sources: from glycerol formed in the breakdown of fats, from monosaccharides—hexoses or pentoses—and from the breakdown products of amino-acids after deamination. This process of *gluconeogenesis* is promoted by certain hormones: the adrenal hormone cortisone, thyroxine and a pancreatic hormone, glucagon.

The liver both synthesizes lipids, whether arriving from fat depots or from the food via the portal circulation, and breaks them down, the long chain fatty acids finally being oxidized to water and CO_2. If this breakdown is impaired, however, ketone bodies are formed—acetoacetic acid, hydroxybutyric acid and acetone—giving rise to the ketosis or acidosis we have previously noted. Ketosis is really an index that predominantly fat metabolism is going on, as may occur in starvation, in prolonged fever, after long exertion or in uncontrolled diabetes.

Cholesterol is synthesized in the liver (and elsewhere) and excreted in the bile to be reabsorbed from the intestine, though some is lost in the faeces. The liver is also important in the formation of lipoproteins.

Nine-tenths of ingested alcohol is oxidized in the liver.

The bile salts are formed in the liver from cholesterol.

As for the bile pigments, the most important of these, *bilirubin*, is formed by the continuous breakdown of the haemoglobin of the red corpuscles in the blood. This is brought to the liver and there conjugated in the diglucuronide form; it is then excreted in the bile to enter the bowel. In the bowel, processes induced by bacteria convert it into *stercobilin*; some of this serves to colour the faeces and is lost with these, while some returns to the liver and is either circulated again and again between bowel and liver or excreted by the kidneys as *urobilin*, which confers the characteristic yellow colour on urine. *Jaundice* occurs when the normal level of bilirubin in the plasma is exceeded for any reason. Mild degrees of jaundice may be present without the patient being obviously yellow, but can be detected biochemically. It can be seen that jaundice may be due to overproduction of bilirubin, and its accumulation in the plasma, to disease of the liver itself, or to obstruction to the outflow of bile from the liver. The first type occurs when red cells are being destroyed in unusually large numbers, as they are in some anaemias or in malaria; liver disease will prevent the handling and transport of normal amounts of bilirubin; and obstruction to outflow of bile, as from a stone or cancer in the bile ducts, will dam back the pigment. In most forms of jaundice, bilirubin appears in the urine.

The liver also metabolizes and breaks down certain of the hormones, such as cortisone and the sex hormones. It also metabolizes many drugs and toxic agents; and if it fails to do so effectively or if the agent is toxic, liver damage and failure may result, one of the features of which is impairment of fat metabolism and fatty degeneration.

The liver also constitutes the chief depot of the fat-soluble vitamins A, D, E and K, and holds enough of the first two of these to prevent deficiency disease appearing for a long time, even if they are absent in the diet. Vitamin K is used for the synthesis of important clotting factors. Water-soluble vitamins are also stored, and the liver contains enzyme systems which can convert dietary tryptophan into nicotinic acid (page 342).

Because of its wide range of biochemical functions, there are a number of tests of *liver function*. These include measurement of plasma bilirubin or serum albumin, the first being raised and the second lowered in liver inefficiency; a galactose-tolerance test, designed to assess the organ's ability to convert galactose to glucose; and a test for the presence of escaped liver enzymes in the serum.

16 Blood, lymph and the reticulo-endothelial system

The blood

The blood is a complex fluid medium pervading all parts of the body—except cartilage and the cornea—that conveys food and oxygen to the tissues and carries away wastes to be excreted. It varies in colour between the purple of venous blood and the bright red found in the arteries according to its degree of oxygenation. Microscopic examination shows that its uniform texture is only apparent, for there are myriads of *corpuscles* of different kinds suspended in the fluid component, or *plasma*. Corpuscles and plasma occupy an approximately equal volume and can be separated by standing or centrifuging. The blood cells consist of red cells or *erythrocytes*, white cells or *leucocytes* and platelets or *thrombocytes*. In centrifuged blood the deposit consists of red cells; above this is a thin layer of white cells and platelets, the *buffy coat*; and above this is the clear, yellowish supernatant plasma.

The plasma is a clear, pale yellow fluid containing some 10% of solids—most of this being protein—together with a certain quantity of salts, particularly sodium chloride. There are also some sodium bicarbonate, phosphates, potassium, etc., and also representative products of various stages of metabolism of all the different classes of foodstuffs—glucose, urea, amino-acids, fatty acids and so on. The proteins of the plasma are coagulable, and one of them, fibrinogen, is concerned exclusively with the clotting of blood. The others, albumin and globulin, are responsible for the considerable osmotic pressure of blood.

Blood cells

The red cells. The *red corpuscles* are by far the most numerous, amounting to some five and a half million per cubic millimetre. They are cells that have lost their nuclei—biconcave discs that appear narrow-waisted when seen in profile. They are little more than envelopes containing the respiratory pigment, haemoglobin, which gives the individual cell a yellow colour and large aggregations the

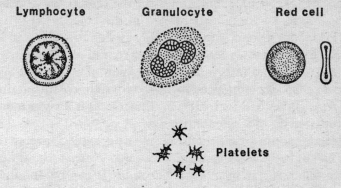

Fig. 164. The different types of blood corpuscles.

typical red colour. The splitting of haemoglobin into innumerable little packets vastly increases the surface area at which oxygen interchange can occur, for the total area of the red cells is between 1000 and 2000 times that of the body itself. The red cells are peculiarly sensitive to changes in the osmotic pressure of the medium in which they float. An increase in osmotic pressure makes them shrivel up by withdrawing water, while a weaker solution allows them to swell up and burst, when their pigment is set free and the blood is said to be *haemolysed*. The natural fragility of red cells is pretty constant, except in certain individuals where they rupture very easily, and these are prone to jaundice from the liberation into the circulation of an excessive amount of pigment.

Before birth the red cells are formed in the bone marrow,

liver, spleen and lymph glands. After birth the marrow is
their only source; in the first few years of life the bones are
full of red marrow, but later this retreats to the bone ends,
leaving the shaft occupied by yellow fatty marrow. The
survival time of the adult red cell in the circulation is some
120 days, after which it fragments and is absorbed. The
corpuscles represent the end of a line of development from
specialized cells which eventually lose their nuclei; but an
occasional intensive demand on production, in response to
haemorrhage or anaemia, results in the temporary appearance of abnormal nucleated forms in the blood.

If blood treated with an anticoagulant is allowed to
stand in a narrow vertical tube, the red cells sediment
slowly, clumping into rouleaux-like piles of coins. The *rate
of sedimentation* is influenced by the composition of the plasma
proteins; it is often increased in disease and is a good index
of the activity of a disease process.

Haemoglobin itself consists of a complex protein—
globin—linked to a chemical grouping—*haem*—containing
iron, which is responsible for its colour. It can be obtained
in crystalline form, and its particular value lies in the ease
with which it combines loosely (and reversibly) with oxygen
in the lungs to form bright red oxyhaemoglobin and with
which this oxygen is given up to the tissues, leaving purple
reduced haemoglobin. It may be convenient to add here
that carbon dioxide is not carried to any extent in the red
corpuscles but is dissolved in the plasma.

A large number of chemical factors are essential for the
proper maturation of red cells and for filling them with the
right amount of haemoglobin, and lack of any one of these
factors may give rise to a particular type of anaemia. Amino-
acids are required for the synthesis of globin and the red
cell envelope, and iron is essential because of its central
position in the haemoglobin molecule. Women need more
iron than men because of blood loss in menstruation, and
the growing child needs more iron proportionately than the
adult. So, anaemia due to iron-deficiency is common in

women and children but uncommon in adult males unless they are losing blood from, say, a stomach ulcer or a blood-destroying parasite. We have referred previously to the importance of vitamin B_{12} and folic acid, and to the gastric intrinsic factor essential to the proper utilization of B_{12} (page 303); lack of these factors, or any one of them, is the cause of *pernicious anaemia*. Other essential factors include thyroid hormone, copper and vitamin C, and a rather mysterious kidney secretion called *erythropoietin*.

In health a red cell of normal shape and size is termed *normocytic*, and, if it has a normal complement of haemoglobin, *normochromic*. Smaller or larger cells than normal are microcytic or macrocytic; cells with a reduced haemoglobin are hypochromic. Now, the normal haemoglobin content of whole blood is from 13·5–18·0 g/100 ml in men and 11·5–16·5 g/100 ml in women. *Anaemia* exists when the level is lower than this, either because there are fewer red cells per millilitre of blood or because their haemoglobin content is reduced, or both. Numerical expression can be given to these factors by relating haemoglobin and red cell count to *packed cell volume* (PCV), the proportion of blood volume occupied by packed red cells after centrifuging, some 40%–55%. From these we can obtain the *mean cell volume* and the *mean cell haemoglobin concentration*:

$$MCV = \frac{PCV\%}{\text{Red cell count} \times 10^6 \text{ mm}^3} \times 10$$

$$MCHC = \frac{\text{Haemoglobin (g/100 ml)}}{PCV\%} \times 100$$

Ordinary anaemia is due to a lack of iron, where the MCV and MCHC are both reduced, i.e. the red cells are smaller and paler than usual; the haemoglobin may fall by as much as a half, and this is known as hypochromic anaemia. In pernicious anaemia, on the other hand, the red cells, though fewer than usual, are larger than normal and have their usual complement of haemoglobin. The obvious results of anaemia are pallor of the skin and mucous membranes—which is, however, difficult to estimate casually—and breathlessness and easy fatiguability due to the reduced oxygen-carrying capacity of the blood, which makes extra demands on the heart and may result in cardiac failure. The body is capable of considerable adjustment to even severe anaemia if its onset is gradual; it may be possible to walk about in comparative comfort with a third of the normal level of haemoglobin if the cause is a very chronic one, whereas a rapid fall to a much higher level may produce severe breathlessness. The listlessness and inertia of many tropical races is often due to severe chronic anaemia produced by chronic parasitic diseases or dietary deficiencies, or both.

The infant at birth has (proportionately) more haemoglobin and a greater packed cell volume than in later life. Sex differences in these values appear to be entirely due to the effects of menstruation, as they are not present before this begins. The red cell content of the peripheral blood increases during exercise and under emotional stress. In individuals living at high altitudes, where barometric pressure and available oxygen in the air are reduced, there is a stimulation of red cell production, and both haemoglobin content and PCV of the blood are increased as a compensatory mechanism.

The white cells. There are three varieties of these: polymorphonuclears, lymphocytes and monocytes (Fig. 164, page 347). The total white count lies in the range 4000–10 000/mm³. If this is markedly increased, as in infective disease, we speak of a *leucocytosis*; if decreased, as in toxic bone marrow failure, this is a *leucopenia*. No specific factor is required for the formation and maturation of white cells comparable with the role of iron in red cell production.

The *polymorphs* (granulocytes) have a typical segmented or

lobed nucleus and account for some two-thirds of all the white cells. They also have a granular cytoplasm. 90% are neutrophils (which refers to their reaction to biological stains); a small proportion are either eosinophils or basophils, according to their affinity for acid or basic dyes respectively.

The *lymphocyte* has a large, central basophil nucleus occupying most of the cell; nearly a third of all white cells are lymphocytes. The *monocyte* is the largest of all the white cells and the scarcest—some 3%–4%. It has abundant cytoplasm and an indented or lobed nucleus.

The polymorphs are produced only in the red marrow and only a small proportion are to be found in the blood at any one time, i.e. there is a large reserve pool in the marrow to meet emergencies. Their life span is one to two weeks. Polymorphs are always being lost in carrying out defence operations (see below), and are also shed into the lungs, and bowel and body cavities.

The lymphocytes are formed in lymph nodes, the tonsils, the spleen, lymphoid patches in the small intestine and also in the bone marrow. There is a large reserve store in the lymphoid tissue of the body. There are two types of lymphocytes, large and small; and this cell is capable of reversion to more primitive types.

Monocytes are produced mainly in the spleen and lymph nodes.

The leucocytes play an essential part in the body's defence reactions to infection. They destroy and remove foreign particles and bacteria, particularly the polymorphs, while the lymphocytes have to do with the production of antibodies. An important property of the polymorphs is their power of *phagocytosis*, the ingestion or swallowing-up of foreign material. At the site of injury or infection, these cells ooze through the walls of the blood capillaries in large numbers and assemble in the affected part to overcome the micro-organisms and remove waste material—including dead and damaged tissue cells—and the debris of repair.

If, instead of incorporating and digesting the bacteria, they are themselves killed, their dead bodies form a large part of the pus that results. The most active phagocytes are the monocytes and neutrophil polymorphs. They are actively mobile within the blood and tissues, and they seem to be attracted to sites of damage or infection by *chemotaxis*, i.e. a chemical attraction by locally released substances. Either they flow round the particle and engulf it in a streaming amoeboid movement or the particle sinks into their cytoplasm. The other two types of polymorph are not so active in phagocytosis, and lymphocytes are not phagocytic at all.

The *platelets* are tiny refractile bodies lacking a nucleus and number from 150 000–350 000/mm^3. They have radiating spicular processes that give them some resemblance to an asterisk in form. They are not true cells but fragments of large bone marrow cells that have been shed into the circulation. Their main function appears to be to assist in the process of blood clotting.

Clotting

Normal blood clots a few minutes after being shed, becoming completely solid. The jelly-like clot is composed of a meshwork of threads of an insoluble substance called fibrin, with blood cells trapped therein, derived from the soluble fibrinogen protein of the plasma. This soon contracts to a smaller bulk, squeezing out a straw-coloured fluid, or *serum*, which is what is left of the plasma after the loss of its fibrinogen. Serum and plasma, therefore, are largely, but not entirely, identical.

BLOOD (PLASMA + CORPUSCLES)

$\rightarrow$ CLOT (FIBRIN + CORPUSCLES) + SERUM

Clotting is a complex process requiring the presence of calcium salts. The fibrinogen-fibrin conversion is activated by an enzyme, thrombin. However, thrombin is not present in normal blood and is formed from an inactive blood protein precursor, prothrombin:

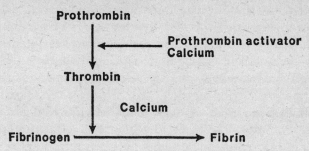

Prothrombin

Prothrombin activator
Calcium

Thrombin

Calcium

Fibrinogen ——————————————→ Fibrin

But, in fact, the process is far more complicated than outlined above, for there are a large number of accessory and adjuvant factors which will not be described here. Clotting may take place in the injured blood vessels or in shed blood in the tissues. The former is more important. When a blood vessel is damaged or divided, the injured end contracts and retracts; platelets then become attracted to the damaged lining endothelium and aggregate into a mass, sealing the cut end of the vessel. It is after this initial phase of platelet plugging that the secondary phase of clotting described above occurs. Of course, it is essential that the coagulation process should not spread back through the vascular tree and jeopardize free circulation elsewhere; so there is a complementary process of fibrinolysis, in which the fibrin is dissolved, except where locally needed. It is probable that in the tissues minor microscopic foci of injury, platelet adhesion, coagulation, and ultimate fibrinolysis and resolution occur at many sites every day. The process of clotting is facilitated by substances derived from injured tissues, and if blood is collected with a minimum of damage and received into a perfectly clean or waxed vessel it remains fluid for a considerable time. Certain other facilitating clotting agents are necessary, including vitamin K stored in the liver; the well-known haemorrhagic tendencies in liver disease and jaundice are due to its absence.

Blood may fail to clot normally within the body under

certain conditions. Thus an essential clotting factor is constitutionally absent in 'bleeders' or haemophiliacs, victims of a disease that is transmitted to male descendants by female members of the family, themselves being immune. And there are certain chemical substances that greatly delay or even abolish coagulation. One of these is *heparin*, isolated from the liver; another is *dicoumarol*, originally extracted from spoiled clover and analogous to the 'warfarin' used to poison rats. These agents have proved useful in surgery to prevent post-operative vein thrombosis and in the repair of injured blood vessels, where it is important to keep the new channel free from clot.

Blood groups

Human tissues and body fluids are sensitive to foreign proteins and react, sometimes violently, against them. In solution this reaction causes their precipitation. This applies as much to blood as to any other tissue. Thus human plasma mixed with animal blood causes the red cells of the latter to clump together in sticky, granular masses like grains of pepper, a process known as *agglutination*. It is for this reason that animal blood is useless and dangerous for transfusion, the agglutinated transfused cells blocking the recipient's blood vessels, while the debris after their destruction suppresses the secretion of urine by the kidneys, perhaps fatally.

Unfortunately, this same sensitivity exists not only between man and animals but also between one individual and another, if they happen to belong to different blood groups, and may produce the same disastrous results. Within the same group, transfusion is quite safe; outside this, only certain combinations are safe and these must be rigidly adhered to. Human blood groups are inherited characters, with antigens (page 359) borne on the surface of the red cells. There are four main groups, designated as O, A, B and AB. Their mutual relations from the aspect of compatibility in giving or receiving a transfusion are indicated in the following table; it will be noted that these are complex by virtue of the fact that it may be quite possible for a member of one group to donate blood to another, though it might be fatal for him to receive blood from the latter.

Group	Can give blood to	Can receive blood from
O	O, A, B, AB	O
A	A, AB	O, A
B	B, AB	O, B
AB	AB	O, A, B, AB

It will be noted that Group O is safe for everybody and therefore a universal donor; and, fortunately, this is found in some 40% of the population. Group AB, safe for donation to its own group only, is capable of being a universal recipient, but it occurs only in some 2% of individuals.

Blood groups are transmitted to children by the ordinary laws of inheritance and are therefore of value in cases of disputed paternity, though the evidence they provide is only negative—i.e. the only certainty being when it is possible to say that someone could *not* have been the father of a child when the child's blood contains a special factor not present in that of the accused or of the mother. The racial distribution of blood groups differs in different parts of the world, and this is of some anthropological value in analysing populations and tracing racial migrations.

There is a second important blood group system based on the discovery that the red cells of some individuals possess a component analogous to one in the blood of the Rhesus monkey, known as the *rhesus factor*. Most people (about 85%) have this factor; they are known as Rh positive and can marry and have children normally. But if an Rh positive man marries a woman of the Rh negative minority and she becomes pregnant, the Rh factor inherited by the foetus from its father excites a serious antagonistic reaction in the mother. Substances are formed in her blood that enter the womb and thus the child's circulation, causing extensive destruction of the latter's red corpuscles, with the result that the infant may be born dead, or with serious anaemia or jaundice, or spastic because of haemorrhage in the brain. It is possible to save some of these children—if the condition is suspected by tests on the parents—by replacing the infant's blood completely with that of a safe group as soon as possible after birth. This haemolytic disease of the newborn is rare in a first pregnancy, but the risk increases with each successive pregnancy as the mother's Rh antibodies build up.

Blood volume and pH

The total volume of blood in the body is easily estimated by injecting a known amount of a dye into the circulation and noting the extent to which it becomes diluted; the average blood volume is or 4–5 litres, varying with the size of the individual. After a severe haemorrhage, the fluid loss is rapidly made good by entrance of water into the circulation from the body reserves of extracellular fluid; as a result, there is considerable dilution and a lower red corpuscle count than normal. The reverse occurs in severe shock or dehydration, where plasma leaks out into the tissues, leaving a thick and viscous concentration of corpuscles in the blood vessels. The blood is slightly alkaline in reaction, with a pH of 7·4. But if we try to neutralize it with acid, considerably more is required than would be expected from this modest degree of alkalinity. This is because there is a considerable loosely-tied alkali reserve of sodium bicarbonate which, together with the plasma proteins, acts as a buffer agent, resisting any change in hydrogen-ion concentration. Only when this reserve is greatly reduced, as in ketosis, can true acidosis develop, and this is exceedingly rare.

Lymph

Lymph is a pale yellow fluid, similar in composition to interstitial (extracellular) fluid, which clots on standing like blood and contains a number of lymphocytes similar to those of the blood and rather less than 1% of protein. We have seen that the lymphatic (lacteal) system of the bowel plays the main part in absorption of fat, so that lymph obtained from the thoracic duct after a meal is milky and opalescent with fat globules. (For details of the *lymphatic system* see page 362.)

Reticuloendothelial system

This system is not physically localized, like the heart or brain, but is a tissue widely dispersed throughout other tissues and organs of the body. It consists (Fig. 165) of a

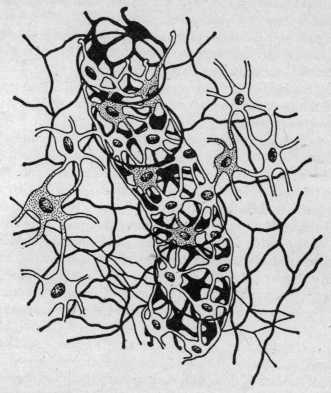

Fig. 165. Reticuloendothelium. Note the three-dimensional network of reticulin, the reticulum cells and the endothelial-like lining which they form on the reticulin framework of the fluid channels or sinusoids. (From *A Companion to Medical Studies, Vol. 1.*)

meshwork of fine fibres supporting large, irregular stellate cells and permeated by vascular or lymphatic channels, or sinusoids, lined by an endothelium composed of these cells.

These cells are characteristically phagocytic—i.e. they are able to swallow up and retain foreign particles, such as dyes—but they include the debris of tissue cells, such as red corpuscles, and some of them are able to wander actively about in the tissue as scavengers. The monocytes of the blood form part of this system; and phagocytosis, in blood or tissues, is an essential defence mechanism in combating invasion by bacteria or parasites. The human body is liable to attack from three sources:

(1) *viruses*, like those of infantile paralysis and chicken-pox, which are too small to be seen under the light micro-scope and can traverse even very fine filters;

(2) *bacteria*, visible microscopic organisms like those of plague and typhoid;

(3) larger parasites, which are single cells but still of microscopic size, the *protozoa* of malaria, syphilis and other diseases.

The primary defence against these is, of course, the outer coverings and internal linings of the body, the normally in-tact 'epithelial envelope'. The horny layer of the skin is a physical obstacle. The mucous membranes are more sensi-tive, but even here organisms become entangled in sticky mucus or are wafted away by the waving hair-like processes (cilia) of the cells in certain regions (as the lining of the air passages). Alternatively, the secretion of certain glands may itself be actively bactericidal, as with the tears and saliva. An intact epithelial lining is normally proof against inva-sion, except in certain sites and with certain organisms, notably the gonococcus and spirochaete, which cause gonorrhea and syphilis and can traverse the intact mucous membrane of the genital tract.

However, once organisms have gained access to the tissues, the first line of defence lies in their being attacked and swallowed by the phagocytes of the reticuloendothelial system that wander to the part and devour the debris of destruction and repair, as well as the organisms themselves. In minor injuries and infections the process can be completed by the phagocytes locally available, without any changes in the circulation. But it is usually necessary to bring up reinforcements by increasing the local blood supply to provide a much greater number of phagocytes from the polymorphs and monocytes of the blood. All the capillaries dilate widely and are occupied by a hurrying central stream of red cells, with white cells clinging to the vessel walls. The granulocytes move about independently and make their way through the capillary walls to the site of infection, where they accumulate in crowds. As soon as the infection has been overcome and the debris removed, the injured tissue is repaired by an overgrowth of fibrous tissue in the depths of the wound and an ingrowth of surface epithelium to cover any superficial breach.

Although the reactions to injury and infection are essentially similar, the latter is characterized by the much greater number of phagocytes required. In an infective focus that is still unresolved after some forty-eight hours, *pus* begins to form at the centre, this being a fluid collection of dead phagocytes and bacteria. On the other hand, pus formation is entirely absent in wound healing, unless a secondary infection develops.

This process of *inflammation* is marked by redness and warmth of the part due to its increased vascularity, by swelling due to exudation of fluid from the capillaries and by pain due to stretching of nerve endings—the four classical signs of *rubor, calor, turgor* and *dolor*.

But, in addition to this front-line defence against the actual physical presence of invading organisms, there is a defence in depth against the foreign substances produced by the invaders. This defence operates throughout the entire body and is known as the *immune response*. This is a reaction to the presence of foreign substances, usually proteins, known as *antigens*. And the response lies in the synthesis of gammaglobulins in the serum, known as *antibodies*, which interact specifically, like lock and key, with these foreign antigens, neutralizing their toxic effects, destroying bacterial cells or facilitating their phagocytosis. This defence is made necessary because bacteria, in addition to the local damage they cause, may affect the body as a whole, either by invading and spreading through the blood and lymph streams, and multiplying in the tissues, or by forming a poisonous toxin which circulates in the blood, or both. Some toxins, such as those of diphtheria and tetanus, are extremely virulent and may cause death, while the bacteria themselves remain localized to a small site in the throat or a wound.

Different species of animals exhibit great variation in their susceptibility to toxins. Some are quite unaffected by certain organisms, e.g. the dog by typhoid; they possess a *natural immunity*. In most cases, the primary immune response by production of antibodies is not immediate; but, once appropriate antibodies have been formed, the individual is said to have developed an *acquired immunity*, or secondary immune response, since antibodies either remain or can be restored very quickly to cope with second or subsequent attacks. And, since the individual owes this immunity to having suffered an attack of the disease and having overcome it by manufacturing specific protective antibodies in his own tissues, this is also described as acquired *active immunity*. Such active immunization may also be achieved by giving small doses of toxin or dead bacteria by inoculation or injection; this acts by producing a minor and much modified form of the disease, although the active response of the body is as great as that after a real attack and guards against any subsequent infection.

Now, it is possible to produce artificially an actively acquired immunity against, say, tetanus by injecting into a horse first minute and then increasing doses of the toxin until eventually the animal remains unaffected by an enormous dose which previously would have been rapidly fatal. Its blood serum now contains a large amount of antitoxin, capable of destroying and neutralizing the tetanus toxin. The injection of such serum into a man confers on him what is known as an acquired *passive immunity*—passive since his own tissues have played no part in elaborating the protective agent. This protection is only temporary, unlike the relative permanence of active immunization, and soon passes off. So it is only of value in treating early established disease, or in trying to forestall it by giving antitoxin as soon as possible after a wound that may have been contaminated with tetanus organisms; it is of no value whatsoever for long-term protection, and even its immediate use has waned since the development of powerful antibiotics.

It is of interest to note here that babies possess a congenital immunity against infections such as measles; this is essentially passive, since they have derived the protective agents while in the womb from the bloodstream of the mother, who has a permanent active immunity of her own dating from a childhood attack. Because the newborn infant's resistance is passive and borrowed, it lasts only a few months and the child is then susceptible.

The production of antibodies which circulate in the blood is known as the *humoral* type of immune response and is dependent on soluble protein agents. There is another specific immune mechanism, known as the *cellular* response, which is effected by small lymphocytes that recognize and respond specifically to certain antigens, and migrate from the blood to localized concentrations of such antigens to protect the body against their effects—i.e. the overall immune response is partly *humoral*, effected by circulating antibodies, and partly *cellular*, effected by small lymphocytes. One or other component of this reaction may predominate, but usually both are operant.

Now, the function of the reticuloendothelial system is to mediate both the early phagocytic response and the long-term immune reaction. It produces both antibody and lymphocytes in lymphoid organs, such as lymph nodes, tonsils and spleen, and also in the reticulum of the bone marrow; in the latter, however, the reticulum cells are the precursors of red cells and polymorph white cells, i.e. the reticulum cell is a multipotent cell, capable of developing into a lymphocyte, polymorph or red cell according to its situation.

The *lymphatic system* consists of lymphatic vessels and lymph glands, or nodes; the spleen is the largest mass of lymphoid tissue. The lymphatics form plexuses in all the body tissues (except the central nervous system) (Fig. 44, page 112). They drain into large trunks, which have one-way valves, allowing fluid to pass only towards the heart, and which accompany the major blood vessels. We have already noted the lacteal drainage from the villi of the bowel into a large lymph sac, or *cisterna chyli*, just below the diaphragm, and this continues in the chest as the thoracic duct to empty into the great veins of the neck. In the limbs the nodes are situated mainly at the bends of elbow and knee, and at the groin and armpit. Other nodes are sited close to all the main viscera in the abdomen, lying in the mesentery, and along the course of the aorta and inferior vena cava; others still are situated at the root of each lung and at the bifurcation of the trachea, these last often being stained black with carbon particles trapped from the lung lymphatics, which have acquired them from polluted inspired air.

An individual lymph node is pinkish-grey, bean-shaped and of varying size. Its main features are shown in section in Fig. 166. Afferent lymphatics drain into it and efferent vessels leave it, carrying lymph to which lymphocytes and antibodies have been added. New lymphocytes are constantly being formed in the germinal centres of the node, according to need, and entering the blood or lymph draining the node; they also constantly re-enter the node from the circulating fluids.

The lymph node functions as a filter or trap, where afferent lymph brings cellular debris or bacteria for ex-

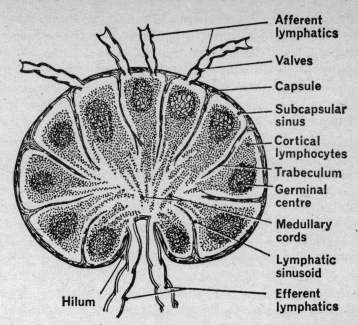

Fig. 166. Diagram of lymph node. (From *A Companion to Medical Studies, Vol. 1.*)

posure to phagocytosis by reticuloendothelial cells. It also receives antigen and reacts by a combined humoral and cellular response. And it is the site of production of new lymphocytes and a staging centre for old lymphocytes.

The *spleen* (see page 215 for gross anatomy) can be regarded as a very large and specialized lymph node. Its reticuloendothelial content carries on the phagocytosis of effete red cells and other cellular debris from the bloodstream. In certain conditions, such as severe anaemia, red cells are, abnormally, produced there. The organ can contract and expel contained blood into the general circulation to meet emergency demands, as in exertion, though this mechanism is less effective in man than in some animals. It is also a source of lymphocytes and antibodies.

In its phagocytic role, the spleen removes from the circulation any wandering bacteria, tissue breakdown

debris, and worn-out red and white blood cells and platelets. It therefore contains a good deal of iron pigment.
Whenever red cells are increasingly destroyed, the spleen
enlarges to cope with the situation, e.g. in malaria where
the parasite inhabits the red cells. Splenic enlargement is
almost the rule in countries where malaria is endemic.

Despite its important functions, the spleen is not essential
to life. It is often damaged in road accidents and can be
removed without obvious harm, partly because its various
functions are taken over by other organs of the reticuloendothelial system.

As already indicated, the combined lymphatic and
reticuloendothelial system traps antigens and responds with
an immune reaction. This response is partly humoral—
antibodies are formed in spleen and lymph nodes, and released into the bloodstream to be freed for chemical
neutralization of the antigen all over the body. It is also a
cellular response—cells, chiefly lymphocytes, are endowed,
as it were, with the capacity to recognize the antigen and to
move towards and attack it. In the course of the immune
response, there is an intense stimulation of the production
of lymphocytes in the organs concerned; also, of two other
important types of cell: the macrophage and the plasma cell.

The macrophage is a generic name for a large connective
tissue cell with the capacity for movement and phagocytosis; it is identical to the monocyte of the peripheral
blood. It can swallow up both non-antigenic particulate
substances, like dyes and carbon black, and antigenic
structures such as bacteria. This phagocytosis is very much
enhanced when specific antibody is present, for this seems
to render antigen much more attractive and devourable to
the monocyte; this is sometimes called the *opsonic* action of
antibody. The macrophage itself produces no antibody.

The *plasma cell* is a specialized cell found in the blood,
tissues, lymph nodes, spleen and bone marrow, and has a
great capacity for producing antibody. On the other hand,
it is not phagocytic.

The lymphocytes themselves manufacture little or no antibody, but in response to antigenic stimulus they become, as it were, immunologically primed and endowed with the capacity to 'recognize' antigen, and to crowd round it at local sites. In addition, the lymphocyte can transform itself into a larger, more primitive cell that does form antibody.

Lymphocytic immunity plays an important part in the reaction of a host individual against grafts from other individuals. The chemical make-up of any person is so specific that tissue material from any other person, except an identical twin, is recognized as foreign and is cast off. This is the case, for instance, with skin-grafts; a badly burned child may be skin-grafted from its mother or father with apparent early success, but the grafts will be sloughed off after two to three weeks, and this is due to local accumulation and attack by lymphocytes at the grafted sites. It depends on the fact that any one individual contains antigenic substances not present in any other person. Similar considerations apply to the transplantation of whole organs, such as the heart or kidney; and recently attempts have been made to combat lymphocytic attack on these organs by giving an antilymphocytic serum, prepared by the preliminary administration of lymphocytes to some animal that develops antigens against the lymphocytes themselves. All this relates to what is called the *host versus graft* reaction. But there is also a possible *graft versus host* reaction; certain cell transfusions may react against host antigens and produce severe damage to the liver, lungs, kidneys and lymphoid system. This is most frequently noted in connection with transfusion of blood or blood fractions.

Thymus

The thymus is a large lobed structure situated behind the upper part of the breast bone, covering the front of the great vessels arising from the heart and the upper part of the pericardium (Fig. 167). It is at its largest in the young

child and then shrinks until it may be unrecognizable in
the adult, a mere fibro-fatty remnant. It is a lymphoid
organ but does not share in the general lymphatic circulation. A great many lymphocytes are formed in the organ,
but most are short-lived and never leave the gland, though
a small proportion do enter the peripheral blood.

In the foetus and the young infant, the thymus seems to
control the normal development of lymph nodes and other

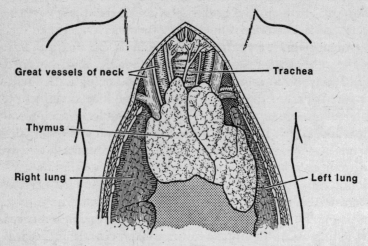

Fig. 167. The thymus in the new-born child. (After *Gray*.)

lymphoid tissue generally. If it is absent these tissues fail to
develop, lymphocytes are few and death results from infection because a failure of immune processes is usual. In the
adult the thymus is a source of supply of fresh lymphocytes
for immunological imprinting, but now it is not the only
source of these cells and its removal causes little or no harm.
The organ is sometimes said to be responsible, in part at
least, for the maintenance of *immunological competence*, and
although its removal may not be serious there is a general
diminution of the body's capacity for cellular immune response. It is interesting that, after heavy irradiation, the
activities of bone marrow and lymphoid tissue are seriously

depressed and sometimes temporarily abolished; but the thymus is more resistant, and controls the gradual restoration of lymphocyte production and cellular immunity. The clinical importance of this is that irradiation is often deliberately used to suppress an unwanted immune response, e.g. after a transplant.

The thymus seems to play an important part in what is known as 'immunological surveillance', i.e. it provides the body with the capacity to recognize certain antigens as foreign and to react against them. The converse of this is 'immunological tolerance'. For instance, an individual is tolerant of his own cells and tissues, and does not normally react against them. He is also tolerant of foreign antigens encountered in the foetal or neo-natal period. An animal whose thymus has been removed will not react against foreign antigens, though such reaction may be biologically necessary; the capacity so to respond can be restored by giving it a thymic graft. Now, in certain circumstances, the immune response can be stimulated by an individual's own cell antigens and can mount a destructive attack on his own tissues, resulting in various forms of disease such as thyroiditis or rheumatoid arthritis. The thymus is actively involved in this process, in which a self-antigen, perhaps modified by some other cause, is no longer tolerated but comes to be regarded as hostile.

Finally, we may add that sometimes the essentially protective antibody reaction may occur with such violence as to endanger life. This happens when the interval between the first and second dose of a sensitizing agent is rather a long one. Then the neutralization reaction occurs not in the bloodstream but in the tissues, and severe shock and collapse—even death—may result. Although easy to produce in animals, this reaction, known as *anaphylaxis*, is rare in man; this, however, has not prevented it from being employed as the ingenious basis of detective stories.

17   The heart and circulation

The circulation is essentially a closed circle of tubing, round which the blood is propelled by the contractions of the muscular heart. Blood is driven into the *arteries*, thick-walled elastic tubes which by their recoil aid the distribution of the blood to all parts. The arteries divide into smaller and smaller branches in their course to the organs and limbs, and finally break up into a meshwork of fine *capillaries*—microscopic thin-walled vessels that pervade every tissue of the body except the cornea of the eye, the outermost layer of the skin and articular cartilage. This meshwork joins up again to form small *veins*, which become larger venous trunks as they travel centrally towards the heart. The veins are thin-walled, have no pulse and contain valves to prevent any backward flow of blood.

The smaller branches of arteries are called *arterioles*. In general, arterioles communicate freely with each other by cross-branches called *anastomoses*, which ensure an adequate blood supply to any part, even when one vessel is obstructed. In a few situations there are no anastomoses and the local artery is an *end-artery*, which provides the only blood supply locally; the retinal artery is an end-artery and its occlusion causes blindness.

The circulation is basically a device for transporting materials to maintain the internal equilibrium of body tissues and fluids, and some sort of circulation is necessary in most animals larger than microscopic size. Its functions are to transport oxygen from the lungs to the body cells and carbon dioxide from cells to lungs; to transport nutrients to the cells and metabolites, and wastes from cells to kidneys; and to transport excess heat to the skin surface, there to be

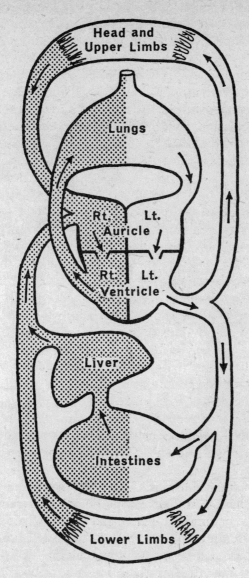

Fig. 168. A diagram of the circulation of the blood.

lost to the exterior, so as to keep the body temperature constant whatever its activities, resting or at exercise. In addition, the blood is a fluid vehicle for a number of control systems, e.g. the feedback mechanism between blood glucose level and insulin production, or between state of dilution and the activity of the pituitary gland (page 470).

Now there are, in fact, two quite separate circulations: a *systemic* circulation, concerned with the body as a whole and

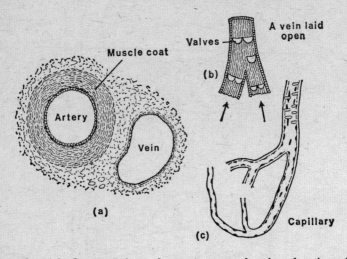

Fig. 169. (a) Cross-section of an artery and vein, showing the much thicker muscular coat of the former. (b) Vein slit open longitudinally, showing the valves with their pockets directed towards the heart. (c) A capillary with its delicate walls.

driven by the left side of the heart; and a *pulmonary* circulation, concerned with the passage of blood through the lungs and driven by the right side of the heart. The two sides of the heart are quite shut off from one another, and each side has an upper chamber or atrium receiving blood from the great veins and a lower chamber or *ventricle* discharging blood into the great arteries. Stale venous blood from the body enters the right auricle, passes to the right ventricle and is expelled through the pulmonary artery to traverse the capillaries of the lungs. Here it is aerated, receiving fresh oxygen from the air in the air sacs and giving up carbon dioxide to be exhaled. The fresh blood returns from the lungs in the pulmonary veins to the left auricle of the heart,

Venule

Arteriole

Fig. 170. Diagram of a capillary network and its supplying and draining vessels. (From *A Companion to Medical Studies, Vol. 1.*)

then down to the left ventricle and is discharged into the great artery of the body, the aorta, which supplies the head, trunk and limbs through its branches. In the body tissues the blood is rendered dark and venous, and is collected up ultimately into two great veins: the *superior vena cava* draining the head and arms, and the *inferior vena cava* draining trunk and legs.

Note that, whereas the arteries of the body contain bright blood and the veins dark blood, the reverse is inevitably the case for the pulmonary vessels because the lungs are concerned with reversing these chemical states of the blood.

Then there is a special arrangement of the abdominal vessels that must be noted. Whereas the veins leaving most structures pass directly to the heart, those from the stomach and intestines enter another organ, the liver, where they break up into a second set of capillaries so that the blood is filtered through the liver before reaching the heart. This, as we have seen, is to ensure that the liver utilizes and stores foodstuffs absorbed from the bowel, and the arrangement is known as the *portal system*.

Physiological features of the general circulation

The circulation is everywhere a closed system and, although the calibre of the blood vessels varies enormously, the blood as a whole never escapes into free contact with the tissues. The motive power of the circulation is the pumping action of the heart, which acts as a boosting mechanism set in the middle of a pipeline, for it receives blood from the veins at very low pressure and transmits it to the arteries at a very high one. When the blood eventually leaves the arteries to circulate through the tissues in delicate thin-walled capillaries, the plasma can diffuse through to mix with the tissue juices, which bathe the cells and form the lymphatic fluid.

There is, therefore, a pressure gradient in the circulatory system, from the high level in the great arteries of the body to an intermediate point in the capillaries and still lower to the pressure in the veins, which may be zero; in fact, the circulation in the veins is largely dependent on such outside assistance as the sucking action of respiration, which draws blood into the chest and heart, and on the contraction of the muscles that surround the veins in the limbs. Thus, on walking, the leg veins are constantly being emptied by the milking action of the muscles, and the simple valves in the veins prevent any reflux and direct the flow towards the heart.

This difference between the veins and arteries is reflected in their structure; the veins are thin-walled with little muscular or elastic tissue, whereas the arteries have thick coats with a good deal of muscular tissue and elastic fibres. This allows the arterial side of the circulation to act as a distensible and elastic distributing system with a recoil to the heartbeat, which is transmitted away from the heart as far as the capillaries to produce the pulse as we see it at the wrist or temple. Moreover, the contractile power of the arteries serves to maintain a higher blood pressure at a greater distance from the central pump than would a system of simple inelastic tubes. And, in addition, the smaller arteries and arterioles can exert a selective action on the local circulation in particular regions by constricting or dilating to shut off or increase the blood flow to a certain organ, muscle or area of skin. This is in marked contrast to the veins, which form an inelastic reservoir which may become a stagnant pool if their valves become inefficient or the heart begins to fail.

The flow in the veins is slower than in the arteries because of their greater total cross-sectional area, though, of course, the total volume of blood transmitted per minute must be the same. In the arteries particularly, there is friction between the blood and the vessel wall, and also between outer and inner streams of blood, so that flow is most rapid in the centre of the stream; in addition, the blood cells tend to cling to the periphery, leaving a fast-moving central axial stream of fluid blood.

As the arteries divide, the cross-sectional area of their branches is greater than that of the parent stem until, when we reach the capillaries, the total sectional area or 'capillary bed' may be almost 1000 times that of the aorta, the great artery, leaving the heart. This enormous difference in area is another aspect of the progressive fall in blood pressure as the arterioles branch into the capillaries; and, when all the capillaries of the body are widely dilated, as in shock after injuries, the general level of blood pressure falls very seriously, in the great arteries, as well as the smaller vessels.

The heart

Although we have compared the heart to a pump, it is far more efficient than any manufactured by man, for it continues to beat regularly and without interruption, rarely missing a beat, for on average seventy years with a rest interval never longer than a fraction of a second. This power of rhythmic contraction is inherent in the organ itself and independent of nervous control, for it continues when the animal organ is experimentally removed from the body and kept under suitable conditions; chicken hearts have been kept contracting in fluid culture media for many years.

The heart beats at the rate of some seventy times a minute at rest in the adult, more rapidly in children and as fast as 150 times a minute in the developing child in the womb. Contraction of the organ is known as *systole*, relaxation as *diastole*. Each beat begins as a simultaneous contraction of both auricles expelling blood into the ventricles; this is followed by ventricular contraction, which throws the blood into the great arteries. This regular cycle lasts eight-tenths of a second, of which auricular contraction makes up one-tenth, ventricular contraction three-tenths and complete relaxation or diastole the remaining four-tenths. During diastole the auricles act as passive receptacles, draining the blood from the great veins; and since they contract a little before the ventricles, while the latter are still relaxed and offering no resistance to the influx of blood, the auricles do not have a great deal of work to do and are relatively thin walled. On the other hand, the ventricles have the whole resistance of the peripheral circulation to overcome, notably the elasticity of the arteries, and are

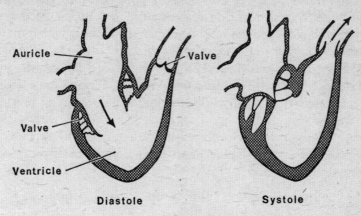

Diastole   Systole

Fig. 171. In relaxation or *diastole* the blood flows into the ventricle from the auricle through the open valve between. In contraction or *systole* the valve is slammed shut as the ventricle expels its blood into aorta or pulmonary artery.

correspondingly thick-walled and powerful. Since the right ventricle is only concerned with driving the blood through the lungs, it has rather less work to do than the left, which deals with the circulation of the whole body, and its wall is therefore only a quarter as thick.

The regular and rhythmical sequence of events in the cardiac cycle is due to the existence within the organ of two little foci of primitive and sensitive connective tissue which act as the switches that fire off each contraction. These *nodal points* are stationed one at the junction of great veins and auricles, one at the auriculoventricular junction. They are connected to the heart muscles as a whole by a special cable of fibres which runs down the length of the septum between the two halves of the organ, giving a main branch to right and left side. The heart is quite soft and flaccid in the relaxed period of diastole; when it contracts it becomes thick and cone-shaped, and transmits an impulse which may be seen and felt to the front of the chest on the left side. This is the *apex beat* and is found a little below the left nipple, in the space between the fifth and sixth ribs. If one listens with a stethoscope, or even with the unaided ear, over this point, it will be found that two characteristic sounds go with each beat, coming very closely after each other and repeated after the diastolic pause. These may be graphically represented as follows:

... lubb, dupp ... lubb, dupp ... lubb, dupp ...

and are both associated with ventricular contraction. The first is the slamming shut of the parachute-like valves that prevent any reflux from ventricles into auricles. The second occurs at the end of ventricular contraction, when the valves at the mouths of the great arteries fall together to prevent any backflow into the ventricles.

In heart disease, which affects the efficiency of the shutting of the valves, these sounds are correspondingly altered; and, although each sound is made up by the closure of a *pair* of valves, one on each side of the heart, it is usually possible to specify which one is out of order by listening at different points over the organ.

It is essential for the contraction of the heart chambers to be both effective and co-ordinated. A haphazard unco-ordinated contraction is known as fibrillation, which tends to prevent any proper pumping action. Atrial fibrillation is common; it is not so serious and is well controllable by the drug digitalis. Ventricular fibrillation is rapidly fatal if not treated promptly by administering an electric shock to the heart or by some other means.

It is important to realize that, although the two halves of

the heart are welded together as a single organ, they are really performing quite separate mechanical functions. The auricle and ventricle on each side constitute a pair of chambers concerned with the pumping of blood through a particular territory; and, although the blood traverses each pump in turn, it is quite possible to imagine the right and left halves as two separate organs fixed together for convenience like the crossed vinegar bottles in an old-fashioned cruet (Fig. 172). The output of the heart, however, must

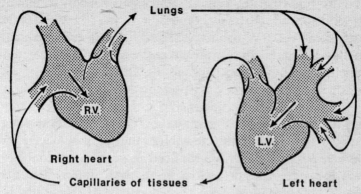

Fig. 172. The two sides of the heart represented as separate organs.

obviously be the same on both sides; the flow from the left ventricle into the aorta must equal that from the right ventricle into the pulmonary artery, or otherwise blood would be dammed up on one or other side. This output, at rest in the adult, is some 3 litres a minute—in other words, the whole of the blood in the body is passed through the heart in something under two minutes.

The *stroke volume* is the amount of blood ejected in systole from the left ventricle into the aorta, some 60–100 ml/min. The output per minute is obviously equal to this quantity multiplied by the pulse rate.

The control of the heart

The heart's activity is varied by a number of factors; these act by altering either the *heart rate* or its power of contraction, i.e. its *output*. The most important factor influencing output is the inflow of blood from the great veins, for output must be exactly equal to this, whatever changes may occur in the resistance offered by the general circulation, or blood will be dammed up in the venous system; this is what actually happens in heart failure. The venous inflow in its turn is greatly influenced by the sucking action of respiration. Breathing in produces a negative pressure within the cage of the chest which draws blood into the great veins and right auricle from the venous trunks of the head, abdomen and limbs. And, therefore, the filling of the heart is largely a rhythmic result of ordinary breathing; a deep breath increases the intake, the output must rise correspondingly and the pulse can be felt to quicken at the wrist. Conversely, if the pressure within the chest is raised by violent straining exertion, the great veins are flattened out, the cardiac output falls and the pulse weakens, or may even be made to disappear.

From the mechanical viewpoint, the work achieved by the heart, i.e. the energy produced in ventricular contraction, is used up in two ways. First, the column of blood expelled acquires a definite velocity and therefore a momentum or kinetic energy; and, secondly, the elastic resistance of the arterial tree has to be overcome, the stretching of the arterial walls and the resulting tendency to recoil providing a store of potential energy like a compressed spring. In actual fact, the greater part of the work done is in overcoming arterial resistance, since the velocity of the blood is unimportant except during violent exercise. It has been stated that, during the course of a normal life, the work done by the heart is equivalent to the lifting of a weight of 10 tonnes through a height of 16 km.

The circulatory resistance that has to be overcome is, of course, exactly equal to the arterial blood pressure, and the heart adapts itself wonderfully well to expel exactly its intake of blood however the arterial pressure changes, or, conversely, to get rid of an intake varying between wide limits against a constant resistance. It is comparable to a bath that remains filled to a constant level with the tap running and the outflow pipe open, even though the tap may be turned up or down and the outflow pipe made narrower or wider, or even though both are varying at the same time. The essence of this adaptability lies in the power of heart muscle fibres to contract more forcibly the more they are stretched when an increased intake of blood dilates the organ in the relaxed period of diastole, and vice versa. This is a phenomenon corresponding closely to Hooke's famous physical law applied to the stretching of elastic non-living matter—*Ut tensio sic vis.** Obviously there are limits to the heart's power of meeting an increased demand by dilating, with increased contraction and hypertrophy of its fibres. One limit is that the fibrous bag, or pericardium, containing the organ cannot stretch; the other is the increasing mechanical disadvantage at which the stretched fibres have to work. In medical terminology, this property of overcoming handicaps, such as a leaking valve or a grossly-raised blood pressure, is known as *compensation*, and it results in the development of an organ that is permanently larger and thicker than normal. But when the limits of adjustment are surpassed, cardiac *decompensation* results; output can no longer keep pace with influx, and the veins of the lungs and body become stagnant reservoirs.

The blood pressure within the chambers of the heart can be measured directly by inserting a cardiac catheter, a hollow tube passed up from an artery or vein of the arm or leg into the organ itself; this also yields blood for gas analysis. The pressure in the systemic circulation is five times greater than that on the pulmonary side. The stabilization of blood pressure is important, particularly with regard to the functioning of the kidneys and brain. The kidneys stop secreting urine if the systolic pressure falls much below 50 mm of mercury; and man, as an erect animal, cannot survive for more than a few minutes if his brain is deprived of arterial blood.

* 'As is the tension, so is the force.'

The nervous control of the heart

Apart from the self-regulating mechanism just described, there is a nervous control designed to adapt the organ to the needs of the whole organism. And this is effected by branches of the two constituents of the independent autonomic or vegetative nervous system—the sympathetic and parasympathetic. As we have seen in connection with the bowels and bladder, the two have mutually antagonistic actions but work in collaboration to achieve a balanced control (Fig. 173).

The *parasympathetic* fibres are derived from a nerve called the *vagus*, running down from the base of the brain through the neck and chest. They exert a constant depressing action on the heart, slowing down its rate and weakening its force of contraction. If the nerve is stimulated experimentally, the heart may stop altogether; similarly, firm pressure on the eyeball excites the vagus reflexly and slows the pulse. Vagus action is achieved ultimately by the liberation at its nerve endings in the heart muscle of a particular chemical substance—*acetylcholine*.

The *sympathetic* fibres are derived from the sympathetic chain, which runs the length of each side of the spinal column, and their action is to accelerate the heartbeat and increase the force of contraction, which they do by the liberation of *adrenaline* at their terminals. The heart is very sensitive to this substance, which is also, and indeed mainly, produced by the adrenal glands. Adrenaline from the adrenals reaches the heart via the circulation at times of anxiety and stress, producing a rapid, forceful beat.

The ultimate cell-stations that all these fibres, both sympathetic and parasympathetic, are derived from are found in the hindbrain, or medulla, where they form two centres, one inhibiting and the other accelerating cardiac function. These act as central exchanges, controlling the heartbeat on the basis of information received from various parts of the body and, in particular, from stretch or pressure receptors situated in the heart itself or in the great arteries—

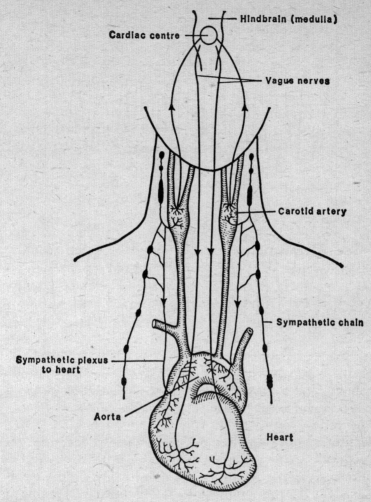

Fig. 173. The nervous control of the heart; anatomical arrangements. (Adapted from Starling's *Principles of Human Physiology*.)

aorta and carotids—nearby. These messages indicate the degree of filling of the heart, the blood pressure and the extent of cardiac dilatation; and the control centres react so as to secure that the heart beats faster and more forcefully if its venous influx increases or the arterial pressure falls,

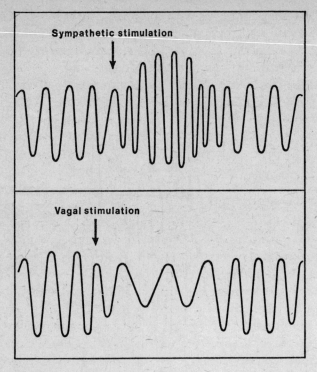

Fig. 174. The nervous control of the heart. Sympathetic stimulation produces a rapid forceful beat, parasympathetic action a slower beat of less amplitude.

and slows and beats less strongly if these changes are reversed.

It is necessary, finally, to say a few words on the metabolism of the heart muscle itself. The heart gets its blood supply from two tiny but all-important vessels, the coronary arteries, which spring from the aorta at its commencement and ramify out over the surface of the organ. These little vessels are very sensitive to nervous control and changes in oxygenation of the blood, and respond to secure the necessary increase in blood supply to the wall of the heart when extra demands are made on it. Any coronary disease (atherosclerosis) or obstruction by clot (coronary thrombosis) gravely interferes with the heart's capacity to respond to demand and, in consequence, there is a risk of fatal results on exertion. From the biochemical aspect, the great energy output of the heart is obtained not so much from glucose—as with ordinary skeletal muscle—as from free fatty acids, keto acids and lactate circulating in the blood, and the heart muscle is peculiarly sensitive to changes in local concentration of sodium, potassium and calcium ions.

The circulation

We have already mentioned that there is a *pressure gradient* throughout the circulation, falling as we pass from the left side of the heart through the arteries and capillaries until it becomes very low or reaches zero in the veins. This pressure is usually measured in the length of a counter-balancing column of mercury. It is much the same in all the arteries, whether in the aorta or at the wrist, i.e. some 120 mm of mercury. But it falls to 20 mm in the capillaries of the tissues, and to only a few millimetres when these have rejoined to form the main veins. In fact, in the large veins of the neck and chest the pressure is often negative because of the sucking action of respiration. When cut across, arteries bleed from the end nearer the heart, in forcible spurts derived from the impetus of left ventricular contraction. Veins bleed from the cut end farthest from the heart in a steady, gentle flow. And wounds of the larger veins of the neck often actually suck in air from outside, so low is their internal pressure; this is a great danger to the circulation, for the air gets churned up into foam in the heart and obstructs the smooth flow of blood.

It must be realized that the blood pressure in the arteries is alternately high and low, corresponding to the contraction and relaxation of the left ventricle. The *systolic pressure* in man is some 130 mm of mercury, the *diastolic pressure* only 70 mm, and the fluctuation between is spoken of as the *pulse pressure*, i.e. 60 mm. In the veins and capillaries, on the other hand, there is no pulsation and the pressure is a steady one. The term 'blood pressure' in common medical and popular usage always refers to arterial pressure and is expressed in double figures as 130/70, though there is a considerable normal range of variation in different individuals.

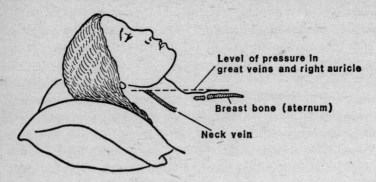

Fig. 175. The basic level of venous pressure in the reservoir of the right auricle of the heart is shown by the extent of the visible column of blood in the superficial veins of the neck.

Since the higher systolic figure depends on the force with which the left ventricle contracts, and since this in turn is greatly influenced by emotion and exertion, it is the relatively steady diastolic pressure that provides a constant base line and gives a clue to any abnormal changes in circulatory disease.

A basic feature of the circulation is that the intermittent flow from the heart into the great arteries is converted into a constant flow by the time the tissue capillaries are reached, and this is due to the elastic recoil of the arteries and arterioles, which continues to provide a driving force while the heart itself is relaxed in diastole.

The time taken for the blood to pass from any given point in the circulation, say an arm vein, through the heart and lungs and back to the same point is known as the *circulation time*, and is of the order of 25 seconds at rest, though speeded up in exercise. It is easily measured by injecting a substance that is identifiable because of its colour or radioactivity and noting the time before its reappearance. It does not, of course, imply that the actual speed of the blood in its passage is constant, as this varies with the heartbeat and the size of the blood vessel; at its maximum, in the great arteries springing from the heart, the velocity is rather more than 1·6 km/h. Nor must we confuse the velocity of the blood flow with that of the pulse wave, for this is a pressure wave transmitted through the column of blood contained in the arteries with great rapidity and is quite unrelated to any forward movement of the fluid itself. It is, in fact, transmitted at a speed of some 24 km/h.

We have seen that the main arteries divide and subdivide into a network of tiny arterioles and capillaries of a total cross-section many times greater than that of the aorta; it is this capillary bed that provides the main resistance to the thrust of the left ventricle. But it is not a fixed resistance, for the arterioles have muscular coats and the capillaries contractile cells in their walls, which are intimately subject to control by nervous and chemical agents. It is as if a hosepipe were divided into innumerable smaller branches that were not only elastic but could also be altered in calibre; we speak of *vasoconstriction* when these are narrowed and *vasodilatation* when they are opened up. This provides a means for adjusting the dynamics of the local circulation in different parts of the body; such an arrangement is absolutely necessary to meet changes of circumstance—for example, the changing effects that gravity exerts in different postures.

In standing upright the whole of the blood in the body would collect in the small vessels of the abdomen and legs (and could easily be accommodated there) were these not shut down to a considerable extent by a vasoconstriction that prevents such an influx. It is obvious that such a widespread narrowing must be reflected in a general rise of blood pressure throughout the body, as the remaining vessels are overstretched; conversely, an extensive vasodilatation leads to stagnation of blood and a fall of blood pressure. In fact, the enormous vascular area provided by the abdominal organs forms what is known as the *splanchnic pool*, in contrast to the *somatic* part of the capillary bed in the skin, muscles and skeleton. When the splanchnic vessels are widely dilated, as after a heavy meal, the blood pressure falls and the blood supply to other parts, particularly to the brain at the highest level, is decreased. We become sleepy and tend to lie down because there is a temporary cerebral anaemia and the circulation can be maintained more easily in the horizontal position. Fainting or *syncope* is an acute change of exactly the same nature after an emotional shock which produces splanchnic vasodilatation.

Often, in aged persons, this control mechanism is inadequate and there is a sudden drop in blood pressure, or postural hypotension, on standing, which may cause falls or fainting attacks. And severe emotional shock may also cause fainting, even in healthy young people while they are in the upright position. The faints that may occur in guardsmen, who are compelled to stand motionless for long periods, are also due to postural pooling and may be obviated by regular contraction of the leg muscles. And in what is known as *surgical shock*, which develops in the hours following severe injuries or burns, there may be such widespread expansion of the capillaries that the body 'bleeds to death into its own blood vessels', the blood pressure falling profoundly and too little blood being available for the heart to work on efficiently or the lungs to aërate.

The addition of fluid to the blood volume, as by a rapid intravenous infusion of some pints, has little effect if the fluid is crystalloid; the excess water and salts rapidly pass out into the tissue spaces. If the fluid has a colloid content, however, this cannot happen and the vessels must dilate to accommodate the extra fluid without causing any rise in pressure; meanwhile, the kidneys are stimulated to pass more urine. The loss of blood, as in blood donation of, say, 600 ml, is compensated so rapidly and efficiently that the circulatory effects are negligible. But the sudden loss of several pints, as from haemorrhage, causes a severe fall in blood pressure which must be compensated. There is general vasoconstriction, the heart beats faster, sweating occurs, the skin becomes cold and moist, and the kidney blood flow is reduced and urinary output falls. At the same time, fluid enters the circulation from the tissue spaces. Thus both blood pressure and blood volume are restored, although the blood itself will be more dilute, i.e. it will contain less red cells and haemoglobin.

Thus, in some ways, the circulation is to be compared not to a main railway line but to a whole railway system with numerous sidings but only a constant number of trains. If some of these are shunted off into the sidings to remain idle, the main line traffic will be much less heavy, i.e. the blood pressure will fall. If all the trains are in use and the sidings empty, the traffic is intense, i.e. the blood pressure rises. Such a regulation occurs in all parts of the body; any inflammation in the skin or elsewhere sets up a local vasodilatation which flushes the part with blood. And in these circumstances the arterioles may be so widely opened that they communicate their pulse to the normally pulseless capillaries and the whole part throbs.

The mechanism controlling the small vessels of the body corresponds to that of the heart and is a double one—partly nervous, partly chemical. The sympathetic fibres that accelerate the heart also constrict the vessels, and are thus fundamentally concerned with preparing the body for emergency and violent action, for they divert the blood from the skin and non-essential organs to the muscles, and to the blood vessels of the brain.

The parasympathetic fibres promote general relaxation and fall of blood pressure. However, the two systems are collaborators, and the vessels are never completely shut down and never completely dilated (except in severe shock) but possess a normal *tone*, which varies between wide limits.

On the chemical side, the arterioles and capillaries are

very sensitive to circulating adrenaline, which has the same effect as the sympathetic nerves, and are dilated by certain substances liberated from the tissues when they are bruised and injured—e.g. the substance *histamine*, which is thought to be partially responsible for the phenomenon of shock after wounding. And again, as in the case of the heart, the ultimate controlling (vasoconstrictor and vasodilator) centres are situated in the medulla. To a certain extent it is true to say that the prime function of these centres is to ensure that, whatever else happens, the supply of blood to the brain itself is safeguarded; for, although other tissues can survive deprivation of oxygen for many minutes, perhaps an hour more, the brain cells must die after more than a few minutes of anoxaemia as the oxygen consumption of the brain is 20% of the whole body resting consumption.

A beautiful example of this mechanism is seen after blows or falls on the head severe enough to cause progressive bleeding within the rigid and unyielding bony skull; such bleeding steadily compresses the brain and makes it more difficult for blood to be pumped in from the heart. The vasoconstrictor centre in the medulla reacts to this by so shutting down the vessels all over the body that the general blood pressure rises, increasing to enormous heights; this enables blood to be forced into the brain-box, at any rate until the whole process of adjustment breaks down.

Exercise

There is one particular point in connection with the regulation of the circulation that, though found in animals, is most developed in man. Although the heart and vessels are quite capable in themselves of adjusting to meet the needs of a particular situation, they are anticipated by mental factors which produce the same changes before they are strictly necessary. Thus the emotion implicit in waiting to start running a race produces a rise in blood pressure and pulse rate long before these are made physically necessary by the demands of active muscular contraction.

When strenuous exercise is actually in progress, the oxygen demands of the muscles may be ten times those of the resting stage, and this must be met by a corresponding multiplication of cardiac output. This is because arterial blood is already saturated with oxygen and cannot dissolve more; the only way to bring more oxygen to the tissues is to rush the blood round the circulation at a faster rate, loading oxygen in the lungs and unloading it in the tissues more times in a minute. This is obviously comparable to increasing the food supply to this country during a period of emergency by a quicker turn-round at the ports on each side; there is faster loading and unloading at the docks, though the number of ships available remains the same.

But the bodily changes in exertion are very complex. Muscular contraction speeds the return of blood to the heart by milking it upwards in the veins of the limbs; and this increase in venous inflow must, as we have seen, imply an equivalent increase in cardiac output if the heart is not to fail. The total amount of blood in the circulation cannot be increased, but more can be diverted to the muscles by shutting down the vessels of the skin and the splanchnic area, and by the contraction of the spleen, an organ that can squeeze out quite a large amount of blood into the general circulation when necessary. The exaggerated respiratory movements suck more blood into the great veins of the chest, as well as more air into the lungs. The increased activity of the sympathetic system, aided by the pouring out of adrenaline into the blood that accompanies any excited state, increases the heart rate and the force of each beat, and mobilizes the liver glycogen as sugar to be used by the muscles. And the waste products of muscular contraction— carbon dioxide and lactic acid—stimulate the brain centres controlling the heart, the vessels, and the rapidity and depth of respiration to continue the effects that emotion has initiated.

18 Respiration

The respiratory system

Air inhaled through the nose or mouth enters the pharynx behind and then travels down through the air passage proper. The first part of this is the *larynx*, which is also the organ of voice; and this leads on to the windpipe (*trachea*), which divides in the upper part of the chest into a right and left *bronchus* for the right and left lung. Each bronchus sub-divides within the lung to form numerous branching *bronchioles* which end in clusters of tiny *air sacs*; and it is in the walls of the latter that the actual interchange between the gases dissolved in the blood and those of the inhaled air occurs (Plate 15).

The cavity of the chest is divided into right and left halves by a massive partition in which the heart lies embedded; the two halves are quite separate and contain the right and left lungs. Each lung cavity is lined by a smooth, glistening membrane, the *pleura*, which facilitates the gliding of the lung on the chest wall; and each pleural space is a closed sac, for the membrane is reflected from the deep aspect of the chest wall onto the lung root and covers the surface of that organ. Normally, there is really no actual pleural cavity, for the two layers of pleura are everywhere in con-tact, each lung entirely filling its side of the chest. But the lung is a very elastic structure, constantly tending to shrink down and expel its contained air; since in health it cannot do this, there is always a negative pressure in the potential pleural space. The moment that air is admitted to the latter, whether by wound from without or by perforation of the lung from within, the lung immediately collapses into a small, solid, airless mass. This was made use of in the

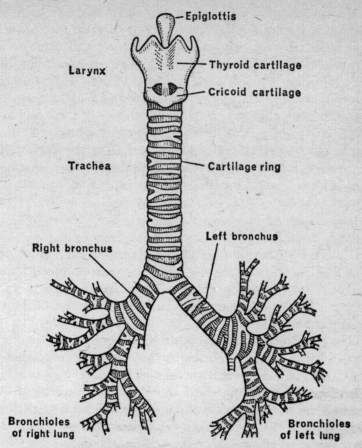

Fig. 176. The respiratory passages.

treatment of pulmonary tuberculosis, where a diseased lung used to be rested by deliberately letting in air from the outside, a process known as 'artificial pneumothorax' treatment.

Respiration

The purpose of respiration is the entry and exit of air to and from the lungs. In *inspiration* the chest cavity is enlarged and air enters; in *expiration* the reverse occurs. Respiratory movements may be thoracic or abdominal, or both. In thoracic inspiration the breast bone is lifted up and forward, and the ribs are simultaneously elevated and come to lie more horizontally; thus the chest is increased in diameter both from side to side and from front to back. In abdominal inspiration the diaphragm contracts and flattens down on the abdominal organs, bulging out the abdominal wall and increasing the vertical height of the chest. Thus inspiration is an active process, due to muscular exertion. Expiration, however, is passive; the chest wall subsides, the abdominal muscles recoil and the diaphragm relaxes, and air is driven out of the lungs.

Lung ventilation

During normal quiet breathing, about 600 cm³ of air are taken in and breathed out at each respiration, and this steady flow is known as the *tidal air*. By means of a very deep inspiration, some 2000 cm³ more can be taken in, the *complemental air*; and, on the other hand, after a normal expiration it is possible by breathing out very forcibly to expel a further 1200 cm³, the *supplemental air*. The sum of these figures, which is the total possible flow in one direction, is known as the *vital capacity* of the individual and represents the maximum volume of air that can be expelled after as deep an inspiration as possible, i.e.:

$$
\begin{array}{lll}
\text{Complemental air} & = & 2000 \ \text{cm}^3 \\
\text{Tidal air} & = & 600 \ \text{cm}^3 \\
\text{Supplemental air} & = & 1200 \ \text{cm}^3 \\
\hline
\text{Vital capacity} & = & 3800 \ \text{cm}^3 \\
\hline
\end{array}
$$

The vital capacity obviously varies a good deal from one individual to another, depending on such factors as physical build, and state of fitness and training; it is also greatest when standing and least when lying down, and is obviously diminished when the ribs and chest are fixed by arthritic disease, a state of affairs predisposing to lung infections. But no expiration, however forcible, can empty the lungs of air; an irreducible minimum is left occupying the bronchi and air sacs, the *residual air*, a volume of some 1500 cm³. Finally, in normal breathing it is obvious that not all the 600 cm³ of tidal air will reach the air sacs, for some is needed to fill the air passages themselves, a *dead space* of some 150 cm³.

We must not allow ourselves to be led into regarding these air passages as rigid tubes of standard cross-section. In fact, they have a strong muscular coat to their walls which is capable of altering considerably their calibre by constriction or relaxation, so varying the volume of the dead space and the resistance to the flow of air. Asthma is an intense spasm of the bronchial muscle and makes the expiratory act a long-drawn-out effort; inspiration is less hindered, for it is produced by active contraction of the rib muscles and diaphragm, whereas expiration is merely a passive relaxation without the impetus to overcome the bronchial spasm. Like the smooth muscle of the blood vessels, that of the bronchi is under the control of the parasympathetic and sympathetic nerves; it is the latter, in exertion and fright, that open the air passages widely so as to achieve the maximum possible influx of air to the lungs. Consequently, adrenaline, which stimulates the sympathetic nerve endings, may be injected to overcome an attack of asthma.

Chemical aspects of respiration

In the tissues of the average man, some 400 cm³ of oxygen is used every minute to burn up food material, and this results in the production of a considerable quantity of carbon dioxide. The oxygen is taken by the tissues from the blood in the capillaries and the carbon dioxide returned in exchange, so that the venous blood leaving any part contains less oxygen and more carbon dioxide than the arterial blood that entered it. The lungs exist to reverse this state of affairs in the venous blood, restoring its oxygen quota and expelling its excess carbon dioxide by gaseous interchange with the air in the air sacs. Thus the venous blood is rendered arterial in its passage through the lung capillaries, and thus the changes in the tissues are ultimately reflected in changes in the composition of the tidal air, which, when breathed out, contains less oxygen and more carbon dioxide than on inspiration. This is made clear by analysis of the composition of inspired and expired air:

	Oxygen	*Carbon dioxide*	*Nitrogen*
Inspired air	20·95%	·05%	79%
Expired air	16·5%	4·0%	79·5%

The content of nitrogen, a gas inert from the respiratory point of view, is unaltered, in fact, but appears to be in slightly greater concentration in the expired air because slightly less air is breathed out than is breathed in. And the reason for this is that the volume of carbon dioxide exhaled is slightly less than that of the oxygen for which it is exchanged. This ratio:

$$\frac{\text{carbon dioxide output}}{\text{oxygen consumption}}$$

is always less than unity and usually about 0·85. As we have seen at page 290, this ratio or *respiratory quotient* depends on the exact nature of the foodstuffs being burnt at the time in the body.

The respiratory gases are carried between the lungs and the tissues dissolved in the blood, and measurement shows that the total gas that can be expelled from a given amount of blood is some 70 volumes per cent. Of this, the relatively insoluble nitrogen forms only a small proportion. And, as we should expect, analysis of venous and arterial blood reveals their different gaseous content:

	Oxygen	*Carbon dioxide*	*Nitrogen*
Arterial blood	19	50	1
Venous blood	10·5	58	1

expressed in volumes per cent

But the problem is not a simple one of relative solubilities of oxygen and carbon dioxide, for it is affected by the following factors:

(1) The amount that can be dissolved is proportional to the pressure of the particular gas in the surrounding medium, i.e. in the lungs or tissues.

(2) The gases are not held in simple solution but in loose chemical combination—the oxygen with the haemoglobin of the red blood corpuscles, the carbon dioxide with the carbonic acid and sodium bicarbonate of the plasma.

(3) The two gases, oxygen and carbon dioxide, can be regarded as opponents that tend to displace each other from their combinations with corpuscles or plasma.

Taking each in turn, the processes at work are as follows:

Oxygen: When venous blood, with its reduced content of oxygen at low pressure, reaches the lungs, it is exposed to a higher pressure of oxygen in the alveolar air contained in the air sacs. The pressure gradient is towards the blood, and oxygen is taken up by combining loosely with the haemoglobin of the red cells to form oxyhaemoglobin. When the blood has been swept through the left side of the heart and round to the capillaries in the tissues, it has arrived at a region where oxygen has been used up in combustion and is at a lower pressure than in the blood; the gradient sets the other way, and oxygen is set free from the oxyhaemoglobin and diffuses out into the tissue fluid. But, at the same time, this dissociation of oxyhaemoglobin is accelerated or catalysed by the presence in the tissues of the excess carbon dioxide formed by combustion. In other words, the very waste products that have been formed with the using-up of oxygen act as stimulants to the liberation of oxygen from its bound state in the blood so as to make more available to the tissues.

Carbon dioxide: The pressure of this gas is high in the tissues, where it accumulates as a waste product—much higher than in the fresh blood brought to the part by the arteries.

The pressure gradient therefore favours the passage of the gas from the tissue fluid into the capillaries, where it enters into loose combination as carbonic acid and sodium bicarbonate in the plasma. This is carried with the venous blood through the right heart and round to the capillaries of the lungs, where it is exposed to the freshly inhaled air in the alveoli; this air has a very low carbon dioxide content, the gradient sets the other way from the blood and the gas is expelled with the exhaled air. At the same time, this transference is assisted by the simultaneous saturation with oxygen that is occurring in the lungs, as a high oxygen content in blood tends to break up the loose carbonic compounds.

Tissue respiration

What we have so far been describing is sometimes referred to as 'external respiration', as it is concerned only with gaseous interchange in the lungs and tissue capillaries and not at all with the vital processes in the tissue cells themselves. This latter and less obvious process is called 'internal respiration', and is the source and motive of all the obvious and complex apparatus of heart, lungs and circulating blood. Every tissue uses up oxygen and liberates carbon dioxide to obtain its energy requirements, and the amounts involved depend on the state of its functional activity; obviously they are greatest in the intensively active tissues of the heart, brain, liver and contracting muscles. The process can be studied experimentally by removing thin slices of living organs, or by making ground-up cellular suspensions and keeping them in a warmed fluid medium containing dissolved food and oxygen. In actual fact, we know little about the fundamental processes in the living cell that intervene between the intake of oxygen and the ultimate formation of carbon dioxide. There is a complex system of enzymes and catalysts at work, known generally as *oxidases*, and these are very susceptible to modification by toxic substances. Thus the poisonous action of the cyanides is due to their wide-spread inhibition of tissue respiration. Finally, it should be stated that this method of respiration, involving the use of oxygen, is known as *aërobic*. Many bacteria and other organisms are capable of an indirect respiration without using oxygen, though still forming carbon dioxide, an *anaërobic* process; but this, though possible to a certain extent in human tissues, is not of any great practical importance.

The regulation of respiration

The respiratory act is a complex one, involving the muscles of the nostrils, the larynx, chest wall, diaphragm and abdomen, so that accurate co-ordination at a higher level is necessary. This control is achieved by a *respiratory centre* of nerve cells situated in the medulla or hindbrain; it is one of the group of vital centres in this region that includes the cardiac and vasomotor centres. And it consists of two parts, one controlling inspiration and one expiration. It is obviously important that the centre should be subject to constant influence by nervous information from the lungs and other parts, and that it should be sensitive to changes in the oxygen and carbon dioxide content of the blood circulating through the cerebral vessels so that it may react to alter the character of respiration as needed. It is essentially automatic and unconscious, though we are capable of controlling our breathing at will at any time by a superadded voluntary effort. And it is one of the most fundamental activities; in progressively deep anaesthesia or poisoning, respiration is one of the last bodily functions to stop, though it usually ceases before the heart makes its last beat.

The *chemical control* of respiration is effected by the response of the centre to changes in the oxygen and carbon dioxide content of the blood; of these two, the carbon dioxide is more important, for it is of prime importance to get rid of any excess of waste products, whereas the oxygen content of the air is amply sufficient for respiration within wide limits of variation. The slightest rise of carbon dioxide increases the rate and depth of respiration, and an increase in the content of the gas in the air to only 1·5% multiplies the lung ventilation by half. In progressive asphyxia, with accumulation of carbon dioxide, respiratory convulsions develop, followed by collapse and death. On the other hand, there is much less sensitivity to oxygen lack, and the atmospheric level can fall from its normal 21% to 13% without noticeable effects; even then, although respiration is augmented and the victim becomes blue, there is no great distress and quiet collapse follows without warning—as happens to unmasked pilots flying at high altitudes. These chemical stimuli also act indirectly, but with the same results, on sensory nerve endings in the aorta and carotid artery.

The *nervous* or *reflex control* is effected by the great vagus nerve on each side of the body, which connects the vital centres of the medulla with the heart, lungs and abdominal organs. On the sensory side, these provide the respiratory centre with information on the state of distension of the lungs, and the blood pressure and gaseous content of the great arteries. On the motor side, they carry down reflex responses, modifying the action of the respiratory muscles accordingly. Gross examples of this mechanism are the inhibition of respiration following a blow on the larynx or solar plexus and the violent expiratory efforts (coughing) caused by food entering the air passages.

Of the two parts of the respiratory centre, the inspiratory one is dominant, and the respiratory act is simply an active inspiration due to messages sent along the vagus nerves, with expiration succeeding as a passive relaxation owing to falling-off in these stimuli. The ultimate co-ordination of the two is a very pretty example of reflex action. When inspiration is at its height, the lungs are so inflated that their sensory nerve endings are greatly stretched; they stimulate the expiratory part of the nervous centre, the vagus action is inhibited and expiration follows, to cease in its turn when deflation is accomplished. In other words, the very act of inspiration produces the necessity and the stimulus for the succeeding expiration; it is as if a pair of bellows were fitted with an automatic electric control which made it absolutely certain that opening would be followed by closing and that, since the tendency from outside was always to open the bellows, a rhythmic opening and closing would be normal and automatic.

Thus the nervous control of respiration affects its rhythm, while the chemical control affects its depth, and these are normally in collaboration, for obviously if increased carbon dioxide increases the volume of inspiration it must also accelerate the stretch-response and thus the nervous control of rhythm. In exercise the rising of the level of carbon dioxide in the blood as a waste product of muscular activity reaches the brain and at first stimulates respiration. But, as the effort increases, carbon dioxide is actually washed out of the blood by forced breathing, and the stimulus now comes from the lactic acid formed in muscular contraction and from the adrenaline secreted in excited states. It is this steady succeeding stage that is known as 'getting one's second wind'. Deliberate, forced overbreathing at rest will so empty the blood of carbon dioxide that the normal stimulus to respiration is entirely removed; if one breathes in and out as forcibly as possible for a couple of minutes, it will be found that there is no wish to breathe and no actual breathing during the next couple of minutes, but the experiment is a little dangerous.

Oxygen deficiency

A lack of oxygen available to the tissues may occur in many ways: by a lowered oxygen content of the atmosphere; by obstruction to the air passages impeding the free entry of air; by lung disease preventing gaseous interchange; by inability of the blood to carry its full quota of oxygen, as in haemorrahge or anaemia, or when the haemoglobin is put out of action by combination with a poison such as *carbon monoxide*, or when the circulation is slowed as in heart failure;

and by inability of the tissues in poisoning to utilize oxygen. All these conditions have the same result, an overbreathing resembling that due to exertion but occurring at rest, in an effort to make up for the deficiency by increasing the intake of air, so that the cardiac invalid in his chair may be panting away as strenuously as a man who has just run a mile race. In all these states the blueness of the insufficiently oxidized haemoglobin is evident in the lips and under the nails—the condition known as *cyanosis*.

Heights and depths

Similar changes occur at increasing altitudes, on mountains or in aircraft; the oxygen available decreases as atmospheric pressure falls and reaches the danger point of 11% at 4877 m, above which an oxygen mask is desirable. The immediate compensation is by overbreathing and by an increase in the rapidity of the circulation, though, of course, both these mechanisms fail at high levels. On the other hand, a more permanent adaptation can be made by mountain dwellers, who can acclimatize themselves at up to 4572 m by two means: an increase in the number of red blood corpuscles, which may reach 8 000 000 per cubic millimetre, and an habituation of the tissues to a lowered oxygen content.

In working at depths, as with divers, or in compressed air, two factors come into play. The normal pressure of carbon dioxide is increased in the compressed atmosphere proportionately to the pressure and produces its usual toxic effects. At the same time, the nitrogen, normally poorly soluble in blood, is driven into solution; and, if the pressure is suddenly released, nitrogen is liberated in the blood as bubbles, which may block the heart or nervous system and cause death or paralysis. Hence the necessity for gradual decompression in these cases, and divers suffering from the 'bends' after being brought up too quickly must be treated by immediate recompression in a pressure chamber.

19 Excretion

The gross anatomical structure of the urinary tract and kidneys is set out on pages 219–221. Microscopically (Plate 13), each kidney contains over a million secretory units, or *nephrons*, each nephron comprising a tuft of blood capillaries, or glomerulus, invaginated into a capsule lead-

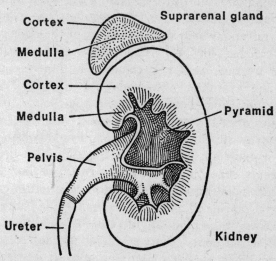

Fig. 177. Kidney and suprarenal gland, longitudinal section.

ing to a convoluted tubular system which eventually discharges with others into a larger collecting tubule, draining at the apex of a pyramid into the renal pelvis. Such is the functional reserve here, as elsewhere in the body, that this provision is more than adequate, and a mere two-thirds of one kidney will suffice to carry out the total excretory function necessary to life.

Each tubular system is closely enveloped by a plexus of

blood vessels derived from a branch of the renal artery. An afferent arteriole supplies the glomerulus and an efferent arteriole leaves the glomerulus to break up again into a network surrounding the tubule, and from this the blood is gathered up into a branch of the renal vein, which removes the blood after its purification.

The general function of the kidney is to maintain bodily homeostasis, i.e. to maintain a constant internal environment, whatever the changes in the outside world. It is a very efficient mechanism for preserving a constant chemical and physical composition of both the blood plasma and the extracellular fluid in the tissues. The kidneys regulate water content, pH and osmotic pressure so as to maintain electrolytic and water equilibrium. The blood and tissue levels of mineral ions, organic metabolites and wastes (such as urea) must be kept between narrow limits to preserve health and even life. Removal of the kidneys causes a piling-up of urinary constituents in the blood (uraemia), with death after two to three weeks.

Practically all the waste products of metabolism are got rid of, or excreted, in the urine, together with drugs and toxic agents. The only exceptions are (*a*) carbon dioxide, which is exhaled from the lungs, and (*b*) those breakdown products of bile and other digestive juices which are expelled in the faeces. If the blood should contain an abnormal non-colloidal constituent, or an excess of normal constituents such as water and salts, the kidneys excrete these to restore normal composition. The kidneys are the only means for eliminating the wastes that are the end-products of protein metabolism. But it must be realized that they do not themselves form any of the wastes they excrete—they merely rid the blood of substances formed in other parts of the body. In renal failure or disease this process is inefficient, the blood levels of urea and other wastes rise enormously, and the tissues become saturated with and poisoned by these metabolic products. The only exception to this rule is the kidney's ability to manufacture ammonia and hippuric acid.

To fulfil these functions the kidneys possess a relatively huge blood supply; they may be regarded as a pair of filters through which approximately 1 litre of blood circulates each minute, i.e. the whole of the blood in the body passes through them in five or six minutes, or 1800 litres a day, which is 400 times the blood volume.

Urine

The water content of the urine is derived partly from ingested food and fluids, and partly from oxidation processes in the tissues. So the exact composition of the urine will vary with the kind of food, the fluid intake and the fluid loss by other channels. Nevertheless, it remains fairly constant under average conditions.

It has a clear yellow colour—paler when diluted by excessive drinking, darker when concentrated by excessive sweating. It contains a certain amount of mucus, derived from the lining of the urinary channels through which it has passed. It is usually moderately acid in reaction but can be made alkaline by a vegetarian diet or by taking large amounts of bicarbonate. The total output in the twenty-four hours varies with fluid intake and outside temperature —it averages 1·5 litres, with a range of 1–2·5 litres—and its formation by the kidneys is at its lowest in sleep and maximal during the day or accompanying exertion. Its specific gravity varies around 1·01–1·03; a fixed low specific gravity is an index of chronic renal disease.

The urine may be turbid from calcium phosphate. It may, on standing, deposit mucus, urates, crystals of calcium oxalate or phosphates. Microscopically, deposits include occasional casts shed from the lining of the renal tubules, sometimes red cells after exertion. Urine is frothy when shaken due to the presence of bile salts.

The individual constituents may be grouped as organic and inorganic. The *inorganic* group includes chlorides, phosphates, ammonia, sulphates, traces of magnesium, calcium and iron. We have seen that the ammonia is derived from protein breakdown. Sulphates result from oxidation of sulphur in protein; phosphates derive partly from the food and partly from oxidation of organic phosphates in the tissues. The *organic* components are mainly nitrogenous end-products of protein metabolism—urea, uric acid, creatinine, etc. Urea is the most important of these, and in urine allowed to stand it is converted by bacteria into sharply-smelling ammonia, the familiar change that occurs in babies' napkins. The urea content is an important index of the state of protein metabolism. There is an excess of uric acid in *gout*; this is a breakdown product of nucleoprotein. There are also various organic *pigments*, yellow urochrome and greenish urobilin—both derived from bile pigments—and pink uroerythrin. The non-nitrogenous organic constituents include oxalic acid from the food (e.g. rhubarb), lactic acid (especially after exercise), and derivatives of the water-soluble hormones and vitamins.

Certain *abnormal* constituents appear in the urine at times. If the kidneys are damaged by disease, some of the normal serum proteins will enter the urine, which will now coagulate on warming. Normally, the urine only contains very small amounts of protein and this only under certain conditions: after exertion, in pregnancy and after long standing in certain healthy young people. Certain unusual proteins appear in diseases of the bone marrow. Glucose is found in diabetes, lactose in nursing mothers and ketone bodies in severe diabetes or in starvation. The urine may contain haemoglobin, as after an incompatible blood transfusion or in malignant malaria, or a great increase of bile salts and pigments, as in liver disease and jaundice. Finally, it is interesting to note that in certain disorders due to inborn errors of metabolism, where particular enzyme systems are lacking, unchanged amino-acids may appear in the urine. Thus there is phenylketonuria, where phenylalanine is not converted normally to tyrosine but to phenylpyruvic acid; this is associated with mental retardation and may be prevented in infants by eliminating phenylalanine from the diet.

Filtration

From the physical viewpoint, the formation of urine entails the passage of water and crystalloids from the blood capillaries of the glomerulus into the nephron capsule and tubules. Colloids, such as the proteins of blood, do not appear in the urine except in kidney disease. The crystalloids involved are inorganic mineral salts, and organic compounds like urea and uric acid. Now, the essential feature of renal activity is that these substances are *concentrated* in the urine. The water content of urine and blood is much the same, but there is sixty times as much urea in the urine as in the blood, fifteen times as much uric acid, forty times as much phosphate, seven times the amount of potassium, twice as much chloride and so on:

	Plasma g/100 ml	*Urine* g/100 ml	*Concentration factor*
Water	90–93	95	—
Proteins and other colloids	7–8·5	—	—
Urea	0·03	2	×60
Uric acid	0·002	0·03	×15
Glucose	0·1	—	—
Creatinine	0·001	0·1	×100
Sodium	0·32	0·6	×2
Potassium	0·02	0·15	×7
Calcium	0·01	0·015	×1·5
Magnesium	0·0025	0·01	×4
Chloride	0·37	0·6	×2
Phosphate	0·003	0·12	×40
Sulphate	0·003	0·18	×60
Ammonia	0·0001	0·05	×500

Table 3. Relative composition of plasma and urine in normal men.

The basic processes concerned are those of *filtration* in the glomeruli, and *reabsorption* of water and certain substances in the tubules. The energy required for filtration is considerable, for the osmotic pressure of the plasma proteins tends to hold the excreted substances back in the circulation. The driving force that overcomes this inertia is the blood pressure in the glomerular arterioles and is therefore ultimately derived from the pumping action of the heart. Should the blood pressure fall below the osmotic pressure of the blood proteins (some 40–50 mm of mercury), as it may do in severe shock, the formation of urine comes to a standstill. Thus urine formation begins as a process of ultrafiltration of a large volume of blood plasma from the glomerular capillaries into the capsular space, colloids such as proteins being held back while crystalloids pass through. The only difference between plasma and this initial filtrate lies in the absence from the latter of molecules above a certain borderline size. The filtrate contains all the substances present in the blood, and in the same concentration, except the colloids and fats.

Tubular absorption

Urinary secretion is not, however, a matter of simple physical filtration, with reabsorption of water back into the blood in the tubules—though this latter certainly occurs. There is a *selective* action which reflects the living processes in the tubule cells and accounts for the fact that, while there is sixty times as much urea in the urine as in the blood, there is only twice as much chloride. What happens is that the filtrate from the glomeruli enters the tubules, and there most of the water and some of the electrolytes are reabsorbed into the bloodstream, while wastes such as urea are selectively retained in the urine. The evidence for tubular reabsorption is (*a*) that the faster the circulation, and therefore filtration, the more nearly does the urine approximate to the blood plasma in composition, because it is hurried along the tubules without allowing time for reabsorption of substances that properly belong to the blood, and (*b*) the same holds if tubular function is reduced or temporarily abolished by cooling or poisoning.

80%–85% of the water and sodium in the glomerular filtrate are reabsorbed in the tubules. There is, for each particular constituent of the blood, a threshold value of concentration which must be exceeded before that constituent appears in the urine. Thus sugar is normally never excreted by the kidneys because its threshold is rarely exceeded, save in diabetes; or, in other words, all the glucose arriving in the filtrate is reabsorbed. On the other hand, reabsorption of chlorides is much less significant, and so chlorides are normal constituents of urine and only disappear in gross dehydration. Further, the urine may be acid or alkaline and so compensate for an excessive intake of alkalis in the food or overproduction of acids in the body.

The pituitary and diuresis
The final concentration of the urine by water absorption back into the circulation occurs in the main collecting ducts. This process is intimately controlled by an antidiuretic hormone—ADH—secreted by the posterior lobe of the pituitary gland (page 470). This hormone increases the permeability of the duct lining to water, and it is secreted, or its secretion inhibited, as part of a feedback mechanism sensitive to the osmotic pressure of plasma and tissue fluids. This is mediated by sensitive osmoreceptors situated in the base of the brain. Thus, at normal blood osmolarity, there is a steady receptor discharge and a steady ADH output. Suppose the plasma becomes hypertonic from ingestion of excess sodium chloride; receptor discharge increases, ADH output rises, the duct walls become more permeable, more water is reabsorbed from the urine in the ducts and a smaller volume of more concentrated urine is passed, i.e. there is an *antidiuretic* effect, the object of which is to retain water in the system to dilute the excess sodium chloride. On the other hand, an excess ingestion of water dilutes the body

fluids, ADH secretion is reduced, the duct linings become less permeable and there is a water *diuresis* resulting in the passage of a larger quantity of paler, more dilute urine, i.e. the mechanism has functioned to rid the body of its excess water.

Sometimes the mechanism gets out of hand; disease of the pituitary may lead to the output of great quantities of pale dilute urine, a condition known as *diabetes insipidus*, not to be confused with sugar diabetes, *diabetes mellitus*.

The bladder and micturition

Urine is propelled down along the ureters from the kidneys to the bladder by successive waves of contraction in the muscular walls of these channels. It is discharged from the ureteric orifices into the bladder in intermittent jets, every twenty seconds or so.

The bladder resembles the general plan of the internal hollow organs in being a muscular sac whose outlet below is normally closed by a tight ring of muscle, or sphincter. Bladder wall and sphincter must obviously be reciprocal in their contraction; in emptying the bladder contracts and the sphincter relaxes to allow the efflux of urine, and in normal resting phases the bladder is relaxed to allow its gradual distension as urine collects in it, while the sphincter remains closed. *Micturition* is the expulsion of urine from the bladder to the exterior along the urethra. The control of the bladder in this respect is both voluntary and involuntary. Basically, the organ is unconsciously and automatically self-regulating, and empties itself once the internal pressure has reached a certain level. This is well shown in individuals paralysed below the waist after injuries to the spinal cord; in them, the bladder continues to fill and empty because it is controlled by a nervous centre in the lowest part of the cord. When, later, we come to discuss the vegetative or autonomic nervous system (the independent regulator of such automatic bodily functions as the heart-beat and bowel movements, which are normally unconscious and independent of the central nervous system), we shall see that it consists everywhere of two parts, a *sympathetic* and a *parasympathetic* system, with precisely opposite actions on glands and muscle but always working in reciprocal co-ordination. This is so with the bladder also. The parasympathetic nerves empty it by contracting the organ and relaxing its sphincter; the sympathetic allow it to fill by exerting the reverse action

But there is in man, acquired in infantile development, a voluntary control of the process due to the action of the brain which can inhibit or modify the basic reflexes. There is a second sphincter surrounding the urethra itself, and this is under the control of the will, so that micturition is usually a deliberate act and one aided by voluntary contraction of the abdominal muscles so as to increase the pressure expelling the urine. Nevertheless, although micturition is usually initiated voluntarily, the process once begun continues to completion under automatic control.

The skin and temperature regulation

The skin

In all animals the skin functions as a protective and a sensory agent. But in the warm-blooded creature it has also come to play an essential role in the regulation of body temperature, for heat loss occurs almost entirely via the skin surface, with the exception of the little that occurs with the exhaled air and excreta.

There are two kinds of water loss from the skin: ordinary perspiration and sweating. Perspiration is the insensible loss of water vapour; it is continuous, uses heat because of the latent heat of vaporization and produces a loss of weight of almost 0·5 kg in twenty-four hours. A good deal of this is from the palms of the hands and soles of the feet, which lose water thrice as rapidly as other parts. Vaporization is a simple physical transpiration through the skin; it is not a function of the sweat glands.

The sweat itself is an acid, acrid secretion, mostly water—a weak solution of sodium chloride. The amount of salt

lost in a day from profuse sweating may be as much as
113 g, and this loss of chloride is an important factor in the
condition known as *heat-exhaustion*, which occurs, for instance, in miners working underground at high temperatures. Here there is a collapse with a normal or even subnormal body temperature, and the condition may be prevented by adding a little salt to the water drunk during
work. Note that this is quite different from *heat-stroke*,
which will be discussed in a moment.

The sweat also contains some potassium, sugar, lactate
and urea. The urea loss by this route is not normally of
much value in excretion, but occasionally in severe kidney
failure the alternative excretion of urea by this route covers
the skin with a hoar frost of urea crystals.

The action of the sweat glands is controlled by the
nervous system, and this is set to regulate not the amount of
water lost (for this is taken care of by the kidneys) but the
body temperature. That is, it depends on the difference between the internal and external temperatures, on the rate
of heat production in metabolism and also on the humidity
of the outside air, which may act as a serious limiting factor
for, unless continuous evaporation of the liquid sweat is
possible, the process will be interfered with. A moving current of air is another factor that encourages sweating. In
practice, the stimulus to sweating is a minimal rise of body
temperature by a third or half a degree, which sets the
mechanism in action. Strangely enough, the palms and
soles are not stimulated by a rise in internal body temperature, but they are intimately affected by the general nervous
and emotional state—the so-called 'psychic sweating'.

Temperature regulation

This is a convenient place to consider the general question
of the regulation of body temperature. In cold-blooded
animals the body temperature closely parallels that of the
environment. In warm-blooded creatures the body is maintained at a constant temperature, though the outside level

may vary enormously—a particular instance of the general physiological principle of the maintenance of a constant internal milieu. The body can be kept warmer than cold surroundings by reason of the heat produced in the oxidative process of metabolism and colder than very hot surroundings by the power of controlled heat loss in sweating. Normally, therefore, a constant body temperature implies an exact balance between heat production and loss; the latter is a function of the surface area and is thus relatively much greater in children, whose surface is so much larger than in adults in proportion to body weight.

The average body temperature taken in the rectum is 37·5°C, in the mouth 36·5°C and in the armpit 35·5°C. But it must be understood that the idea of a standard internal temperature is only conventional, for two reasons. First, there is a daily fluctuation over one or two degrees, highest in the early evening and lowest in the small hours of the morning; and it is interesting that this is reversed if the individual is made to work at night and sleep by day. This fluctuation has superimposed on it, in women, characteristic variations at different times of the menstrual cycle. Secondly, organs like active muscles are hotter than others such as the skin, which is losing heat, so that there is really a temperature gradient rising from the surface to the interior over several degrees.

The regulation of body temperature may obviously be effected by changes in the rate of heat production or the rate of heat loss; usually both are simultaneous.

Regulation of heat production. Although all the tissues produce heat, the muscles predominate because there is always some contraction occurring, as a part of the machinery for maintaining posture, and the liver is also an important heat-forming organ as a result of its tremendous metabolic activity. With any change in the temperature of the environment, heat production is increased or decreased to compensate. Thus in response to cold there is more muscular activity, which may amount to violent shivering; and, at the same time, liberation of adrenaline from the adrenal glands stimulates metabolism generally and sets free glycogen from the liver to be burnt in the tissues. External warmth reverses these changes.

It must be realized that these changes are due to the control of the nervous system, which plays on the muscles, and of chemical messengers such as adrenaline. Left to themselves the tissues would behave like those of cold-blooded animals, increasing their activity and producing *more* heat as the outside temperature rises; and this is what actually happens, with possibly fatal results, when the central regulating mechanism fails, as in heat-stroke.

The regulation of heat loss. Heat may be lost by *conduction*, *convection* or *radiation*, or by *evaporation* from the body surface. Of these, *conduction* is of little importance, since the skin and subcutaneous fat are such good insulators. *Convection* is more important; the layer of air immediately adjacent to the skin is warmed and then rises, and each warmed layer of air is successively replaced by colder layers, which are warmed in their turn; and this process is obviously greatly accelerated by external air currents. *Radiation* is also of significance since the body surface happens to be, from the physical viewpoint, a very efficient radiator indeed.

Finally, *evaporation* varies greatly with circumstances. At low temperatures there is only the insensible vapour loss; sweating begins at an air temperature of 29–30°C, and if this reaches body temperature sweating and evaporation become the only method of heat loss, since convection and radiation must obviously come to a standstill when there is no longer any temperature gradient between the skin surface and the environment. The cooling effect of evaporation is due to the fact that a considerable amount of energy is needed to convert the water of sweat into water vapour, and this energy is abstracted from the heat of the body.

Under average conditions, a clothed adult effects about

two-thirds of his heat loss by convection and conduction, and rather less than a third by evaporation. In addition to these purely physical considerations, heat loss is controllable to some extent by certain physiological activities. Most important of these is the possible fluctuation in the calibre of the small blood capillaries of the skin. These may open up widely to produce a flushing, or *vasodilatation*, which increases the surface temperature and so accelerates heat loss because of the steeper gradient to the outside level. (It also improves the actual conducting power of the skin by raising its fluid content.) Or they may shut down to produce a cold pallid skin surface—*vasoconstriction*—with a reduced rate of heat loss. This fluctuation may be enormous; the maximum blood supply to a patch of skin may be fifty times the minimum. And these changes are under the remote control of the central nervous system, via the sympathetic (vasoconstrictor) and parasympathetic (vasodilator) nerves already referred to.

However, this mechanism, sensitive as it is, ceases to operate when the outside temperature reaches the level at which sweating and evaporation become significant. And the mechanism of sweating and evaporation is so efficient that a man can easily sustain exposure to outside temperatures that will cook a piece of meat in a few minutes, provided the atmosphere be *dry* so that evaporation is unhindered. If the air is saturated with water vapour, or if—which amounts to the same thing—the individual is born without sweat glands and is unable to sweat, his temperature will rise rapidly, with possible death from heat-stroke. In this condition, which is not necessarily due to direct exposure to the sun, there is actual damage to the vital centres of the brain; the whole heat-regulating mechanism breaks down and the body temperature continues to rise until death occurs in coma or convulsions. This is quite separate from heat-exhaustion, which we have already referred to as following excessive sweating, with chloride loss, *fall* of temperature and circulatory failure.

The ultimate regulation of body temperature is fundamentally nervous. Although sweating and skin blood supply are immediately managed by the autonomic nervous system, the ultimate control is from special centres in the middle and hind parts of the brain. These act in two ways. They are very sensitive to minute changes in the temperature of the blood brought to them by the circulation and so react as to restore it to its previous level. They are also sensitive to the *nervous* stimuli that a hot or cold application produces on any part of the body surface, for the skin is provided with special heat and cold receptors which are greatly modified nerve endings, those for cold being rather the more sensitive.

The heat-regulating centres effect their results by the ways we have already discussed: by producing shivering, sweating, more or less heat production and muscular activity, changes in metabolism, vasoconstriction and vasodilatation, and so on. The important point is that, like the governor of a machine, the mechanism is usually *set* at a certain level and reacts to maintain that level under all changes of outside circumstance. But if it is shifted by some general upset the setting may change, and this, for instance, is what happens in fevers where the gearing is adjusted upwards a few degrees.

The effect of *clothing* is mainly to protect from both extremes of temperature by providing a good insulating layer whose efficiency is increased by the amount of air entangled in the interstices of the material and between its layers; the effect is therefore greater with several light garments than with one thicker one. But clothing considerably cuts down the radiating power of the body, and its capacity for permeation with moisture is responsible for the very chilling effect of a cold, damp climate. It must also be remembered that different races are helped to solve their environmental heat problems both by the way they are made and by their social behaviour. Genetic considerations play a part. Thus the long, thin limbs of tropical Africans favour heat loss because of their proportionately greater surface area. And Finns and Eskimos manage to exist in what is almost a subtropical micro-climate by heating their houses and wearing heavy, layered, insulating clothing. The Bedouin, on the other hand, wears voluminous, loose clothing, which allows sweating to occur beneath it and which is white to act as a maximal reflector of heat.

All protoplasm is sensitive to stimulation, or *excitable*, and *conducts* such excitation. This property is highly developed in nervous tissue. Nerve cells are particularly sensitive, and the fibres that spring from these cells are specialized for the transmission of impulses from one part of the cell to another part, from one nerve cell to another and from nerves to muscles or glands. The nervous system of man can be regarded as a network pervading the entire body, having a two-way connection with central control and enabling the individual to make a co-ordinated response to any stimulus from outside or from within his own body.

The nervous system is divided into main parts: the *central* nervous system (brain and spinal cord) and the *peripheral* nervous system, long bundles of fibres attached to central cells. The *receptor* side of this system conveys such information as sight and touch from the outer world by *afferent* impulses along nerve fibres to the controlling nerve centres, but it also provides information as to the state of the body and internal organs. Some of these messages enter consciousness, giving us our picture of the universe; others, notably those from within us, remain largely unconscious, though they do help to regulate the normal working of the body. The *effector* side of the system carries *efferent* messages out from the nerve centres to the effector organs—muscles and glands—and this response is meant to deal with the situation provoking the original sensory stimuli. In fact, from the purely physiological view (though not necessarily the philosophical), the brain is nothing but the centre of a glorified reflex arc, to which, at the apex, consciousness has been added, perhaps as an incidental by-product of the whole complicated arrangement. For the brain does nothing independently, only in response to stimulation from outside; so-called voluntary movement is merely the result of some provocation of which we may be unaware.

Basic structure

The basic elements of the nervous system consist of:
 (1) the actual nerve cells or *neurones*;
 (2) the supporting cells within the central nervous system, or *neuroglia*;
 (3) *connective tissue*, forming the investing membranes of brain and cord, and the sheaths of nerve fibres and of feeding blood vessels, though there is no connective tissue in actual brain or spinal cord substance.

The neurone

Each nerve cell has several branching processes, or *dendrites*, interlocking with those of adjacent cells. One process is elongated as the main axon for transmission of stimuli; often this is enormously long compared with its microscopic origin, traversing the length of the limbs or spinal cord. Nerve cells are sometimes multipolar with many processes, sometimes bipolar with one small dendrite plus one axon and occasionally unipolar with an axon only. Cell, dendrites and axon constitute the nerve unit, or *neurone*, and the nervous system is built up of millions of such units in close and complex interrelation.

There are two kinds of peripheral nerve fibre: myelinated and unmyelinated. The larger, white or myelinated fibres have a fatty sheath formed by investing Schwann cells, and this sheath is indented at intervals at the nodes of Ranvier. These fibres are also found in the *white matter* of the central nervous system. The unmyelinated fibres lack this coating; they are finer, and are found in the autonomic nervous system and in the *grey matter* of the brain and cord. The nerve bundles that constitute a peripheral nerve contain both kinds of fibre; the bundles are bound together by con-

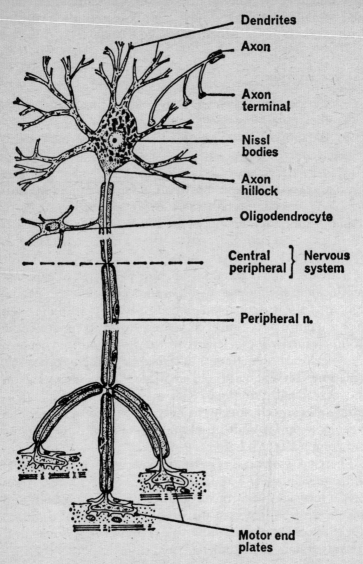

Dendrites

Axon

Axon
terminal

Nissl
bodies

Axon
hillock

Oligodendrocyte

Central } Nervous
peripheral } system

Peripheral n.

Motor end
plates

Fig. 178. Diagram of a ventral horn cell from the spinal cord.
(From *A Companion to Medical Studies, Vol. 1.*)

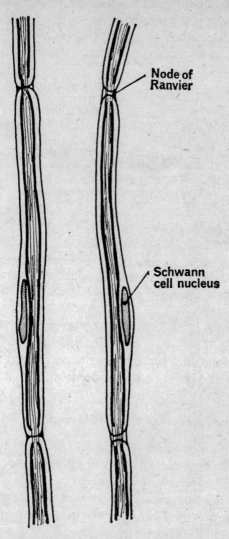

Fig. 179. Internodal segments of two nerves from a young animal. Stained with osmic acid. (From *A Companion to Medical Studies, Vol. 1.*)

nective tissue sheaths—the *perineurium*—and the sheath of
the entire nerve is the *epineurium*.

Damage to nervous tissue is irreparable insofar as the
cells are incapable of reproduction; but the function of

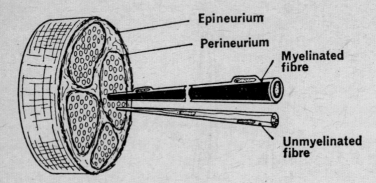

Fig. 180. Diagram of the structure of a peripheral nerve containing myelinated and unmyelinated nerve fibres, to show the neural
sheaths. (From *A Companion to Medical Studies, Vol. 1.*)

destroyed areas in the cerebral cortex may sometimes be
taken over by other parts of the brain. Injured fibres may,
however, grow again, but only in the peripheral nerves and
never within the brain or spinal cord.

The nature of nervous transmission

Although the nerve fibres are sensitive to direct stimulation,
they normally only *conduct* stimuli, either those passing out
from the central nervous system to muscles or glands, or
those travelling inward from skin and sense organs to the
brain and cord. These impulses originate at one terminus of
the fibres, and fibres normally conduct in only one direction
because of their particular connections, but this is not an
essential property. If we graft a portion of sensory nerve
into a gap in a motor nerve, the graft will conduct impulses
quite efficiently in the reverse direction from that which it
used to do. Thus nerves are comparable to telephone wires

(or, more strictly, considering their physical make-up, to coaxial cables); the nature of the conversation conveyed depends only on who happens to be connected at either end. Thus we associate the optic nerve with sight and the olfactory nerve with smell. But they subserve these particular functions only because they link the appropriate sense organ with the corresponding area of the brain; in themselves, the impulses transmitted are identical to those carried by nerves in general.

Nerve fibres are readily stimulated experimentally—electrically or by pinching—and stimulation sets up a wave of depolarization or negative electric potential, the *action potential*, travelling along the nerve at many feet per second. This electrical wave is not, in fact, the nerve impulse itself; it is merely the measurable index of an underlying disturbance, itself obscure, which seems to be of the nature of a sudden change in an unstable local membrane permeability to ions.

Nervous action is essentially a stream of such small, discrete, separate impulses, repeated at very short impulses and sounding like a burst of machine-gun fire when magnified through an amplifier. There is no such thing as a continuous flow of nervous excitation. And there is a maximum rate for such a stream, since each impulse is followed by a short *refractory period* during which the nerve will not transmit a succeeding stimulus. Further, a single nerve fibre exhibits the *all-or-none* phenomenon; if sufficiently stimulated it will convey an impulse of fixed magnitude, but it cannot convey a larger or smaller impulse. In its activities nervous tissue undergoes metabolic changes in exactly the same way as the body tissues generally, using up oxygen and generating carbon dioxide with the evolution of heat. The quantities involved are minute compared with those concerned with, say, muscular activity, but are nonetheless essential. Nervous tissue, particularly the cells of the brain, cannot tolerate oxygen deprivation for more than a minute or two. There is a distinction between the larger, sheathed motor

and sensory fibres of the central nervous system, which convey impulses at a velocity of some 90 m/s, and the finer
fibres of the autonomic system, where the speed is only some
1·5–15 m/s. These differences are related to the necessity for
immediate response to an external situation, whereas the
regulation of the internal organs can afford to be a more
leisurely process.

As far as a single nerve cell and its fibres are concerned,
the nervous impulse is conducted smoothly and, for each
action potential, uninterruptedly; but there is an inevitable break when the impulse is relayed by another cell
unit. Such relays, or *synapses*, are an integral part of the
nervous system. Motor impulses descending from the brain
are relayed by fresh cells in the spinal cord, whose fibres run
out to the muscles, and sensory impulses from the skin are
relayed by other cord cells before being transmitted up to the
brain to reach consciousness. Every relay is associated with
a breach in anatomical continuity. The incoming fibre ends
in a network of branches, which embrace those of the adjacent
cell but do not blend with it; there is physical discontinuity.
Because of this breach, messages are delayed a little at the
synapses—just as a call made through a telephone and its
extended wire is delayed at the exchange while the connection is made to its destination.

The bridging of the synapse gap between one neurone
and another is effected by chemical means; the nerve endings liberate chemical substances which stimulate the
adjacent cell to start off a fresh impulse along its own fibre.
Thus nervous activity is a complex blend of electrical conduction, associated with altered ionic membrane permeability, along the fibres and chemical excitation at the
synapse bridges. A similar gap must be crossed when a
motor nerve fibre ends in a muscle, and this again is
achieved by liberation of a chemical agent. Although nerve
fibres can conduct in either direction, transmission at the
synapses is strictly one-way and this sets the pattern for the
directional flow of impulses in the central nervous system;

the same applies to the nerve-muscle junction, which is designed to permit only a flow of impulses into the muscle fibres.

These *chemical mediators* are largely either acetylcholine or noradrenaline, though some other substances are also involved in brain function. Acetylcholine is liberated at motor end-plates in muscle. Although in the central nervous system these substances are perhaps subsidiary to neuronic control, they are very potent and are particularly closely associated with the working of the autonomic system, whose ultimate fibres to glands and viscera are labelled cholinergic or adrenergic as the case may be (page 447). When we come to consider the autonomic system, we shall see that chemical agencies play a very large part in regulating the behaviour of the internal organs and blood vessels, and that there is an accessory arrangement by which certain endocrine glands (e.g. the adrenals) can secrete the same substances directly into the blood as hormones to secure a rapid response in an emergency.

Anatomy of the central nervous system

The brain and spinal cord, enclosed in the cranium and spinal canal, and continuous with each other at the foramen magnum in the skull base, make up the *central nervous system*. The twelve pairs of cranial nerves arising from the brain and the thirty-one pairs of spinal nerves arising from the cord constitute the *peripheral nervous system*. Together, the central and peripheral arrangements form the cerebrospinal or voluntary nervous system, which is mainly concerned with the control and sensation of the somatic structures of the body wall—skin, muscle, bones and joints—though many of its activities are far from conscious.

In contradistinction to this, the semi-independent *autonomic* or *vegetative nervous system* deals with the automatic functioning of the splanchnic structures—viscera, glands and vessels. Nevertheless, the two systems are closely interconnected, as we shall see later.

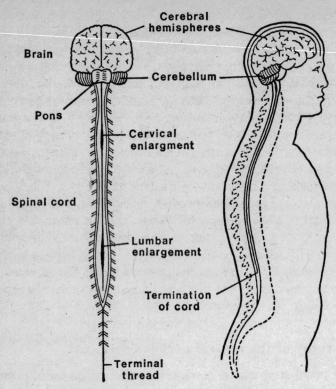

Brain
Cerebral hemispheres
Cerebellum
Pons
Cervical enlargment
Spinal cord
Lumbar enlargement
Termination of cord
Terminal thread

Fig. 181. The central nervous system.

Cerebrospinal system

Membranes, cerebrospinal fluid

Brain and cord have three enveloping membranes, which are prolonged as sheaths along the nerve roots which pierce them on their way out. The outermost layer is the *dura mater*, a tough, protective envelope loosely applied and not entering the interstices of the cord or following the convolutions of the brain; in the cranium it also forms the inner lining periosteum of the skull bones. The innermost layer is the *pia mater*, a fine vascular membrane closely applied to the outer surfaces of brain and cord, entering and exactly following every cleft and crevice, and carrying in with it the fine blood vessels. Intermediate is the *arachnoid layer*; this fits closely inside the dura, but there is a considerable *sub-arachnoid space* separating it from the pia, a space filled with cerebrospinal fluid and traversed by spidery meshes of connective tissue (Fig. 182).

The cerebral and spinal membranes are continuous with each other, the fluid freely bathing the outer surfaces of brain and cord. The fluid also occupies the hollow chambers

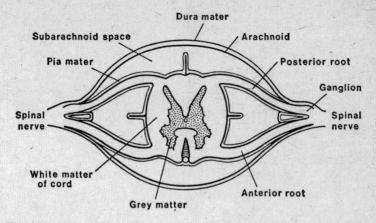

Fig. 182. Cross-section of the spinal cord and its membranes. (After *Gray*.)

or ventricles of the brain which communicate with the small central canal of the cord. It is formed as the secretion of the vascular *choroid plexus* lining the ventricles, circulates inside the brain and cord, escapes through the roof of the hind-brain to the outer subarachnoid space and is finally re-absorbed into the bloodstream via the venous sinuses of the cranium. Thus there is a continuous circulation of fluid from the blood vessels in the ventricles, round the brain and cord, and back into the bloodstream again.

Spinal cord

The spinal cord is elongated and cylindrical with two fusi-form swellings, the *cervical* and *lumbar enlargements*, which are

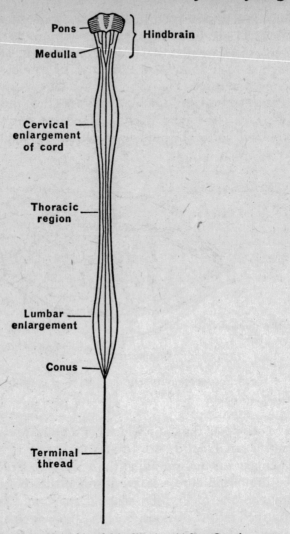

Fig. 183. Spinal cord and hindbrain. (After *Gray*.)

the sites of origin of the roots of the brachial and lumbar nerve plexuses for upper and lower limbs.

In the foetus the cord occupies the whole length of the spinal canal. But it fails to keep pace with the growth of the spine until, in the adult, it ends at the first lumbar vertebra in a tapering conical extremity. And, since the nerve roots must still emerge at the correct levels from the pairs of intervertebral foramina, they have to run more and more

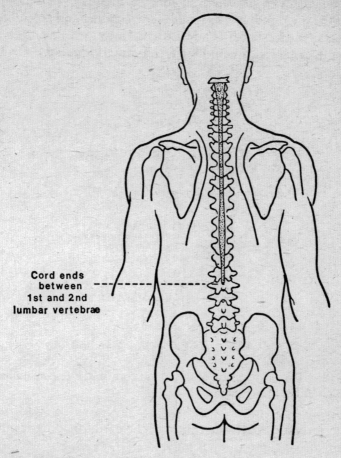

Cord ends
between
1st and 2nd
lumbar vertebræ

Fig. 184. Surface marking of the spinal cord. (After *Gray*.)

obliquely downward and out as they arise lower in the cord; so the spinal canal below the termination of the cord is filled with a mass of roots—the cauda equina or 'horse's tail'—descending almost vertically to their appropriate lumbar and sacral exits. This relative recession of the cord with growth is responsible for the lumbar enlargement being actually in the lower thoracic region.

Cross-section of the cord shows the inner H-shaped arrangement of *grey matter*, composed mainly of nerve cells, and the surrounding *white matter*—descending and ascending tracts of nerve fibres (Plate 12(b)). In the very centre is a minute canal. In the midline a cleft anteriorly and a fissure

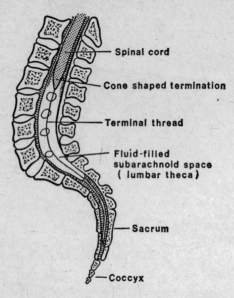

Fig. 185. The termination of the spinal cord in the upper lumbar region. (After *Gray*.)

posteriorly approach the centre of the cord, dividing it into symmetrical right and left halves.

On each side the grey matter projects in front and behind as the *anterior* and *posterior horns*. The *anterior horn* contains the motor nerve cells, whose fibres are destined for the stimulation of muscles, and these leave the cord in a bundle as the anterior or motor nerve root.

The *posterior horn* contains sensory cells; fibres enter it from the posterior or sensory nerve root and are mediated by a group of cells forming a knob or ganglion on this root just outside the cord, a sort of relay station.

Spinal nerves. The two roots on each side join just beyond the ganglion to form the spinal nerve proper, the junction lying just within the intervertebral foramen before the nerve

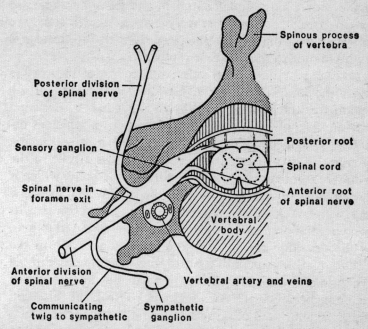

Fig. 186. Cross-section showing the spinal cord, its issuing roots and the appropriate spinal nerve, as they lie in relation to the bony spine. The section is in the cervical region. Note the communicating twig to the sympathetic ganglion. (After *Gray*.)

leaves the spine. Thus the spinal nerve contains mixed sensory and motor fibres, and these are not separated out again until, in the later distribution of the nerve and its branches in limb or trunk, specific motor and sensory twigs are given off to particular muscles and areas of skin.

The peripheral nerves follow the simplest pattern in the thoracic region, the spinal nerve on each side encircling the chest in its appropriate intercostal space without connection with its neighbour above and below. But, as we have seen, the lower cervical and the lumbar and sacral nerves form complex networks as soon as they have left the spinal column. From these brachial and lumbosacral plexuses emerge the actual peripheral nerves of the limbs, e.g. media and ulnar in the arm, sciatic and femoral in the leg; and these are usually mixed motor and sensory nerves, though some are almost entirely one or the other, e.g. the radial in the arm is mainly motor to the extensor muscles.

It should be noted that our cross-section of the cord is much the same in appearance whatever the level of the section, i.e. the spinal cord may be regarded as consisting of a number of identical *segments*, to each of which is attached a pair of spinal nerves, and each spinal nerve emerges by two roots on either side. This is a relic of the primitive segmentation pattern of the whole body, which is very obvious in such an animal as the earthworm, in which each segment exactly reduplicates the next and contains its own essential organs. This pattern, though obscured in man by the development of head and limbs, lingers on in the repetitive arrangement of the ribs, the vertebrae and in the structure of the cord itself. It is of particular importance in the latter because each cord segment controls the sensation of a particular area of skin surface and the movement of a particular group of muscles; together these make up the corresponding body segment, still retaining its fundamental connection with a particular segment of the nervous system, though the pattern is grossly distorted by evolution. Thus the sensation of the little finger is always supplied by the first thoracic segment of the spinal cord, the muscles straightening the knee always derive their motor nerve fibres ultimately from the third and fourth lumbar segments, and so on for every part of the body.

This constancy in innervation is of the greatest value in helping to locate the precise site of injury or disease of the central nervous system because we can deduce from the particular muscles paralysed, or areas of skin surface devoid of sensation, exactly what the level of the disease process in the cord must be.

The reflex arc. It is obvious that the spinal cord provides a relatively simple means for the performance of certain primitive and essential actions without the intervention of the brain, though many reflexes are mediated through higher parts of the central nervous system. Thus an irritation of the skin sends a sensory impulse through the posterior root to the posterior horn of the appropriate segment; this is relayed to the anterior horn cells of the same or related segments of the cord and becomes a motor impulse travelling out through the anterior root to activate muscles, which remove the affected part from the noxious agent. This is the reflex arc at its simplest, the so-called withdrawal reflex; in some cases, scratching may be the appropriate motor effect. It is the only pattern of behaviour in the lower invertebrates and is independent of the brain, though the brain is aware of its results and can to a certain extent modify them. Another simple reflex arc is involved in the *tendon-jerk*. The simplest example of this is the knee-jerk; when one leg is dangling crossed over the other, a smart tap on the patellar tendon (Fig. 86) causes the quadriceps muscle—i.e. the muscle whose tendon is involved—to contract and straighten the knee. There are many other similar reflexes, and their basis lies in the stimulation of specialized stretch receptors within the muscle.

The reflex arc has certain characteristics. It is slower than one would expect from the normal rate of nerve conduction; it tends to continue to produce motor behaviour after the stimulus has ceased—the 'after discharge'; it is fatiguable; and it can be inhibited or facilitated by simultaneous stimuli elsewhere.

On a more complex level, but basically similar, are the reflexes of micturition, defaecation, sexual orgasm and childbirth, all controlled by a particular segment or group of seg-

ments of the spinal cord. All these activities remain possible,
even when the cord-brain link has been severed by injury, so
that the body is completely paralysed as far as conscious
sensation and voluntary movement are concerned.

The tracts of the cord. The grey matter is composed mainly of
cells, with short fibre connections at the same transverse
level. The white matter is made up of fibres running longitudinally in the cord in cable fashion, and these are grouped
in a constant pattern of individual tracts, some of which are
sensory and some motor. A few tracts run the length of
several cord segments only, but most run from cord to brain
and link the anterior or posterior horn cells with the appropriate controlling or receptive centres of the brain. Thus
from that part of the motor area of the cerebral cortex concerned with arm movement fibres run to the anterior horn
cells of the seventh cervical segment of the cord, and from
these cells fibres leave to take part in the brachial plexus,
enter the median nerve and actuate the flexors of the wrist.
Again, sensory impulses from the knee travel up in the
femoral nerve, through the lumbar plexus to the posterior
horn cells of the fourth lumbar segment of the cord and are
then relayed up the long sensory tracts of the latter to the
sensory portion of the cerebral cortex.

In other cases, other parts of the brain are the termini—
cerebellum or midbrain—but an invariable principle is that
of crossing or decussation; at some point, usually in the
brain stem, the fibre tracts cross completely from one side
to the other so that the sensory and motor regions of the
brain control the opposite halves of the body. Some of the
more important tracts of the cord are shown in Plate 12.

The brain
The brain is the greatly expanded and elaborated upper end
of the cerebrospinal nervous axis. It has the same enveloping
membranes as the cord, it almost entirely fills the cranial
cavity and it makes indentations on the inner aspect of the

cranial bones. The arrangement of grey and white matter is the reverse of that in the cord, for the grey matter is now on the surface and the white fibre tracts within. The main portions of the brain are, from above down, as follows:

(1) The *forebrain* is the great overhanging pair of *cerebral hemispheres*—the bulk of the organ—symmetrical, rounded masses of convoluted nervous tissue which completely hide the lower parts of the brain when viewed from above.

The two hemispheres are separated by a deep median fissure, which contains a partition of the dura mater, but are

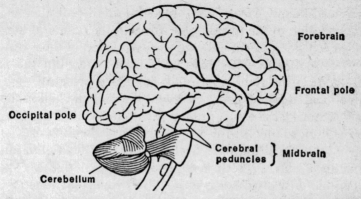

Fig. 187. The brain in side view; diagrammatic. (After *Gray*.)

connected at the base of this cleft by a great bridge of fibres from side to side. The superficial grey matter, the cerebral cortex, has an enormous area owing to the intricate pattern of convolutions. This produces a number of elevated ridges, or *gyri*, separated by narrow valleys, or *sulci*, and the increasing complexity of this pattern in evolution through the vertebrates to man is a measure of the increase in intelligence and consciousness.

Each hemisphere is composed of several *lobes* which are not clearly demarcated from each other. In side view, the *frontal lobe* is seen at the anterior end of the hemisphere, occupying the anterior cranial fossa, its tip the anterior pole of the brain. Behind is the *occipital lobe*, lying in the posterior cranial fossa, its tip the posterior pole. The *temporal lobe* lies immediately behind the frontal and projects below, occupying the middle cranial fossa, and the *parietal lobe* is an ill-defined area squeezed in between frontal and occipital regions above the temporal lobe.

Plate 12 shows the areas of localization of function in the cerebral cortex, those known to control movement or to act as the main reception centres for sensation. The main motor and sensory areas are placed midway between the poles, and the body is represented in these areas in inverted fashion, i.e. the areas for the legs are uppermost, those for the head below. There is a vaguely localized centre for the appreciation of the meaning of words in the temporal lobe of the left side, concerned with the correlation of the meanings of spoken, written, read and heard words, but this is situated on the right side in left-handed people. All these functional areas are concerned with the opposite half of the body, though the visual area of the occipital cortex is concerned not so much with the opposite eye as with gathering up all the visual impulses from the opposite field of vision that have entered both eyes. It will be seen that a considerable area of cortex is assigned no particular function. These are the psychic or association areas, concerned with the correlation of sensory and motor data, and with the higher levels of consciousness and personality, particularly in the frontal lobes. The latter may be destroyed or removed without interfering with essential activities, though with considerable changes in mood and capacity for self-appraisal.

Cross-section of the forebrain shows the surface pattern of sulci and gyri, the contrast of grey and white matter, and the cleft between the hemispheres. It shows on each side the hollow chamber or *ventricle* in the depths of the hemisphere, filled with cerebrospinal fluid, and great masses of grey matter on each side of the ventricles, the *basal ganglia*, one of which—the thalamus—is intimately connected with the emotions.

Although the account of cerebral localization of function given above suggests a very precise division of labour between the different regions of the forebrain, in fact there is some power of one area to compensate for another lost by disease or injury, though this is never very great in man.

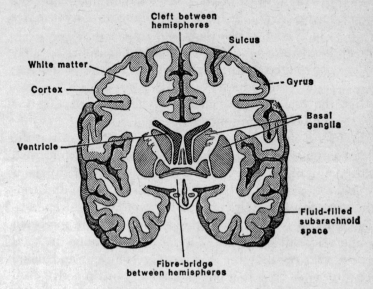

Fig. 188. The forebrain in cross-section.

Man's power of associating linked impressions and ideas represents the highest development of mental processes in the animal kingdom. It takes place mainly in the large areas of cortex to which no particular motor or sensory function can be assigned, and on it depends man's educability and capacity to learn from experience.

(2) The *midbrain* is deeply situated at the base of the hemispheres and consists partly of a squat cerebral *peduncle* on each side, a pillar supporting the corresponding hemisphere and carrying fibres to and from it. The roof of the midbrain contains the important nuclei of the third cranial nerve, which controls the movements of the eyeballs.

The *pituitary body*, an important endocrine gland, projects from the undersurface of the brain just in front of the peduncles.

(3) The *hindbrain* is best seen on the underside of the organ (Figs 183, 187, 189). It includes:

    (i) the *medulla oblongata*, the bulbous expansion formed by the upward prolongation of the spinal cord, the lowest part of the brain, containing the nerve centres which control heartbeat and respiration;

   (ii) the *cerebellum*, a pair of rounded hemispheres, finely convoluted, occupying the deepest part of the posterior cranial fossa;

  (iii) the *pons*, a broad bridge of fibres connecting the cerebellar hemispheres.

From the point of view of evolution, the hindbrain is the oldest and most primitive part of the organ, and is concerned with maintaining the basic activities of the body. It is largely independent of conscious cerebral control and continues to function fairly normally even when sleep, unconsciousness or deep anaesthesia have put the forebrain temporarily out of action. The medulla contains the vital centres that regulate the heartbeat, respiration and blood pressure; and these, in injury or disease, remain active until the last moment before death supervenes. The cerebellum is concerned with the regulation of muscle tone and posture, and does so in response to the changing sensory impulses of position and tension it receives from joints and tendons, and also from information gained as to the body's position in space from the semicircular canals of the ear. That is to say, it is concerned with the essential postural mechanism of the head and trunk, the automatic righting reflexes that restore the upward carriage of the head after an upset of balance.

Fibre tracts of brain. The brain may be regarded as a series of cell-stations situated at higher and lower levels in the grey matter of the cerebral cortex, basal ganglia, midbrain, medulla and cerebellum. To and from these great tracts of nerve fibres ascend and descend within the central white matter of the brain stem, the axial conducting core of medulla and midbrain. These tracts have already been encountered in the cross-section of the spinal cord, and may be classified as motor (descending) and sensory (ascending); at some level, usually in the hindbrain, these fibres all cross to the opposite side, so that functional areas of the brain represent the other half of the body. The principal tracts are as follows (Plate 12(b)):

Motor tracts:

(1) *pyramidal*, running from the cerebral motor cortex to the anterior horn cells of the cord at all levels, thence relayed to the motor roots; concerned with voluntary motion;

(2) *cerebellospinal*, running from the cerebellum to the anterior horn cells, concerned with the automatic regulation of tone and posture.

Sensory tracts:

(1) the posterior bundles of the cord carrying touch and sensation to the cerebral sensory cortex;

(2) the anterior and lateral bundles of the cord, carrying pain and temperature sensation to the sensory cortex;

(3) the spinocerebellar bundle carrying postural sensation to the cerebellum.

All the sensory bundles are relaying impulses entering the cord via the posterior sensory roots and the posterior horn cells. Since the sensory tracts acquire an increasing influx from higher roots as they ascend and since the motor tracts shrink as they descend, because they shed fibres to successive segments of the cord, the bulk of the cord as a whole tapers from above downwards.

The ventricles. We have seen that the brain is hollow and traversed by channels for the cerebrospinal fluid. Within each cerebral hemisphere is a large *lateral ventricle*, and between and below these is a small *third ventricle*, from the floor of which the pituitary body is slung. These communicate and lead on to a tiny channel, the *aqueduct*, which leads backwards through the midbrain to expand as the *fourth ventricle* of the medulla. This is continuous with the central canal of the spinal cord, and from its roof, under cover of the cerebellum, the fluid escapes through these little apertures to bathe the outer surface of the brain and cord in the subarachnoid space. Growing from the smooth linings of the lateral and fourth ventricles is the highly vascular *choroid plexus*, which secretes the fluid; any block to its free flow within the ventricles, or at its exit from the hindbrain, results in a damming-up within the brain, which is slowly ballooned out to produce the condition known as *hydrocephalus* or 'water on the brain'.

The cerebrospinal fluid freely bathes the outer surface of brain and cord, cushioning them within the bony cranium and spinal column against concussion or violent changes of position. It is normally maintained at a constant pressure to support the soft bulk of the central nervous system. This pressure is intimately related to the pressure of blood in the great veins and is raised accordingly when the venous tension is increased by coughing or straining. Any disease process or tumour—i.e. any space-occupying lesion—within the unyielding box of the cranium greatly increases the cerebrospinal fluid pressure as it expands, and this causes headache, vomiting and, ultimately, severe disturbance of cerebral function, ending in coma.

Because the chemical composition of the fluid and the nature and number of its contained cells are greatly affected by disease, examination of the fluid is often of great value in the diagnosis of cerebrospinal disease. Now we have already seen that, while the spinal cord ends below at the level of the first lumbar vertebra, its membranes are continued down as far as the sacrum. The lumbar spine thus encloses a space, or *theca*, filled with cerebrospinal fluid, and this space may readily be entered using a long needle inserted between the spinous processes in the small of the back—the procedure of *lumbar puncture*. Fluid may also be obtained, at greater risk, by inserting a needle in the midline of the upper part of the back of the neck, just at the skull base between occiput and atlas, entering the fluid cistern in the angle between cerebellum and medulla—*cisternal puncture*. Sometimes it is necessary for diagnostic or treatment purposes to insert a needle directly into one of the lateral ventricles; this is done by trephining a disc of skull at the appropriate site.

Vessels and nerves of the brain. The main *arterial* supply of the brain is derived from the two internal carotids, which enter the skull base. They form, with the vertebral arteries which have entered the foramen magnum, a vascular circle around the stalk of the pituitary body. From this circle, *anterior, middle* and *posterior cerebral arteries* are given off to the frontal, parietotemporal and occipital parts of the brain respectively. The hindbrain is supplied directly by the *vertebral artery*.

The main *veins* are grouped as venous sinuses, which run between the layers of the dura mater and eventually drain to the internal jugular; the *superior sagittal sinus* of the vault, which runs the whole length of the brain from front to back between the hemispheres, also drains off the cerebrospinal fluid, which is thus reabsorbed as constantly as it is formed.

The brain substance itself is not supplied with nerves and is quite insensitive; it can be handled and cut at operation without reaction, even in the conscious patient. But its covering membranes are extremely sensitive, particularly to distension, which causes severe headache.

The cranial nerves. Just as the horns of grey matter in the spinal cord give off the roots of the spinal nerves, so in certain parts of the brain there are collections of nerve cells, forming the *nuclei* from which spring the twelve pairs of cranial nerves, best seen on the underside of the brain. These are mostly concerned with movement and sensation in the head and face, including the eye, ear and nose; but one (the

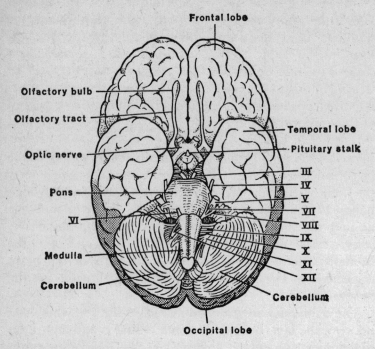

Fig. 189. The underside of the brain, showing the attachments of the twelve pairs of cranial nerves.

eleventh) controls certain neck muscles and another (the tenth or vagus) is distributed widely to the internal organs of the neck, chest and abdomen. They may be listed in order of their origin from the different levels of the brain, from above downwards:

Cerebrum: (I) The first or *olfactory* nerves are those of smell. On each side a group of some twenty rootlets ascends from the nasal mucous membrane, through the perforations in the floor of the anterior cranial fossa, and ends in the olfactory bulb on the undersurface of the frontal lobe. From this, the long olfactory stalk relays the impulses to the cerebral hemisphere. It should be noted that the sense of smell is responsible for a good deal of what is normally regarded as taste, as witnessed by the damping effect of a severe cold on the appreciation of food. These nerves are purely sensory.

(II) The second or *optic* nerve is concerned with sight. Each carries fibres from the nerve cells of the retina and leaves the back of the orbital cavity as the stalk of the eyeball through the optic foramen to enter the middle cranial fossa. The two nerves form an X-shaped crossing or *chiasma* on the underside of the frontal lobes, just in front of the pituitary stalk, the hind limbs of the X being the *optic tracts*. The effect of this crossing is that fibres from the outer half of each retina remain in the optic tract of the same side, while those from the inner halves cross over into the opposite tract. This means that the right tract carries all the fibres from the right halves of both retinae and, therefore, all the sensation from the whole of the left half of the visual field, and *vice versa* for the left tract (Plate 12). This separating out is essential to maintain the principle of one-sided representation of function in each half of the brain, for, though each eye is a one-sided organ, it sees on both sides of the body, and what is needed for efficient co-ordination is that the whole of one side of the *visual field* should be represented as such in the cerebral cortex. The impulses are relayed to the visual area of the occipital cortex, and there is an intimate connection with the nucleus of the third nerve, which controls eye movements. The optic nerve is purely sensory.

Midbrain:

(III) The third or *oculomotor* nerve controls four of the six small muscles moving the eyeball. It is purely motor.

(IV) The fourth or *trochlear* nerve controls another of the ocular muscles. It is purely motor.

Hindbrain:

(V) The fifth or *trigeminal* nerve has a large sensory root carrying a ganglion; this root is divided into three branches:

(i) *ophthalmic*; this conveys sensation from within the orbit other than sight, i.e. ocular pain, and sensation from the forehead and front of the scalp;

(ii) *maxillary*; this conveys sensation from the upper jaw and teeth and overlying facial skin;

(iii) *mandibular*; this conveys sensation from the lower jaw and teeth and overlying skin, and ordinary sensation (but not taste) from the front of the tongue and mouth.

The mandibular division passes down to enter a canal traversing the lower jaw. The ophthalmic division enters the back of the orbit. The maxillary division runs in a canal sandwiched between the orbital floor and the roof of the maxilla to emerge on the front of the cheek.

The fifth nerve also has a small *motor root*, concerned with the muscles of mastication, which leaves the skull with the mandibular division. Thus it will be seen that the fifth nerve is a mixed sensory and motor nerve.

(VI) The sixth or *pathetic* nerve controls the last of the ocular muscles. It is purely motor.

(VII) The seventh or *facial* nerve is mainly concerned with the facial muscles, though it also activates the secretion of some of the salivary glands and conveys taste sensation from the front of the tongue. It leaves the posterior fossa of the skull by entering the internal auditory meatus on the inner aspect of the temporal bone in company with the eighth nerve. After a winding course within the temporal bone, it emerges outside the skull, immediately in front of the mastoid process, and enters the substance of the parotid gland, in which it divides into several twigs supplying the various groups of facial muscles—scalp and eyelids; cheek, nose and mouth; and the platysma in the neck. It is a mixed sensory and motor nerve.

(VIII) The eighth or *auditory* nerve connects the inner ear, deep in the substance of the temporal bone, with the hind-brain; it is really double, serving two distinct functions:

(i) the *acoustic portion*, which carries sensations of sound and pitch from the true organ of hearing, the cochlea; these sensations are relayed to the auditory part of the cerebral cortex;

(ii) the *vestibular portion*, which carries postural sensation from the semicircular canals of the internal ear, the organ of balance, to the cerebellum. It provides constant information as to the changes of position to aid in maintenance of the erect position.

(IX) The ninth or *glossopharyngeal* nerve is almost entirely sensory, though it also activates parotid secretion and some little muscles associated with the pharynx. It conveys sensation from the pharynx, tonsil and back of tongue, including taste.

(X) The tenth nerve, the *vagus*, is very important and complex, extending through the neck and thorax to the abdomen. It is an essential component of the autonomic system controlling the viscera, as we shall see later. Its anatomical course in neck and chest has been referred to previously. Its main branches are:

(i) in the *neck*, for movements of the pharyngeal and laryngeal muscles, sensation of the linings of pharynx and larynx;

(ii) in the *chest* it forms the cardiac plexus controlling heart action and the pulmonary plexuses at the roots of the lungs; and it also supplies the muscle of the oesophagus and bronchi, and their associated glands, and carries sensation from their mucous linings;

(iii) in the *abdomen* it supplies the muscle, glands and mucous membranes of the stomach and duodenum, and sends twigs to the liver, spleen and kidneys.

(XI) The eleventh or *accessory* nerve is purely motor and supplies the muscle of soft palate and pharynx, and the sternomastoid and trapezius.

The ninth, tenth and eleventh nerves all arise from the medulla and leave the skull base through the jugular foramen with the internal jugular vein.

(XII) The twelfth or *hypoglossal* nerve controls the muscle of the tongue and floor of the mouth.

One point, common to both spinal and cranial nerves, must be made here. The fibres of motor cranial nerves, or the motor portions of mixed cranial nerves, and those of the motor roots of the cord arise from their cells in the cerebral nuclei or anterior horns and pass out without a break to their destined muscles. But sensory fibres, passing in from the skin or other sensory surface, end at a group of cells situated outside the brain or cord; these cells make up the *ganglia*, knobby swellings on the posterior spinal roots, or on the sensory cranial nerves as they lie on the skull base. It is invariable that all sensory nerves have such a ganglion, and it is the axons of the ganglion cells, not the original sensory fibres, that ultimately enter the cerebrospinal axis. In other words, motor impulses are direct, sensory impulses are mediated. The outgoing motor pathways are known as efferent, the incoming sensory fibres are afferent.

Autonomic nervous system

Sympathetic and parasympathetic (Fig. 190)

The autonomic nervous system is distributed to the muscular coat of the blood vessels, and to the smooth muscle and glands forming part of the viscera. It controls their movements and secretion, and monitors their state of distension by means of incoming pathways. Because all its nerve fibres are ultimately derived as an outflow from the cerebrospinal axis itself, it is analogous to a local authority in more or less complete charge of sewage and transport arrangements, but only by virtue of powers delegated to it by the government itself. And these arrangements are susceptible to modification from the centre, not through an act

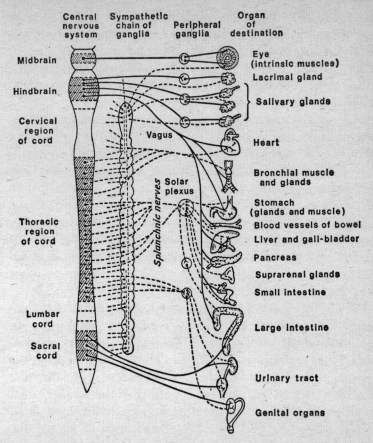

Fig. 190. Plan of the autonomic nervous system. (After *Gray*.)

of will but through emotion, which intimately affects visceral function.

Anatomical arrangements. The *sympathetic system* is a chain of ganglia on each side, running the whole length of the spinal column from atlas to coccyx, where the two meet and fuse in a single ganglion. We have already met this chain lying on the sides of the vertebral bodies at the back of the abdominal and thoracic cavities, where there is one ganglion to each segment and spinal nerve, and in the neck where there are only three. The outflow from the central nervous system to the sympathetic is by a little communicating twig from each spinal nerve in the thoracic and upper lumbar region to the corresponding sympathetic ganglion (Fig. 186, page 429). From this ganglion the peripheral sympathetic system is effected in two ways:

(1) by a returning twig to each spinal nerve, the sympathetic fibres travelling out with that nerve to the somatic structures, mainly blood vessels of the limbs and body wall;

(2) by the formation of *plexuses*, e.g. the cardiac and pulmonary plexuses of the chest, branches of which form a network around the vessels travelling out to heart and lungs; also the splanchnic nerves run down from the lower thoracic chain through the diaphragm to form the solar or coeliac plexus around the upper part of the abdominal aorta, from which branches go to the bowel, and to other abdominal and pelvic viscera.

The *parasympathetic system* is less obviously a distinct structure, for its fibres are merely contained within certain cranial and sacral nerves at either end of the cerebrospinal axis—the craniosacral outflow. The former supply the glands and vessels of the head and the eye, the latter the bladder, rectum and genital organs. But the tenth cranial nerve—the *vagus*—is the great parasympathetic nerve of the body, traversing neck, chest and abdomen to supply the viscera as a whole.

Now there is one major difference between the arrangements of the autonomic and central nervous systems. In the latter, the cell bodies of the neurones all lie within the system; only fibres lie without the brain and spinal cord. In the autonomic, however, there are peripheral ganglia, consisting of masses of cell-stations with synapses. In the case of the sympathetic, these are located close to the vertebral bodies; in the parasympathetic, they are near or actually within the organs to be controlled. So we may speak of preganglionic and postganglionic fibres within the autonomic system. The parasympathetic has long preganglionic fibres within a spinal nerve before synapsing with ganglion cells close to the organ for which they are destined, so the postganglionic fibres will be very short. Conversely, the sympathetic outflow has short preganglionic fibres, but their postganglionic course is often very long.

Chemical transmission in the autonomic system takes place (*a*) at the ganglionic synapse, and (*b*) where the postganglionic fibres enter the target organ. In the former site, both sympathetic and parasympathetic act by liberating acetylcholine. At the peripheral site, the sympathetic liberates acetylcholine where sweat glands and skeletal muscle blood vessels are concerned, but noradrenaline in the heart, smooth visceral muscle, glands and internal blood vessels; the parasympathetic again liberates acetylcholine. The actions of noradrenaline and acetylcholine are completely antagonistic, and noradrenaline may be regarded as the chief sympathetic transmitter at the end-organs.

The effects of sympathetic stimulation are those of great bodily activity and preparation in relation to the primitive needs of fear, fight or flight. Aided by adrenaline secreted by the adrenal gland into the blood, it constricts the blood vessels of the skin, so raising the blood pressure and shunting more blood to heart and brain; speeds and strengthens the heartbeat; dries up glandular secretion; dilates the pupils and the bronchioles; stands the hair on end and initiates sweating to dissipate extra heat production; and inhibits peristalsis, thus relaxing the walls of the hollow viscera. In contrast, the parasympathetic controls relaxed and constructive activities in times of tranquillity, as after a heavy meal. It dilates the peripheral vessels, slows the heart and lowers the blood pressure, and excites secretion and vigorous peristalsis.

However, this generalization of their actions is an oversimplification. The two parts of the system are always in complex and reciprocal control of the internal organs, even though at any given time one or other will predominate, especially in markedly anxious or very placid individuals. Antagonists though they may be, they work together like the ropes of a tiller to set the course of the internal empire they control.

Special sense organs

Sensation in general

We have seen that a large amount of sensory information from the muscles, joints and internal organs is constantly arriving at the brain; but, with the exception of pain, these messages do not enter consciousness. Contrasted with this are the special sense organs: the eye, ear, nose, tongue and skin; and these provide information of a very definite and particular quality, and elicit rather complex responses. In discussing the nerves it was pointed out that the particular nature of a sensation is determined by the receiving organ, not by the stimulus. This means that the same physical cause will produce different effects when applied to different sense organs—e.g. radiation of a certain wavelength is appreciated as light by the eye but as warmth by the skin, a tuning fork is only a repeated touch to the skin but conveys a note to the ear. And, conversely, a sense organ can only react in the one way, whatever the stimulus; the eye not only reacts to light normally but also transmits the sensation of light when the eyeball is pressed vigorously.

Another characteristic phenomenon in the relation between brain, sense organs and the outer world is that of *projection*. Although sensations are only appreciated as the result of changes in the brain cells, they are felt as if taking place more remotely; touch is projected to the skin surface and taste to the mouth, while we regard sight and sound as coming from the surroundings. These are, as it were, conventional delusions established in infancy. For, in fact, our sensations are the mere masks and symbols of reality, which is filtered through our sensory recording apparatus but of whose true nature we can never have direct experience. The situation is very like that of a radar operator who can unfailingly detect a ship as a blob on his screen without necessarily ever getting an idea of what it really looks like.

Finally, all sense organs obey certain physical laws. There is a minimum or threshold value for the stimulus, below which it is ineffective, and this level depends to some extent on physiological factors; it is higher in fatigue or if there are simultaneous distracting stimuli. Further, we do not experience sensations in an intensity directly proportional to the true physical intensity of the stimulus; there is such an enormous range of impressions coming from the outside world that we have, as it were, to summarize them. Thus a light that we feel to be twice as bright as another is, by physical measurement, perhaps ten times as intense, and this relation applies also to sound, i.e. it is a logarithmic one.

The 'code' operative in sensation appears to be the frequency of transmitted nerve impulses, and the important factor in appreciating a stimulus is the change from a previously steady state. The fresh stimulus causes a burst of impulses, which then die away, even though the stimulus persists—the phenomenon known as *adaptation*. This is obviously essential, otherwise we could not tolerate the continued stimulus of sunlight, or even the pressure of our clothes. In fact, some receptors are such fast adaptors; others respond more slowly but continue to respond as long as the stimulus endures.

The skin as sense organ

The skin is the bounding layer that intervenes between the body and the outside world, so that its sensory reactions to stimulation are of major importance. In fact, the special sense organs of eye and ear, and the central nervous system itself, have developed in evolution, and likewise in the embryonic history of the individual, from an infolding of this surface layer. The modalities involved include touch, pressure, pain, temperature, vibration and stereognosis.

The sense of *touch* is transmitted by small organs or end-bulbs of the cutaneous nerves, lying in little bays on the deep surface of the epidermis (Fig. 191). And touch is

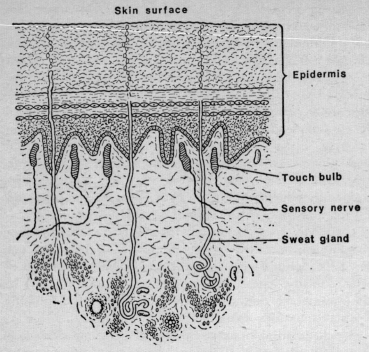

Fig. 191. Section of the skin showing touch bulbs and sweat glands.

essentially a response to distortion of the skin surface, not to pressure alone. The hairs play an important part by acting as levers magnifying the effects of contact, and the sensory bulbs are grouped around their roots in profusion; hence, too, the phenomenon of *tickle*, and the general blunting of sensation that follows the shaving of a hairy surface. The skin is not a uniformly sensitive surface; some points are more sensitive than others. There are a number of scattered *touch spots*, separated by ½ mm or so, and between these there is no sensation. The capacity of discriminating between two adjacent stimuli varies from place to place. On the tongue and fingers, where touch spots are close together, two compass points are felt to be separate if only 1 mm apart, but they must be separated by 6 cm on the thigh before being appreciated as distinct. There are also many fine, unspecialized nerve endings in the skin. There are probably not special fibres for each kind of skin sensation; the different receptors are linked to common fibres, and the information input is unscrambled in the central nervous system.

The skin offers a good instance of how our conventional reading of sensations depends merely on habituation. A pencil point slid between the tips of the middle and index fingers is felt as a single stimulus because these surfaces are normally adjacent and we have learnt to fuse their sensations; but if the fingers are crossed, surfaces that are never normally in contact are brought together and a pencil placed between them is now felt as double.

Pain sense used to be regarded as due to an increase of pressure above the maximum threshold for ordinary touch sensation, but it is now doubtful whether this is really so, or whether there are specific pain receptors and fibres. The two are certainly intermingled; the cornea of the eye is an instance where *any* stimulus is felt as pain and there is no ordinary touch sensation. The object of pain sensation is to protect the individual from danger by provoking an automatic withdrawal reflex, and such a response will persist in animals whose brains have been destroyed—a decerebrate frog will respond just as usual to a drop of acid placed on a hindleg—and in men suffering from paralysis due to loss of the connections between brain and spinal cord. When pain sense is lost in nervous disease, the sufferer may burn himself and remain unaware, and the skin becomes particularly liable to infection and ulceration.

There is also a close link between temperature and pain, and we know that these sensations are transmitted along the same fibre tracts within the central nervous system. The headquarters for pain reception appears to be the thalamus, and 'pain' seems to be the result of analysis of varying stimuli entering the CNS, for one and the same sensation is sometimes painful and sometimes not, and pain is always affected by *attention* and the *emotional state*. Note how the

operation of leucotomy, usually performed for mental illness—the severance of the fibre tracts connecting frontal
cortex and thalamus—may render even severe pain more
tolerable; it is as if the patient becomes indifferent to the
stimulus, and this is why the operation is sometimes performed for severe pain in incurable malignant disease.

Stimuli that would cause great pain if applied to the
skin may not hurt at all in the viscera, which are normally
insensitive to cutting or burning. The substance of the
brain itself, paradoxically enough, appears to be entirely
insensitive. Then there is the phenomen of *referred pain*,
where pain originating from a deeper structure is felt over
the skin surface. Thus pain from the diaphragm is often felt
at the shoulder tip because the phrenic nerve originates
from the same segments of the cervical part of the spinal
cord that supply the skin of the shoulder; and a prolapsed
intervertebral disc in the lumbar region irritates a root of
the sciatic nerve and causes pain in the leg. It is interesting
that the area of skin reference is often hypersensitive to
quite light touch under these circumstances.

Temperature sensation. Feelings of heat and cold are due to a
loss or gain of heat from the object felt. The importance of
this gradient is shown by the familiar trick of chilling one
hand and overheating the other, when tepid water is felt as
hot by one and cold by the other. As with touch, temperature sensation is not distributed uniformly but in discrete
points at the skin surface, and there seem to be separate sets
of receptors for the two kinds of stimulus—heat and cold
spots. The mucous membranes are less sensitive to heat
than the exposed parts, so that one can drink fluids that are
too hot to be held in the hand. It is a matter of common
observation that there is a qualitative difference between
the hot and the merely warm. Very hot water often feels
cold for a short time and this may be due to a total stimulation of all the temperature end-organs. But the problem of
temperature sensation is complicated by the changes that

such stimuli produce in the small blood vessels of the skin. If these open up, the part becomes warm because flushed with blood, while constriction results in pallor and coldness. And a painful stimulus often causes vasodilatation and hence is spoken of as a burning pain, either because heat and pain receptors are being simultaneously excited or because at a certain level the stimulus is undifferentiated.

The skin is also sensitive to *vibration* from something pressing on it—more so if the vibration is transmitted to underlying bone, as from a tuning fork placed on the skull. Vibration sense is closely linked to touch and pressure sensation.

Finally, there is *stereognosis*, the ability to recognize the three-dimensional shapes of solid objects. For this, the object has to be handled, not merely felt, the fingers must be reasonably warm and sensitive, and the receiving cerebral cortex must be able to analyse the information supplied.

The eye

The globe of the eye lies embedded in the orbital fat, embraced by the six small *ocular muscles* that spring from the bony walls of the orbital cavity and move the eyeball in all directions. It is a sphere almost 2·5 cm in all diameters, with a more local convex bulge at the anterior pole formed by the transparent window of the *cornea*. The *optic* nerve is attached behind and runs back in the orbit like a stalk to reach the cranial cavity via the optic foramen.

The eyeball has three coats:

(1) An outer fibrous coat, the tough opaque protective *sclera*, the white of the eye, which surrounds the whole globe except where it merges into the cornea anteriorly.

(2) An intermediate pigmented layer, the *choroid*, which lines the sclera and is modified in front to form the diaphragm of the *iris*. This rests on the lens and has the central aperture of the *pupil*, the size of which is varied by the contraction of the underlying muscular *ciliary body*.

(3) The innermost nervous layer is the *retina*, adapted

for the reception of light stimuli and extending forward no further than the ciliary body. It is pigmented, containing the visual purple which is bleached by exposure to strong light; and consists of a very complex layered pattern of

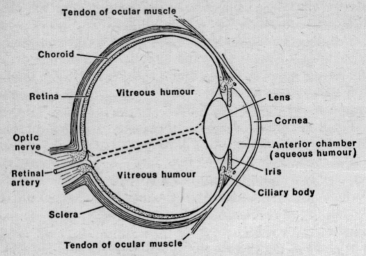

Fig. 192. Section through the globe of the eye.

nerve cells, whose fibres ultimately leave the eye in the common bundle of the optic nerve. The point of attachment of the latter to the retina is the *optic disc* or blind spot, as here there is no room for visual reception; and it is here that the central retinal artery and vein enter the eye by traversing the centre of the optic nerve.

The refracting media of the eye are also three in number:

(1) The watery *aqueous humour* fills the anterior chamber, the space between the cornea and the front of the iris and lens, and also pervades the little posterior chamber, the narrow marginal cleft between lens and back of iris. Like the cerebrospinal fluid, the aqueous is constantly being secreted from the circulation and reabsorbed, and any interference with the latter may cause a dangerous rise of ocular tension—glaucoma.

(2) The translucent *lens* is a solid structure enclosed in a capsule and markedly more convex on its posterior than its anterior aspect. This curvature is modified by changes in the tension of its suspensory ligament to accommodate for near and far vision.

(3) The great bulk of the eye is filled with the thin translucent jelly of the *vitreous humour*, traversed by a central canal running from the entrance of the optic nerve to the back of the lens.

Eyelids and lacrimal apparatus. The supporting substance of each lid is the tarsal plate, a layer of dense connective tissue; the eyelashes are attached at the lid margins, and the inner surfaces are lined by a delicate membrane, the *conjunctiva*, which is reflected over the cornea and sclera in front. These are kept moist by the secretion of tears by the *lacrimal gland*, which is situated in the upper and outer part of the orbital cavity and has several ducts opening under the upper lids. The secretion, which is actively bactericidal, is collected in the *lacrimal sac* at the inner angle of the lids, and from this the *nasolacrimal duct* runs down the sidewall of the nasal cavity to discharge the tears into the nose.

The eye as an optical system. We may compare the eye to a camera, much to the former's advantage. Focusing is automatic, and the lens changes its shape, or *accommodates*, so as to become more curved and powerful for near vision and less so for distant objects, the aim being always to bring the rays of light to convergence exactly on the retina, not in front or behind. A normal eye produces a sharp retinal image of an object at infinity without using any accommodation at all. There is a very wide field of vision, over 200° of a circle, so that objects slightly behind one's eye level are still in sight, and there is little loss of light by reflection at the refracting surfaces.

The cup-shaped retina is everywhere in just the position for the best definition of the image, and it has two different systems of receptor cells: rods and cones. The rods, with their visual purple, are for blacks, whites and greys under twilight conditions; and when this apparatus is deficient, as when lack of vitamin A depletes the supply of visual purple, there results the condition of night-blindness. The cones have other pigments and are designed for colour vision in bright light; when this system is defective, the individual is colour-blind, either partially for reds and greens or completely so. He must use his twilight apparatus for daylight purposes and all objects will appear only different shades of grey. In darkness the retina becomes increasingly sensitive to the available light for the first hour, the phenomenon of *adaptation*.

Refractive errors. Longsightedness is normal in early childhood, for the eyeball is too small for proper convergence. If this condition persists, as *presbyopia*, in adults, light rays are converging behind the retina and correction is therefore by the use of an additional, artificial, convex spectacle lens to supplement the converging power of the natural lens and to focus the rays sharply on the retina. The long sight of old age is due to the degeneration and loss of resilience of the lens at this time so that it cannot assume a more convex shape.

In shortsightedness, or *myopia*, rays are being focused in front of the retina by too powerful refraction and objects are blurred. This is corrected by using a supplementary concave lens of just sufficient power to displace the point of convergence backwards onto the retina.

It will be evident that errors of refraction in either direction may be due not only to inadequacy of the lens system but also to an eyeball that is too short or too long. The short eye and consequent long vision of infancy give place with growth to an eye of normal length and focusing capacity; but too much reading and close work cause the eyeball to go on increasing in size and then the child may become permanently myopic.

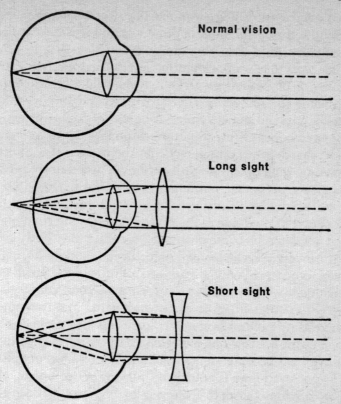

Normal vision

Long sight

Short sight

Fig. 193. The focusing of parallel light rays by the optic lens in normal vision, and in long and short sight. When there is a refractive error an artificial lens has to be used to help make the rays converge on the retina, and the dotted lines show how this is achieved.

Astigmatism results from the lens/cornea system being more curved in one plane than another, i.e. no longer truly spherical. Accommodation will be correct for a given line, but a line at right angles will not be properly focused. Correction is with a cylindrical lens.

Visual acuity is not a measure of the *size* of an object that can be seen, for anything, however small, is visible if it emits enough light. It is measured by estimating the minimum angle subtending two lines that can be appreciated; normally, this is one minute of an arc and it is checked by the familiar rows of letters on a test-card.

Binocular vision is important for several reasons. The combined use of both eyes gives a wide field of vision and eliminates the screening effects of nose and brow, which shut off part of the field from a single eye. And it provides the stereoscopic effect, the appreciation of depth and distance; this depends on the fact that the images of an object formed by each eye are slightly different, and that these two images are presented simultaneously to the brain without appearing double. Each eye, in other words, sees a little way round its own side of the object, and this information is fitted together in consciousness. To secure this both eyes must converge on the focused object, even though the angle of convergence is only slight for distant points. In many individuals with weak eye muscles and poor converging power, the single image is easily dissociated into two; and, if there is a habitual preference for using a better eye, the image of the weaker one is permanently suppressed so that this eye may become virtually blind. The flatness of vision of the one-eyed man is notorious, but it is not complete. He gets a fair idea of depth from shadow contrast, surface reflections and the modifying effect of distance on colour.

The ear

The ear falls naturally into three parts: the *external ear*, the *middle ear* or *tympanic cavity* and the *internal ear* or *labyrinth*.

The *external ear* consists of the *pinna*, the funnel-shaped organ for the collection of sound waves, a thin plate of elastic fibrocartilage covered with skin. This leads inwards to the middle ear along a narrow channel, the *external auditory meatus*, whose walls are 2·5 cm long, cartilaginous at the surface but bony as they approach the skull, and lined by inturned skin possessing wax-secreting glands. Thus the external ear is mostly outside the skull, while the middle and inner compartments are situated actually within the temporal bone. At the bottom of the external meatus, the tense circular *tympanic membrane*—the eardrum—completely shuts off the middle ear cavity, except when previous disease has left a perforation.

The middle ear itself is a roughly cuboidal cavity. Its roof is the floor of the middle cranial fossa of the skull, the eardrum is its outer wall and the bony case of the inner ear

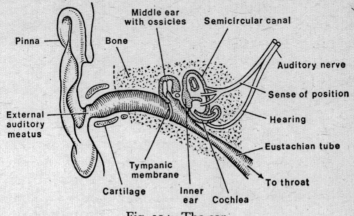

Fig. 194. The ear.

lies medially. A canal, the *Eustachian tube*, connects it with the nasopharynx, providing a means for equalizing air pressure within the chamber and the throat, and also a route for ascending infection. Its function is to help the air pressures identical on either side of the drum; if the atmospheric pressure is suddenly altered, as in an aeroplane ascent, there is temporary deafness until the canal is opened by yawning or swallowing. The cavity is spanned by three tiny bones, the *auditory ossicles*, which link the tympanic membrane with the outer wall of the inner ear, transmitting by bone conduction the vibrations set up by sound impulses in the resonant eardrum.

The *internal ear*, deep within the temporal bone, is a complex organ, the *membranous labyrinth*, which is housed within an exactly-fitting bone chamber. The upper part is concerned only with sense of position and spatial orientation; this consists of three fluid-filled *semicircular canals* arranged in three different planes mutually at right angles. Then there

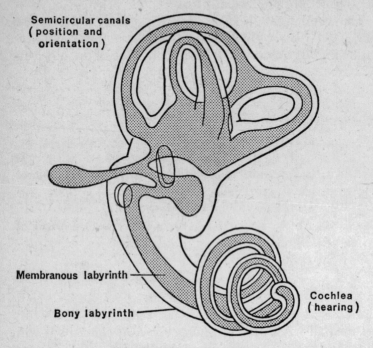

Fig. 195. The internal ear.

is an intermediate portion, the *utricle and saccule*. And inferiorly is the other main part, the *cochlea*, the true organ of hearing, a spirally-coiled structure responding to pitch.

Two little windows lie between the inner and middle ears, but these are closed by membrane and by one of the ossicles. On the inner cranial side an *internal auditory meatus* allows the passage of the seventh and eighth cranial nerves into the temporal bone from the brain.

Gravity and position sense. The semicircular canals, the utricle and the saccule are filled with fluid and lined with hair cells, which respond to changes in position of jelly-like mineralized masses—the *otoliths*. In the utricle and saccule these respond to gravitational changes brought about by changed positions of the head; it is possible in some invertebrates to substitute iron filings for the otoliths, and then the animal can be made to swim upside down by using a magnet. During rotational movements of the head, one or two of the canals are left behind, as it were, and this stimulates the hair cells. Taking both ears together, there are six varieties of detectable spin. It seems that control of accurate movements of the eyes to compensate for head movements depends on information supplied from the canals. If the lateral canal is stimulated by syringing the ear with hot or cold water, this causes *nystagmus*—rapid lateral to-and-fro eye movements, vertigo and impaired sense of object-position in space. The well-known motion sickness is due to overstimulation of the canal system. Impulses from the canals are transmitted along the vestibular portion of the eighth cranial nerve to the cerebellum, which automatically adjusts body posture in response.

Hearing. The tension of the eardrum is not constant but is adjusted by a special muscle, which relaxes when no sound is to be heard but springs into contraction to protect the delicate middle ear mechanism against very loud noises, provided that these are not too sudden. The membrane responds very faithfully to high and low notes of frequencies ranging between 40 and 30 000 cycles per second, and transmits these vibrations to the chain of ossicles in the middle ear; these, in their turn, beat against the bony case of the labyrinth and stimulate the cochlea.

Hearing is very effective over a wide range of both pitch and intensity; it is most sensitive to frequencies around 1000 c/s, and the loudest noise perceptible is a billion times greater than the softest. Hearing is also direction-finding, due partly to the position of the ears at either side of the head and partly to appreciation of even minute differences in the time of arrival of sounds at either ear. Hearing impulses are transmitted along the acoustic portion of the eighth nerve and reach consciousness. Apart from blockage of the external meatus by wax, deafness can originate from either the middle or the inner ear. Middle-ear deafness is caused by loss of bone conduction due to infection which has damaged the drum to lock the ossicles into a rigid chain; but a vibrating tuning fork can be appreciated as well as ever if applied to the skull, for the vibrations bypass the ossicles to reach the inner ear directly. Inner-ear damage causes nerve deafness; a tuning fork cannot now be heard and this condition is basically irremediable, whereas middle-ear deafness can be dealt with by surgery on the drum or ossicles.

Nose

We have seen that the external part of the nose is partly bony at its base, where it is part of the skull; while the tip and much of the body is fibrocartilaginous. The nasal

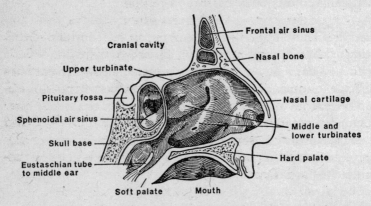

Fig. 196. The nose. The septum has been removed and we are looking at the sidewall of the left nasal cavity. (After *Gray*.)

cavity itself is divided into right and left halves by the nasal septum, cartilaginous in front and bony in the depths. The two cavities open externally on the face at the nostrils or *anterior nares* and behind into the nasopharynx at the *pos-*

terior nares. The side wall of each cavity is marked by three ridges: the superior, middle and inferior *conchae*, the substance of each of which is a fragile scrolled *turbinate bone*; and beneath each overlying turbinate is a corresponding recess or *meatus*. The accessory nasal air sinuses open more or less directly into one or other meatus, the maxillary antrum straight into the middle meatus, and the frontal sinus by a more tortuous channel. The lining nasal mucous membrane is therefore freely continuous with that of the sinuses, which may be secondarily involved by a nasal infection. The actual smell-sensitive region is on the roof of each cavity, where the mucosa contains the olfactory cells, whose fibres make up the first cranial nerve; the remainder of the cavity is purely respiratory in function.

The sense of smell is vestigial in man, varying greatly from one individual to another; nevertheless, it remains one of the most delicate of all the senses. It depends on odoriferous particles actually reaching the nose and there dissolving in the moisture of the mucous membrane. The sensory area is a particular patch of membrane situated at the very summit of the nasal cavities and rather out of the way of direct air currents through the nostril. There is, therefore, a little delay before a new odour is appreciated while the particles diffuse upwards. The nature of the sensation aroused probably depends on the molecular configuration of any particular substance. It should be noted that violent and pungent 'smells', such as that of ammonia, are really stimuli of ordinary sensation within the nose and are transmitted by a different nerve.

Taste

Taste, like smell, is a chemical sense and depends on the entry of foreign particles into solution on the surface of tongue and palate, basically a sampling process to decide on their rejection or acceptance. Special microscopic organs, or *taste buds*, embedded in the mucous membrane serve this function, and in the tongue these are grouped into obvious projections, or *papillae*. Taste buds are also found on the soft palate and epiglottis. Their impulses are collected up in the seventh and ninth cranial nerves. There are four basic tastes: sweet, sour, salt and bitter, and these, when mixed, do not blend like colours but remain distinguishable. The nature of any taste depends not only on the chemical composition of the substance concerned but also on its acidity or alkalinity, and certain parts of the tongue tend to 'specialize' in different tastes. We have already noted that many so-called tastes and flavours are really smells, and are lost when the latter sense is put out of action; food and tobacco are savourless during a severe cold.

There has always existed in the animal kingdom a duality in the methods of responding to stimulation. The most primitive method is by an alteration of the chemical substance of the protoplasm at the stimulated point, with the diffusion of the chemical agent to other parts of the body, where it excites a general reaction. The more sophisticated method is the development in evolution of a nervous system, with ramifying communications and a central control for incoming and outgoing messages.

But the primitive chemical mechanism has been retained, and even developed, in the higher animals. The *endocrine* system is a group of glands without obvious ducts; their secretions are internal, i.e. they are absorbed into the bloodstream and diffused throughout the body, on whose functions they exert an intimate and profound control. These secretions are the *hormones*, or chemical messengers; and these glands obviously differ entirely from the glands of external excretion, such as the liver and salivary glands, which have ducts carrying their juices into a body cavity or onto the skin surface. Nevertheless, some organs are capable at one and the same time of both internal and external secretion; for example, the pancreas secretes digestive juice into the bowel, as well as insulin into the blood, and the ovaries and testicles form ova and spermatozoa, as well as the male and female hormones.

A general characteristic of hormones is the extreme potency of even minute amounts of the secretion. We may recognize three main types of endocrine activity:

(1) a *temporary response* to meet the needs of an emergency, e.g. the liberation of adrenaline from the adrenal glands in alarm;

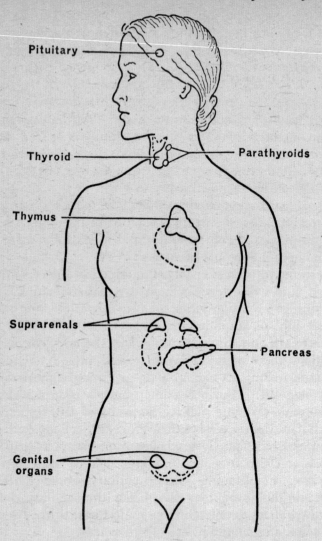

Fig. 197. The endocrine glands.

(2) a *sustained activity*, e.g. the constant stimulating action of the thyroid gland on metabolism;

(3) the secretion of the *sex hormones* by the male and female sex organs.

The endocrine glands are largely, but not entirely, independent of nervous control, which is most important in relation to the adrenal medulla and the posterior pituitary. On the other hand, the glands subtly influence each other's activities through the chemical action of the hormones themselves. *Chemically*, hormones are proteins (or polypeptides), such as the secretions of the anterior lobe of the pituitary; or steroids, such as the secretions of the adrenal cortex; or phenol derivatives, such as thyroxine and adrenaline. The hormone content of a biological fluid or tissue can be estimated, or *assayed*, biologically by noting the effect of an extract on an experimental animal; or chemically by analysis; or immunologically, if they are proteins, by their capacity to excite measurable antigen formation.

Our understanding of the activities of the endocrine glands has grown from appreciating the clinical disturbances that result when their output is deficient—e.g. the cretinism of thyroid deficiency—and from the response of such individuals to treatment with the appropriate gland extracts. Hormones maintain internal homeostasis by controlling such factors as water and electrolyte balance (adrenal cortex), blood sugar level (insulin) and metabolic rate (thyroid). They also influence metabolism and reproduction, growth and development, and the response to mental and physical stresses. We may regard the hormones as having specific 'target' organs or tissues under their control. These may be quite specific, e.g. the thyroid-stimulating hormone of the pituitary affects only the thyroid gland; or they may, like the adrenal corticosteroids, affect the metabolism of most cells of the body. In the target organs and in the liver, hormones are degraded to inert soluble substances, which are excreted by the kidneys.

All the endocrine glands are closely interrelated and affect each other's activity; the precise nature of this balance in the individual is a large factor in the determination of behaviour and personality. They have been compared to an orchestra, of which the leader is the pituitary gland.

Pituitary

The pituitary gland is a pea-sized organ, slung from the undersurface of the brain immediately behind the optic chiasma. It occupies a little bony pocket in the base of the skull between the two middle cranial fossae. It is attached by a short stalk to an important region of the brain called

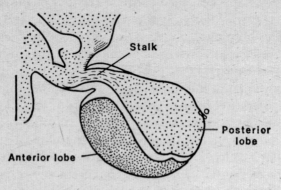

Fig. 198. The pituitary gland, longitudinal section. (After *Gray.*)

the *hypothalamus* in the floor of the third ventricle; hypothalamus and pituitary are in close circulatory communication. Although not essential to life, disease or removal of the whole gland results in arrest of growth, atrophy of the sexual organs, general weakness and premature senility. It consists of two lobes, an anterior and a posterior, each of which produces several types of hormone.

Hormones of the anterior pituitary
These are all proteins or polypeptides.

(1) *Somatotrophic or growth hormone.* This is concerned with growth, particularly that of the skeleton. It is necessary for normal protein synthesis in the body and thus the increase in size of the tissues. Essentially it promotes anabolism, and it also mobilizes depot fat to meet the energy requirements of growth and exercise. It is secreted in irregular bursts and more at night than during the day; this may be the cause

of the nocturnal 'growing pains' common in young children. If there is overproduction of this hormone, as may occur with a tumour of the secreting cells, this causes *giantism* in children due to overgrowth of the long bones while their epiphyses are still patent. In adults, after epiphyseal closure, the corresponding condition is less dramatic and is known as *acromegaly*, an overgrowth of the soft tissues of hands, feet, face and tongue, with coarsening and broadening of the features, giving a characteristic, rather Johnsonian, appearance. The tumour responsible may press on the optic chiasma and cause blindness. Deficiency of growth hormone in childhood produces *pituitary dwarfs*, stunted and often sexually immature individuals who are not, however, mentally retarded; they are like delicate miniature adults.

(2) *Adrenocorticotrophic hormone, ACTH.* This is necessary for the health and activity of the cortex of the adrenal gland (see below), and for the secretion of steroid hormones by that structure. No significant steroid output is possible without such pituitary stimulation, and the close connection of these two fundamental glands is sometimes termed the 'pituitary-adrenal axis'. If the pituitary is diseased or surgically destroyed—as it sometimes is for malignant disease—the patient can be kept in health only by the administration either of the adrenal hormones as such or of ACTH.

(3) *Thyrotrophic hormone.* In a similar way, this is responsible for normal functioning and secretion of the thyroid gland.

(4) *Gonadotrophic hormones.* These control the release of ova and the secretion of female sex hormones by the ovary in the female, i.e. they are responsible for the well-known cyclical changes in the ovary (page 484). In the male they control sperm formation and the secretion of male hormones by the testis.

(5) *Prolactin.* This is essential for the development of the female breast during pregnancy and the secretion of milk by the gland after childbirth.

Pituitary control system. The anterior lobe of the pituitary is stimulated to produce each of the above hormones by a series of specific releasing factors passed down to it from the hypothalamus, and the hypothalamus itself is under the control of higher cerebral centres in response to various stresses—using the word 'stress' in a wide sense. After the appropriate pituitary hormone has reached the target organ—say, the adrenal cortex—the latter is stimulated to secrete its own hormone—in this case, say, cortisol—into the bloodstream. A very ingenious feedback mechanism now comes into play. Once the blood level of cortisol has reached an optimum, this inhibits the further production of ACTH by the pituitary. Then, as the concentration of target gland hormone gradually falls, the pituitary is reactivated and produces more of its trophic hormone. This feedback mechanism applies to all the secretions of the anterior pituitary and is represented in the following diagram:

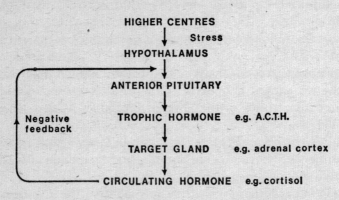

Hormones of the posterior pituitary

(1) *Antidiuretic hormone*, ADH. This is of premier importance in control of water excretion by the kidney (see page 484). It increases the permeability to water of the lining of the terminal ducts of the renal tubules, and so its effect is to promote the reabsorption of water into the circulation and to concentrate the urine. Again, an elegant feedback mechanism comes into play, designed to maintain a con-

stant osmolarity in the blood and extracellular fluid. Thus, should the osmotic pressure of the plasma begin to rise, this stimulates specific osmoreceptors in the brain, more ADH is liberated by the posterior lobe of the pituitary, water is retained in the blood to exert a diluting effect and a small volume of concentrated urine is passed. Should the osmotic pressure of the plasma fall, ADH secretion is inhibited, a large amount of dilute urine is passed and the plasma becomes more concentrated.

This explains how destruction of the posterior lobe of the pituitary may cause the disease known as *diabetes insipidus*, in which the urine is always pale and dilute, and of a fixed low specific gravity and a large volume.

(2) *Oxytocin*. This exerts a stimulating action on smooth muscle, notably that of the uterus and of the milk ducts in the breast. Oxytocin plays an important part in uterine contraction during childbirth and in ejection of milk from the breast.

Thyroid

The anatomy of the thyroid is described on page 273. There are two main secretions:

(1) *Thyroxine*. This is a hormone secreted in response to the stimulus of the thyrotropic hormone of the pituitary. It is a complex glycoprotein containing iodine, and this is the only function of iodine in the body. Dietary iodine is ultimately derived from the soil, and in localities where iodine is deficient the thyroid hypertrophies to try to compensate for this lack and forms a swelling, or *goitre*, in the neck. Goitrous retarded individuals used to be common in certain parts of the world until the condition was prevented by iodizing table salt.

The action of thyroxine is to stimulate metabolism; it increases the rate of tissue build-up and breakdown, and the rate of consumption of oxygen, and it is also essential to normal physical and mental development. Excess of secretion, which may occur spontaneously or after an emotional crisis, causes *thyrotoxicosis* or Graves's disease—overexcitability, rapid pulse, staring eyes, wasting and diarrhoea. Some of these features indicate a sensitization to circulating adrenaline and noradrenaline. Deficient secretion in children causes dwarfism and mental retardation, and in adults the condition of *myxoedema*, an obese slow-wittedness associated with coarsening of the skin, loss of hair, sensitivity to cold and a lowered metabolic rate.

(2) *Calcitonin*. This acts so as to transfer calcium from the plasma into bone. It is therefore beginning to be used in the treatment of conditions characterized by rarefaction of bone.

The *parathyroids* are four small pea-like glands embedded in the thyroid capsule. Their hormone acts together with calcitonin to regulate the level of plasma calcium steadily at 9–11 mg/100 ml. Parathormone mobilizes calcium from bone into the blood and promotes intestinal absorption of calcium. It is secreted whenever there is a tendency of the plasma calcium to fall too low; should the latter rise, calcitonin comes into play. If there is an excess of parathormone, as when there is a parathyroid gland tumour, so much calcium is moved out of the bones that they appear transparent and cystic in the X-ray, and fracture easily. If there is not enough parathormone, as may happen when the glands are inadvertently removed at operations on the thyroid, the plasma calcium falls to low levels, causing twitching, convulsions and spasms, and this may be fatal unless the defect is made good by administering gland extract.

Adrenal (suprarenal) glands

These have been described in the section on the abdomen (page 221) as triangular, cap-like organs sitting on the upper poles of the kidneys. They have an outer rind, or *cortex*, and an inner *medulla*.

The function of the *medulla* is intimately connected with that of the sympathetic system—indeed, the two have developed in common. Its hormones—*adrenaline* and *noradrenaline*—are secreted in response to sympathetic stimulation in emergency situations, such as muscular exertion or emotional stress. Their release into the bloodstream excites the usual adrenergic effects at all the bodily termini of the sympathetic system. There is an increase in the rate and force of the heartbeat, the blood pressure rises, the smooth muscle of bowel and bladder relaxes, the pupils dilate and the hair stands on end; there is also general increased awareness and apprehension as blood is diverted to skeletal muscle, glycogen mobilized from the liver and the available supply of free fatty acids increased to prepare for a fight or flight situation.

The internal secretions of the adrenal cortex are known as corticosteroids and are based on the carbon ring group structure:

There are a number of these, but the most important are cortisol (hydrocortisone) and aldosterone. It is difficult to specify their actions briefly, but they have a profound influence on the metabolism of carbohydrates and electrolytes respectively. *Cortisol* promotes glucose formation and retention, is generally catabolic, favours protein breakdown and inhibits tissue repair; *aldosterone* modifies the rate of transport of sodium and potassium ions across cell membranes, favouring sodium retention and potassium excretion in the renal tubules.

Cortisol formation is increased many times in stress situations—not just those acute emergencies that excite adrenaline release but also those of a more prolonged nature—after operations and injuries, in long exposure to cold, in convalescence and emotional crises. It is produced in response to stimulation of the adrenal cortex via the hypothalamic-pituitary-adrenal axis, and if the cortex is unable to respond the individual may collapse and die. There is a condition known as Addison's disease caused by disease or atrophy of the cortex which is marked by pigmentation, prostration and low blood pressure; this used to be fatal until it became possible to substitute the missing hormones artificially.

The adrenal cortex also has a close connection with the sex glands. It is difficult to say exactly what this connection is under normal circumstances, but some hint is given by the fact that abnormal function of the adrenal cortex can either bring on precocious sexual development in boys and girls or actually tend to *reverse* the sexual characteristics, producing masculinity in women and feminism in men. Such a reversal does not, of course, alter the essential nature of the sexual organs of the individual but changes the external appearances, such as the growth of hair and the body contour, as well as the psychological make-up and the direction of sexual desire. Many unfortunate women looking like men, and with male sexual desires, are the victims of an overacting tumour of the adrenal cortex, and some have been cured by its surgical removal.

Pancreas and insulin

The function of insulin in carbohydrate metabolism has already been discussed. Insulin is the internal secretion of groups of cells in the substance of the pancreas known as the *islets of Langerhans*. It is basically an anabolic hormone that promotes the storage of carbohydrate, protein and fat. Insulin is normally secreted in response to the rise in blood sugar that follows a meal and it acts to lower this level by facilitating the take-up and utilization of glucose in the tissues, and inhibiting the formation of glucose from glycogen in the liver. It also promotes glycogen deposition in skeletal muscle.

The *thymus* may be regarded as an endocrine gland, inasmuch as it secretes a humoral factor that is essential for the development of lymphoid tissue in the lymph nodes and of spleen in the newborn, and the maintenance of these tissues and the production of new immunologically competent lymphocytes in the adult.

The gonads or sex organs

The ovaries and testes, besides their prime function of forming the reproductive cells—ova and spermatozoa— secrete the female and male sex hormones respectively, *oestrogens* and *androgens*. These are steroid hormones, formed in response to gonadotrophic drive from the anterior pituitary, with the usual negative feedback control.

In both cases, these hormones are responsible for the development of what are known as the *secondary sexual characteristics*—the attributes, though not the essentials, of sexuality. These include hairiness, a deep voice, coarse skin, the pubertic enlargement of the external genital organs and a narrow pelvis in men; hairlessness, smooth skin with plenty of subcutaneous fat, the development of the breasts and the broad pelvis of women. And they also include the less definable, but important, differences in male and female personalities and attitudes.

In more detail (and this should be read in conjunction with Chapter 22 on reproduction), the ovarian hormones consist of two main groups: oestrogens and progestagens. The situation is complicated because the pituitary has a direct effect on the development of egg cells and the process of ovulation in the ovary. Oestrogens initiate the cyclic activity of genital tract and breast, promoting active thickening and secretion in the lining membranes, and muscular development in the walls of uterus and vagina. They are more concerned with the first, preparatory or follicular, stage of the menstrual cycle associated with ovulation. The progestagens have to do with the second, receptive or luteal, half of the cycle, in readiness for a possible pregnancy, and their activity persists if actual pregnancy ensues.

Oestrogen also has some general metabolic effects. Its production falls off sharply at the menopause, often resulting in some impairment of emotional stability and some masculinization, both of which can be relieved by giving the hormone.

The most important androgenic hormone is *testosterone*. This stimulates the growth of the male reproductive tract and its proper secretion and functioning. It also has a marked anabolic action on general metabolism, promoting protein synthesis, growth and skeletal development.

Finally, it is of interest that, paradoxically, ovary and testis also secrete small amounts of the *opposite* sex hormones.

22 Reproduction and development

Reproductive organs

The essential sex organs are known as the *gonads*: a pair of *testes* in the male forming the spermatozoa, a pair of *ovaries* in the female forming the ova. The general plan of the reproductive system in both sexes is not dissimilar, but the final adult forms appear very different because, while the ovaries remain in the abdominal cavity, the testes come to lie outside it. And in the female there is a uterus to house the developing embryo, while the penis of the male is represented in the female only by a diminutive organ called the clitoris.

Male organs

The *testes* develop in the abdominal cavity but pass down shortly before birth to enter the loose, corrugated skin pocket of the scrotum externally, where they hang down in the perineum on each side of the root of the penis. There remains an oblique valve-like passage through the abdominal muscles just above the inguinal ligament, the inguinal canal, occupied by the stalk of the testis, or *spermatic cord*, which carries blood vessels to the organ and the sperm duct back to the pelvic cavity (Plate 7(b)).

The testes are ovoid structures, with a tough fibrous capsule made up of lobules containing the fine tubes in which the spermatozoa are formed. Applied to their outer side is a curved organ, the *epididymis*, which receives the sperm from the testis; it is an intricately coiled tube from which issues the main spermatic channel, the *vas deferens*, running up to the abdomen. Testis and epididymis lie vertically in the scrotum, with upper and lower poles, surrounded by a loose

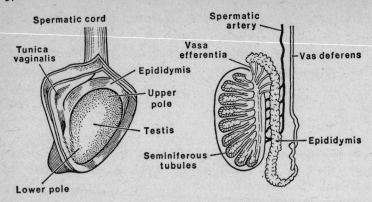

Fig. 199. (a) Testis and epididymis exposed by reflection of their serous covering, the tunica vaginalis. (b) The same in longitudinal section.

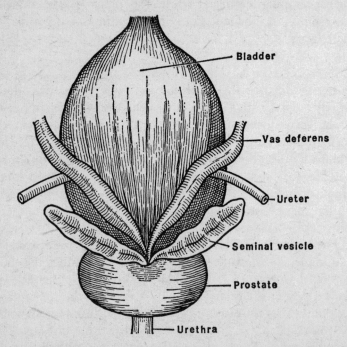

Fig. 200. Bladder and associated organs in the male. Posterior aspect.

serous sac which is sometimes distended with fluid to produce a hydrocoele.

The vas deferens runs down the side wall of the pelvis to
reach the back of the lower part of the bladder, where it
lies on the upper surface of the prostate; here there is an
associated sac attached to its outer side, the *seminal vesicle*,
for the reception and storage of sperm. The ducts of vas
and vesicle open via a common *ejaculatory duct* into the *urethra*,
which lies embedded in the prostate gland. Sperm are continuously formed in the testis, stored in the vesicle and only
enter the urethra in the final ejaculatory act of sexual orgasm, when they are discharged via the urethra and penis
into the female vagina. It should be noted that the ejaculated
fluid is not just the sperm but a complex seminal fluid containing also the secretions of vesicles and prostate. The
prostate is a solid organ of mixed muscular and glandular
structure, of rather the shape and size of a chestnut. It has
its base applied to the neck of the bladder and its apex on
the pelvic floor, and it is traversed by the urethra.

The *penis* consists of a central *bulb* arising from the centre
of the perineum, and traversed by the urethra and two
lateral *crura* springing from the sides of the pubic arch.
These all soon join to form the shaft of the organ, a cross-

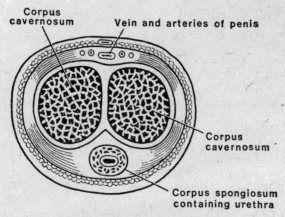

Fig. 201. Cross-section of shaft of penis. (After *Gray*.)

section of which shows the disposition of its constituents (Fig. 201). Here, the part containing the urethra, the continuation of the bulb, is the *corpus spongiosum* below, with the *corpora cavernosa*, continuations of the two crura, above on each side. It is to these latter that the organ owes its remarkable property of rapid increase in length and girth on sexual excitement, becoming rigid and capable of introduction into the vagina. This process of erection, contrasted with the flaccid state, is due to the presence of a system of cavernous spaces which can be rapidly distended with blood from the penile arteries.

Female organs

The *ovaries* are a pair of almond-shaped organs, each lying on the sidewall of the pelvis just below its brim. They are studded with fluid-filled cysts, the ovarian *follicles* in which the egg cells or ova ripen, one coming to maturity every

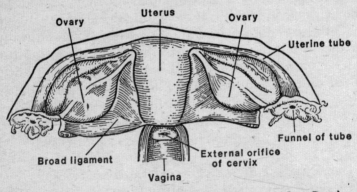

Fig. 202. Internal sex organs of the female. (After *Gray*.)

month. The ovaries are embedded in a membranous sheet, the *broad ligament*, which stretches like a wing from the uterus to either side of the pelvis. And in the upper free edge of this ligament are the *uterine tubes*, attached to the uterus like outstretched arms to the shoulders. These muscular channels open into the uterus medially and have at their outer ends a fringed, funnel-like entrance, which tends to curve towards and embrace the ovaries so as easily to receive the ripe ovum when it is shed.

Uterus. The womb is a hollow organ, with thick muscular walls, lying in the middle of the pelvis between bladder in front and rectum behind. It communicates below with the vagina and at each side with the uterine tubes. The organ is tilted forwards so that the anterior surface rests on the bladder, and the posterior surface faces up as well as back; both surfaces and the dome (or *fundus*) are covered with peritoneum, and the peritoneal pouch between uterus and rectum is the deepest part of the abdominal cavity.

The virgin uterus is 7·5 cm long, its wall 2·5 cm thick. Longitudinal section shows the upper *body*, 5 cm long, with a triangular cavity, and the lower *cervix*, 2·5 cm in length and traversed by a narrow cervical canal opening into the main cavity at its internal orifice and into the vagina at its external orifice. The lining mucous membrane of the body, or *endometrium*, contains numerous mucous glands and undergoes striking regular changes during the menstrual cycle. The organ as a whole becomes enormously enlarged in pregnancy but returns almost to normal after childbirth by a process of involution, though the cavity never quite regains its virgin state. The uterus is small and undeveloped in childhood, and shrinks and atrophies in old age.

The *vagina* is a distensible canal, capable of receiving the penis during intercourse and of allowing the passage of the child in parturition. It extends from the uterus, which it meets at an angle of 90°, to run down and forwards through the pelvic floor and to open externally on the perineum. Part of the cervix actually protrudes into the vaginal vault, the encircling rim of which is known as the fornix. The front wall of the vagina is intimately blended with the back of the bladder and the urethra; the back wall is separated from the rectum by fibrous tissue.

The *external female genitalia* are known collectively as the *vulva*. This includes the *mons pubis*, a fatty hair-covered eminence in front of the pubic symphysis; the skin folds forming the lips of the vaginal orifice, comprising the outer thick *labia majora* and the inner slender *labia minora*; the

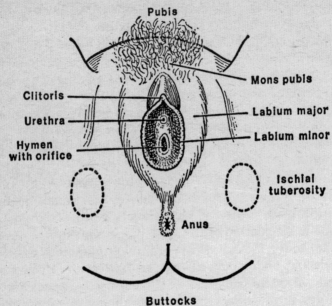

Fig. 203. The female perineum.

clitoris, a diminutive but sensitive erectile equivalent of the penis, lying at the meeting of the labia in front; and the *hymen*, an incomplete partition of mucous membrane stretching across the vaginal orifice in the virgin, which is ruptured by intercourse to leave a few peripheral tags. The external opening of the urethra is just behind the clitoris and immediately in front of the vaginal orifice. There are numerous mucous glands in this region.

There is an inguinal canal in the female as in the male, but this contains no essential part of the genital organs, only a supporting ligament of the uterus terminating in the tissues of the groin.

Reproduction

In animals reproductive activity and sexual desire is usually an intermittent affair, periods of sexual desire alternating with seasons of indifference. Even when the male is continuously active, the female tends to be receptive only in cycles, dependent on a recurring rhythm of activity in the ovaries, and the formation of the sex hormones is correspondingly discontinuous. In man the pressure of sexual feeling is pretty constant, particularly in the male, and the secretion of the sex hormones is on the same level. There is still some periodicity in the female, associated with the monthly phenomenon of *menstruation*, a flow of blood and mucus from the womb which occurs only in the higher apes and man.

We have already seen that the development of the sexual organs before puberty, and their proper functioning thereafter, is dependent on the pituitary gland. This influences the enlargement of the organs in childhood and is responsible at puberty for their beginning to function; but when this time arrives it is the hormones of the sex organs themselves that cause the development of the secondary sexual characteristics, which impart the outward forms of masculinity and femininity—in both sexes the origin of sexual desire, in the girl the swelling of the breasts and the commencement of menstruation, and in the boy the change of voice and enlargement of the external organs. All these changes are dependent on the primary development of the ovaries and testes, and will be prevented if these organs are ineffective because of disease or castration.

The menstrual cycle

This forms a complicated pattern, regulated and controlled by both pituitary and ovarian hormones. It is usually considered in terms of a twenty-eight-day cycle beginning with the first day of menstrual bleeding, though longer and shorter cycles are common.

In the twenty-eight-day cycle *ovulation* occurs at or around the fourteenth day, i.e. an ovarian follicle ripens in the ovary and discharges the contained ovum into the peritoneal cavity. This egg cell is taken up into the open mouth of the uterine tube and propelled along the tube by muscular peristalsis towards the uterus. Here, the lining has become highly vascular, engorged and thickened in preparation for the embedding of a fertilized ovum, for fertilization, if it has occurred, will have taken place in the tube. If, as is usually the case, the ovum has not been fertilized, the mucous membrane of the uterus is eventually shed and this causes the bleeding of menstruation. It lasts three to five days, and then a new lining begins to be built up again and is completed by the fourteenth day of the cycle, midway between two periods, just at the time when ovulation occurs again in the ovary. Thus the whole cycle affects both ovaries and uterus and is divided into the *follicular phase*—the first two weeks—and the *luteal phase*—the second fortnight.

The hormonic aspect of all this is as follows. The ripening of the ovum in its follicle and its discharge at ovulation are effected under the control of the anterior pituitary hormone—the gonadotrophin known as FSH (follicle-stimulating hormone). Once the egg cell has left the ovary, the remains of the follicle become organized into a bright yellow structure known as the *corpus luteum* under the influence of another pituitary gonadotrophin—LH or luteininizing hormone. The follicle itself stimulates thickening of the uterine lining preparatory to a possible conception by means of its own hormone, and the corpus luteum has a hormone controlling the luteal phase which completes these preparations. Should pregnancy supervene, the corpus luteum enlarges and persists as the corpus luteum of pregnancy, otherwise it disintegrates. Thus the cycle involves both ovary and uterus, ovulation and menstruation alternate regularly at fortnightly intervals, and the uterine lining is completed and receptive during the latter half of the cycle, destroyed at menstruation and rebuilt in the first half.

Fertilization

Fertilization, the union of the male spermatozoön with the female ovum, normally occurs in the uterine tube as the egg is moving towards the uterus. It can only occur if, at this time, a living sperm derived from the male by recent intercourse has made its way up the vagina, through the cervix and body of the uterus into the tube. Success therefore depends on a near coincidence of intercourse and ovulation, say within forty-eight hours, so that the likelihood of any single sexual act resulting in pregnancy is therefore a small one. When the male ejaculates during sexual orgasm, a pool of semen is deposited in the vaginal vault around the cervix. The sperm have to penetrate the cervical mucus and progress upward, rather slowly as far as their own motility is concerned—not more than 3 mm/min; their overall progress in the uterus and tube is, in fact, much faster, due to muscular propulsion by these organs. During their climb there is an enormous reduction of sperm count. Around 100 million spermatozoa are deposited in the vagina, but only a million enter the uterus and perhaps only 100 actually reach the ovum. Spermatozoa do not remain active and fertile in the female tract for more than two days. Once the ovum has been penetrated by a spermatozoön, rapid changes ensue. The nuclei of the two cells fuse and the ovum becomes impenetrable to further spermatozoa. Although fertilization takes place in the tube, it is unusual for the fertilized ovum to become embedded in the tube and for pregnancy to develop there. When it does happen, tubal pregnancy is a very serious affair, for the tube cannot support the size of a growing embryo and will at some stage rupture, causing severe haemorrhage. This is an example of ectopic pregnancy, i.e. pregnancy occurring at some site other than in the uterus. It may also occur in the ovary itself, sometimes in the abdominal cavity, and always results in a surgical emergency after a certain stage of development.

The sperm cell has a flattened pear-shaped head containing the nuclear material, connected by a short neck to the

long tail, whose lashing movement gives the cell its motility.
Sperm formation in the testis is a complicated process which
sometimes goes wrong, and a high proportion of spermatozoa
are ill-formed and unsuitable for fertilization. If sperm are

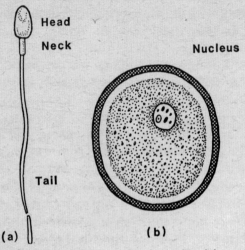

Fig. 204. Male and female reproductive cells, (a) spermatozoön,
(b) ovum.

very few—oligospermia—or even totally absent from the
seminal fluid—azoospermia—this is one possible cause of
infertility. Spermatogenesis is also impaired if the scrotum
is kept too warm or is subject to radiation.

Sterility means the inability of a woman to conceive, i.e.
to house a developing embryo within her uterus, but it is
by no means a purely female problem. For, when we survey
the long line of processes necessary in both sexes for the
proper development of the germ cells and their fusion, it
can be seen that an error at any point in either sex may be
responsible. Thus the sex organs themselves may be abnormal, diseased or absent; the sperm may be few, mal-
formed or non-viable; the vaginal secretion may be lethal
to them; the uterine tubes may be blocked—as after gonor-
rheal infection—and prevent the passage of the ova. And

we have seen that deficiency of vitamin E may possibly cause the death of the very young embryo. Thus no investigation of sterility is complete until both parties have been examined. In some cases, female sterility is due to some failure of proper maturation of ova in the follicles, and this can be treated medically by administering the pituitary follicle-stimulating hormone, though with some risk of producing multiple births.

A point of great importance in the development of mature male and female germ cells is that, at some stage in the formation of these cells from primordial cells, the chromosome count in the cells has to be reduced. A normal body cell contains forty-six chromosomes, which is sometimes called the *diploid* state, and if this applied to ovum and spermatozoön the fertilized egg and its successor cells would contain ninety-two chromosomes. This is obviously not a tenable state of affairs; hence, there is a cell division at a stage in germ cell formation from their precursors when the chromosomes, instead of splitting in two as they usually do during cell division, are simply divided between the two new cells, the final germ cell containing twenty-three chromosomes—the *haploid* state.

Life cycle in male and female

The word adolescence describes a state of physical and social maturity, and cannot be allotted any definite time. In contrast, *puberty* means the capacity for sexual procreation—i.e. in the female that ovulation has begun and in the male that emissions of viable spermatozoa are occurring. The age at which regular uterine bleeding begins is called the *menarché*, and this comes rather late in adolescence at an average age of thirteen years, but ranging between ten and seventeen years. As a rule, proper ovulation does not take place during the first year or two of menstruation. The onset of puberty is a complex hormonal interaction; hypothalamus, anterior pituitary, ovaries and adrenal cortex all play their part.

The eventual cessation of menstruation is known as the *menopause*, which occurs at an average age of forty-eight, and with a range of forty to fifty-five years. Many bodily and mental changes, known collectively as the climacteric, are associated with the menopause. It is marked by the failure of the ovaries to produce either ova or oestogen, and so the genital tract and the breasts atrophy, with widespread physiological repercussions and, frequently, some psychological disturbance.

In the male sperm formation in the testis begins after the age of ten or eleven, and with this occurs enlargement of the penis and scrotum and, a year or two later, the appearance of pubic hair. The voice breaks and the sweat and sebaceous glands overact, which is why acne is so common in boys at puberty. At this time, or earlier, skeletal growth accelerates, and this growth spurt lasts until the closure of the epiphyses of the long bones, which is delayed a year or two later than in the female—hence the greater average height of men. It is difficult to pinpoint the exact onset of puberty in boys because there is no obvious physiological guideline such as menstruation. It is usually between ten and fifteen years of age. The basic mechanism of hypothalamic-pituitary-adrenal gonadal control is similar to that of the female. Once established, testicular function continues for the rest of life and is only slightly diminished in old age. There is no sharp cut-off of gonadal function as occurs in women. Nevertheless, at about the corresponding age many men do suffer some waning of desire and physiological disturbance, suggesting that there is such a thing as a male climacteric

Development of the embryo

The nuclei of the two germ cells fuse, and the egg at once begins to divide or segment to form a rounded mass of cells, whose number is doubled at each division. This process begins before entry into the uterus, the journey along the uterine tube taking several days. Once in the uterine cavity, the ovum adheres like a parasite, excavating for itself a cavity in the mucous membrane, in which it becomes embedded and sealed off by a blood clot.

The early embryo consists of little more than a pair of sacs or vesicles, the *amniotic cavity* and the *yolk sac*, with an intervening plate which is the actual embryonic area at which individual development occurs. The amniotic cavity

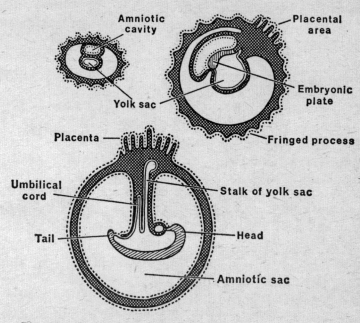

Fig. 205. Stages in the development of the early embryo.

soon greatly expands, coming to line the whole of the uterus, and is filled with the amniotic fluid bathing the embryo. At the actual site of implantation, the amniotic membrane develops a complex system of fringed vascular processes penetrating the uterine wall and bathed freely in the blood spaces of the latter. This constitutes the mass of the after-birth or *placenta*, a thick fleshy disc of 15–20 cm diameter when maximally developed. It is here that maternal and embryonic circulations are in closest proximity; but there is no direct communication, oxygen and food substances having to diffuse across the separating layer of cells. The other embryonic cavity, the yolk sac, becomes squeezed by the growth of the amnion into a narrow stalk, the core of the

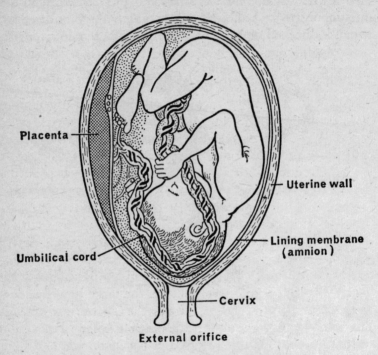

Fig. 206. The foetus *in utero*. (After *Gray*.)

umbilical cord which connects the developing child, floating freely in the amniotic fluid, to the centre of the placenta. The cord is 51 cm long at birth, spirally twisted, and carries in its jelly-like substance the two umbilical arteries and the single umbilical vein of the foetus.

Since the foetus sends its stale blood to the placenta for oxygenation and nutriment, the umbilical vessels resemble the pulmonary trunks, i.e. the arteries contain stale venous blood and the veins fresh arterial blood.

Embryonic layers

The early embryo soon comes to possess three layers of cells responsible for the origin of all body tissues and organs.

The outer layer or *ectoderm* forms the skin, its glands, hair and nails, the nervous system, and the essential parts of the eye, ear and nose. The nervous system originally lies on the surface of the back of the body and is infolded in development as the neural tube, the precursor of the spinal cord, with a bulbous expansion at the head end marking the brain. All the sense organs originate in this outer layer; an infolding of greatly modified skin surface meets an outgrowth from the central nervous system. This arrangement is well shown in the eye, where the retina, an extension from the brain on the stalk of the optic nerve, cups the transparent lens developed from the overlying skin.

The innermost layer or *endoderm* forms the whole of the bowel and its associated glands—such as liver and pancreas —the lining of the respiratory tract and lungs, and the thyroid and parathyroid glands. We have noted that the intestine develops on the back wall of the abdominal cavity and then comes to lie suspended in that cavity by a mesentery still attached to the posterior wall. And, although the small intestine retains this position, the large bowel undergoes a complicated rotation to bring the caecum and ascending colon to the right and the descending colon to the left— a rotation that may be reversed or left incomplete.

The intermediate layer is the *mesoderm*, origin of all the connective tissues of the body; the bone and cartilage of the skeleton; the teeth; all the muscles of both voluntary and involuntary type; the heart and blood vessels; and the urogenital system.

The general features of the development of the limb-buds have already been touched on at page 54. At any point in the unimaginably complex pattern of development things may go wrong, or not go far enough. The spinal cord may be left on the surface of the back as an open layer instead of a tube; or it may develop normally, but the two halves of the vertebral column fail to enclose it completely, leaving a gap behind, usually in the lumbosacral region—the condition known as *spina bifida*. The palate may remain cleft if its two halves fail to meet in the roof of the mouth; the septum of the heart may remain incomplete, allowing a vicious and disabling admixture of venous with arterial blood; a bone or a limb may fail to develop wholly or in part; fingers and thumbs may be too few or too many. What is outstanding in all this is the overwhelming frequency with which the developmental process inches its way to an accurate ending at a constant time, reciprocal organs maturing together. The surprise, indeed, is that any error is a rarity.

Time scale

A note on the times of certain important stages may be given. The two primitive vesicles of the embryo are present some ten days after fertilization and the intervening embryonic area begins to develop in the third week; head and tail folds, neural groove and heart are obvious by the fourth week. In the fifth week there appear the lens of the eye, the rudiments of the face, the gill arches and the stumps of the limb-buds, and the embryo is now 5 mm long. The sixth week sees the body well curved on itself, the head approximating to the long tail, the umbilical cord attached to the belly near the latter; the liver enlarges, and the limb-buds grow out and are demarcated into their segments.

By the end of the eighth week, the embryo is 2·5 cm long and is thenceforward known as the *foetus*; eyes, ears and nostrils are formed, the external genitals differentiated, and the fingers and toes clearly marked out. Fine downy hair appears in the fourth month when the foetus is 20 cm long; in the fifth month, foetal movements begin and the skin becomes covered with a greasy secretion.

By the seventh month, the eyelids have opened and the testicles descended into the scrotum; and though the foetus may now be viable (i.e. capable of living), if born prematurely it is not adequately clothed with subcutaneous fat until the end of the ninth month.

At *birth*, the uterus contracts, rupturing the membranes and expelling first the foetus and amniotic fluid, then the placenta and finally the membranes, turned inside out like an umbrella. Certain important changes occur in the child at and soon after this moment of birth. During foetal life, the lungs have remained unexpanded and airless; they need no blood, and the stream in the pulmonary artery is shunted through a bypass directly into the aorta. With the first cry, the lungs expand, the pulmonary circulation is established and the bypass soon closes off. The portions of the umbilical vessels contained within the foetus also become shrivelled up and obliterated, and the former parasite has become an independent organism.

The breasts and milk

The *breasts* may be described here, as the mammary glands are accessories of the genital system; designed for milk secretion, they occur in a rudimentary form in the male as well as the female. In the latter, they form two large, rounded eminences between the skin and deep fascia on the front of the chest, overlying the pectoral muscles; they extend from the second to the sixth ribs and from beside the sternum out towards the axilla. Small before puberty, they then develop with the uterus, hypertrophy in pregnancy—and especially at the lactation period after delivery—and atrophy in old age. There are minor cyclical changes with menstruation.

The pigmented nipple at the apex of the breast is surrounded by a coloured area of skin, the *areola*; this is studded with sebaceous glands, is pink in the virgin and permanently darkened in the first pregnancy. The nipple is perforated by fifteen to twenty milk ducts opening at its summit; it is sensitive and contains erectile tissue. The gland

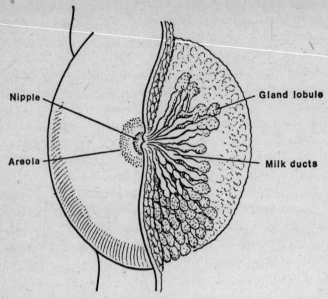

Fig. 207. The female breast, partly dissected to show the lobules.

itself is divided into numerous lobes by fibrous partitions, some strands of which go deeply to fasten the organ to the chest wall. Each lobe contains a branched system of secreting gland spaces supported by much pervading fatty tissue; the larger ducts converge on the nipple and expand just before reaching it as the milk sinuses or reservoirs. The gland spaces are only distended with milk during active functioning, and the organ as a whole is liberally supplied with blood vessels for this purpose.

The breasts develop throughout pregnancy in preparation for lactation. The first few days after delivery of the child they yield only a watery fluid or colostrum containing a few fat globules; only after this does milk secretion begin properly. The adequate secretion of milk is governed by the hormones of pituitary and ovaries, and there is also a reciprocal relation between lactation and menstruation; resumption of the monthly periods after delivery of the child may be long delayed by continued suckling. In addition, the child's sucking at the breast is a powerful stimulus to the proper contraction or involution of the uterus after childbirth, and is therefore of the greatest importance to the mother herself.

The milk of all mammals is essentially similar, a slightly acid fluid, opaque from the presence of myriads of finely dispersed fat globules. Its main constituents are:

(1) *protein* (caseinogen), which is converted into a solid casein clot or curd in the stomach, or when it is allowed to sour; from this can be expressed a liquid whey;

(2) *sugar* (lactose);

(3) *fat*.

In man these are found in the proportions of protein 1·5%, lactose 6·5% and fat 3·5%; and cows' milk differs in that it contains more protein and less sugar, the fat content being much the same. If, therefore, it is necessary to use it for infant feeding, it can be made approximately the same in composition as human milk by adding water to dilute the protein content and making up the sugar; but it is very much a second-best substitute, and the physical benefits natural feeding confers on both mother and child, as well as the inestimable but profound psychological advantages, should always make breast feeding the method of choice.

23 Environmental considerations

Some of these aspects have been considered in previous sections.

Man is able to withstand severe climatic rigours, ranging from subzero to desert conditions, and to maintain his own body temperature and functions fairly constant within narrow limits. This capacity to live and work under widely differing environmental conditions depends partly on clothing, and cultural and social adjustments, e.g. housing. But it is mainly effected by regulating (*a*) the heat produced by muscle, and (*b*) the heat loss from the skin.

The central organs of the body remain at a constant temperature of 37°C, while the temperature of the subcutaneous fat and skin surface varies widely, according to their distance from the heart, and the ambient temperature ranges between 27°C and 34°C. Heat loss is mainly by radiation and convection from the skin surface; there is little loss by conduction, both air and skin being poor conductors. The air temperature alone is not an adequate measure of environmental heat stress as both solar and body radiation must be taken into account.

Although the skin, through changes in calibre of its vessels and by sweating, regulates heat loss, the primary response to thermal stimuli is within the thalamus of the brain, where there are heat receptor cells very sensitive to changes in the temperature of the circulating blood. It is when the control of this centre is damaged by bacteria or viruses that we get the fever characteristic of infections.

Heat stress

The first effect of heat stress is dilatation of the skin vessels; the skin becomes warmer and loses more heat. Secondly, there is sweating, which can stabilize skin temperature at about 35°C. Very high temperatures can be tolerated if the air is absolutely dry. Tolerance lessens with increasing humidity, and air quite saturated with water vapour cannot be borne for long, even if the temperature is only 32°C.

Repeated exposure to high temperatures increases the ability to sweat, and with acclimatization comes a reduction in the salt content of the sweat. In great heat sweat may be lost at the rate of over 4 litres an hour, though this cannot be maintained for long without collapse. Genetic factors are of some importance here as they affect body build; a long, lean trunk and limbs increase the area of skin surface proportionate to body weight, and so the area for heat exchange.

Exposure to cold

The immediate response to cold is by vasoconstriction of the skin vessels, which reduces the blood flow to the skin, whose temperature falls. At the same time, heat production in muscle is stepped up by movement or shivering; the latter appears to be a more efficient source of heat than ordinary voluntary contraction. As far as longer-term acclimatization is concerned, man, as we have already pointed out, tends to evade his problems rather than overcome them by building up a subtropical microclimate within special clothing and buildings.

Survival under conditions of exposure depends on the maintenance of body temperature, body water and supplies of energy—in that order of importance. Failure to maintain one's temperature will kill before dehydration, and dehydration before starvation, though, of course, these factors interact.

If the body temperature falls and remains low, it is surprisingly easy to die of hypothermia in the middle of a civilized community, whether one is a fell walker, an old person living alone in an unheated room in winter or a

young baby kept insufficiently warm. When clothing is
soaked, it loses its insulation value, and a man so clothed
walking against a high cold wind might as well be naked,
for he can only survive while he remains capable of generating a good deal of heat through physical work, e.g. walking or some other exertion. Should he begin to slow down,
his surface temperature will fall; then the central organs
begin to cool, consciousness will be lost at or even above
30–32°C, and death ensues in an hour or two unless these
conditions are reversed by rapid rewarming of the whole
body, as in a hot bath. Impermeable outer clothing is the
great protection against cold and wet.

Immersion

The length of survival during immersion in cold water depends on the water temperature. If this is freezing, death
follows within the hour but may be delayed for up to six
hours at 15°C. At temperatures of 20°C or over, death does
not occur from hypothermia. These facts may seem to be
contradicted at first sight by experience of Channel swim-
ming, where individuals remain in cold water for much
longer periods; but these subjects are exceptionally well
insulated by fat and are trained to maintain a very high rate
of energy production for many hours—and even they may
succumb if they slow down. The ordinary survivor from
a shipwreck does better to remain still because heat loss
from movement exceeds any heat that can be produced by
struggling.

The main problem of the *survivor in an open boat* is that of
dehydration. It is not helpful to drink even small quantities
of sea water, and death comes more quickly to those who do
than to those who refrain, because the concentrating powers
of the kidney are limited to a urine containing the equivalent of no more than 2% of salt. As sea water contains 3%–
4% salt, the drinking of it is bound to raise the salt content
of the tissue fluids, with fatal results. The same considera-

tions apply to drinking urine, for urine is already concentrated to the maximum and cannot provide additional body water. The best course is for survivors not to drink at all for the first twenty-four hours, and then to ration any fresh water that may be available at the rate required to compensate for obligatory water loss in urine, perspiration and expired air, i.e. 1 litre a day. The minimum intake should be of the order of 500 ml/day, even if this only lasts for a week or two, as this maintains the general condition and the ability to co-operate in rescue attempts. Water loss in hot weather can also be reduced by remaining still, shading the body and covering oneself with clothing soaked in sea water. In a cold sea, however, it is important to keep the clothing dry, for there is then a risk both of general body hypothermia and of immersion foot—a condition analogous to trench foot, marked by swelling, insensibility and muscle damage.

Desert conditions

Here the problems are those of heat, dehydration and starvation, in that order of importance. Heat may cause heat-stroke and may also increase water loss to the point of heat-exhaustion. The two conditions are quite dissimilar; the former is due to direct interference with the brain thermoregulatory centres and is associated with fever; the latter is a collapse state, with perhaps quite a low body temperature. Water loss is much increased by heat because the body not only ceases to lose heat through convection and radiation but may even *gain* heat from the environment. Under these conditions, sweating affords the only means of heat loss. Some rise of body temperature, even as high as 39°C, is acceptable. But at 40°C and over, impaired consciousness and collapse begin, while at 42°C, even if death does not ensue, there will be permanent brain injury.

Yet it is possible to survive almost any degree of heat if the victim keeps quite still and has reasonably adequate water supplies, and, preferably, some shade. The direct effects of the sun's rays on the head and neck may be discounted. At night, of course, the desert may be intensely cold; this is the period when some attempt to travel should be made.

Death is usual when 15%–20% of the body water, i.e. 10%–15% of the body weight, has been lost. The body of an average man holds some 40 litres of water and 2–4 litres may be lost without too much ill effect; after 5 litres, mental and physical impairment set in, and death occurs when 6–8 litres have been lost—such are the narrow margins of human existence on this planet.

High altitudes

In addition to the hazards of cold, dehydration (exacerbated by high winds) and starvation, high altitudes carry special risks arising from the reduced barometric pressure and consequent lowered availability of oxygen.

The oxygen content of both inspired air and arterial blood is reduced with consequent circulatory and respiratory changes. There is a stimulation of respiratory rate and volume, and of cardiac output. A more delayed response, taking one or two weeks, is an increased production of red blood cells. Races habitually living at over 3658 m have a rate of red cell production that is 30% higher than normal and as much as 50% more haemoglobin than residents at sea-level.

The oxygen lack at heights may impair judgment; the effects of frostbite may be overlooked, though these can be easily prevented. The winds increase heat loss and more water is lost with the forced respiration, so clothing and shelter become important—even the insulating protection of a snow-drift.

The onset of altitude hazards depends to some extent on

the rate of climb and the possibilities of acclimatization. Mountaineers who ascend gradually to 5486 m may exhibit only slight breathlessness, and below 3048 m a fit young adult usually notices nothing untoward. Sudden exposure is much more serious, as in aircraft when cabin pressurization or oxygen masks fail, or in mountaineering motorists. The features of acute *mountain sickness* include headache, breathlessness, rapid pulse, weakness and sometimes vomiting, but these usually pass off within a few days. *Chronic* mountain sickness is more serious, and is due to reactive thickening and pressure rise in the arteries of the lungs.

Increased barometric pressure

A diver is exposed to increasing pressure on all the tissues of the body at increasing depths. If he is breathing air, the partial pressures of both nitrogen and oxygen in his blood and tissues are increased—but not that of CO_2 because this is produced within the body. This increased nitrogen pressure may cause *nitrogen narcosis* at depths of 46 m and over, with unconsciousness at 117 m.

On return to the surface, the nitrogen may come out of solution as bubbles and cause decompression sickness. The gas blocks small blood vessels and causes joint pains ('the bends'), paralysis of the legs and difficulty in breathing. Within the long bones, large sections of bone may die even two years later. All these effects may also be seen in tunnel workers—'caisson disease'. The treatment for these symptoms is immediate recompression in a pressure chamber, followed by a very slow return to atmospheric pressure. *Prevention* is by staged decompression at various depths on the way up, giving sufficient time at each level for gas to be eliminated without bubble formation. Because this is tedious, divers are sometimes brought up in a submersible pressure chamber at high pressure, which is then gradually reduced on board ship. The dangers of deep diving can be lessened by using oxygen mixed with helium. It is not safe to give pure oxygen, because to breathe this at high pressures risks the development of epileptic fits or severe bronchopneumonia. The use of pure oxygen in *space capsules*, even at not very great pressure, carries a very serious fire risk, as events have shown.

It may be of some interest to mention that oxygen given under increased pressure—about two atmospheres—has proved very useful in the medical treatment of certain conditions, notably gas gangrene, and is also a useful adjunct to the X-ray treatment of cancer, the resuscitation of victims of coal-gas poisoning and the management of peripheral arterial disease or injury.

Starvation

This is less serious than dehydration since survival is possible for as long a period as sixty to seventy days without any food whatever, provided that water supply is adequate and physical work is at a minimum. During starvation, the body lives on its own tissues, mainly on the fat reserve but also to some extent on muscle protein. Characteristic features are apathy and listlessness, giddiness and psychological withdrawal. There is also a fall in body temperature and in basal metabolic rate. In part, these changes are no doubt adaptive. In trying to keep alive during starvation, the important factors are to maintain body temperature, water balance and calorie balance, in that order of importance. As little energy should be expended as possible, but some work is often good for morale. As far as rationing is concerned, the value of even a small amount of food is out of all proportion to its calorie value. It is best not to include protein, so as to avoid overburdening the kidneys and increasing the urine output if there is also a shortage of water. The most suitable ration is carbohydrate; alcohol, if available, can be very useful.

Inherent biological rhythms

In contrast to the variable factors in the environment, the body possesses certain inherent rhythms in the way it functions. These have been called *circadian*, as they are usually based on a twenty-four-hour cycle. Thus body temperature is lowest in the morning and highest in the late afternoon, even in fever. Urine output is lowest at night; urine is more alkaline in the morning. And there are many biochemical variables, as in the level of plasma phosphate or cortisol. This why it is always best to measure such variables at a known time of day and for serial measurements to be made at the same times of day if the readings are to be comparable.

These variations are independent of working or eating patterns and evidently manifest the existence of an underlying twenty-four-hour biological clock, independent of physical or mental activities, or the influence of social pressures. The cycle can be made to adapt by altering circumstances—for instance, by working by night and resting by day, or flying into a different time zone—but the process of adaptation may take weeks. Even in experimental subjects made unknowingly to lead an artificially short or long day, many rhythms retain a twenty-four-hour cycle for weeks. However, in persons exposed to darkness and solitude with absence of stimuli, as in caverns, the cycle gradually drifts into longer and longer periods of twenty-five, twenty-six and twenty-seven hours until, eventually, several days may be 'lost'.

These rhythms are usually beneficial, e.g. wakefulness during working hours, sleepiness at night. The low urine output at night leaves sleep uninterrupted; the higher body temperature by day favours mental and physical work. It is of interest that the daily temperature variation often becomes inverted in night workers, though with a reversion to pattern at weekends. The adrenal cortex secretes more cortisol in the early morning, around waking, due to increased hypothalamic activity and pituitary output of ACTH, i.e. this reflects a change in brain activity and may be associated with the return of alertness after sleep. It is also fascinating that certain parasitic worms that may infest the body also have their rhythms, as noted by their periodic migration from tissues to blood and vice versa. Some favour the night for their appearance in the bloodstream, others the day; and these patterns are reversible if the host inverts his day/night pattern of living.

It is probable that most of the circadian rhythms are acquired during the first years of life under the influence of day and night alternation, and the whole pattern of human activities that is associated with this. The argument for believing that this is so is that they are absent in Eskimos, who are not exposed to such a regular alternation of light and darkness.

We may define man's place in the world of living creatures by successively narrowing our terms. He is an animal, not a plant; multicellular, not unicellular; a chordate—and a vertebrate chordate—not an invertebrate; a vertebrate with jaws, not a jawless lamprey or hagfish; a terrestrial vertebrate, not a fish; a warm-blooded mammal, not a reptile, amphibian or bird; a placental, not an egg-laying or pouched mammal; a primate within the placentals, and an anthropoid within the primates; a man within the anthropoids, and finally a modern man—*Homo sapiens*—as distinct from more primitive forerunners.

Let us discuss these terms in more detail.

Chordates

These are a great group of animals, ranging from degenerate forms such as sea-squirts to every form of vertebrate, but all derived from a simple, free-swimming, marine, fish-like ancestor—an ancestor possessing certain characters separating it from earlier forms and handed down, with modifications, to all subsequent chordates.

The elongated body was clearly differentiated into upper and lower surfaces, and, though this body was segmented, segmentation was confined to the muscle mass along the back, the abdominal organs in the ventral portion of the body remaining as a unified system. This is quite different from the segmentation common in lower forms of life such as the earthworm, where each body segment reduplicates the next and carries a complete set of organs. Chordate segmentation was apparent not only in the dorsal muscle blocks but also in the nerves supplying them, which emerged from a nerve cord running the length of the body. Finally, and characteristic of this group of animals—as its name implies—a stiffening rod, or *notochord*, the forerunner of the vertebral column in vertebrates proper, traversed the body from end to end immediately in front of the nerve cord providing an axis of leverage for the muscles.

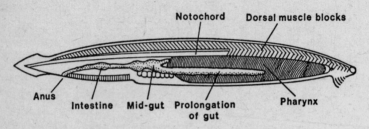

Fig. 208. A simple chordate still in existence, *Amphioxus*, the lancelet. Note the segmentation of the dorsal muscle blocks, partly removed to expose the notochord. The simple intestine, mid-gut and their prolongation occupy a common ventral abdominal cavity and open externally at the anus. The nerve cord is not shown here. (From Young's *Life of Vertebrates*.)

The signs of this primitive organization persist even in man. A relic of the notochord is to be found in the simple jelly of the pulpy central nucleus in each fibrocartilaginous disc between a pair of vertebral bodies—a relic still occupying its old place anterior to a nerve cord which has become the spinal cord. And the segmentation of the muscles of the back, and of the spinal nerves supplying them, is still to be seen—most clearly in the rib spaces—distorted though the pattern has been made by the evolutionary emergence of the limbs.

Biologically, the appearance of the fish-like chordate on the evolutionary scene was important. Simpler forms of life often have a free-swimming intermediate or larval stage in their life cycle, but the typical adult is sedentary on the sea floor. With the chordate we meet an active adult able to

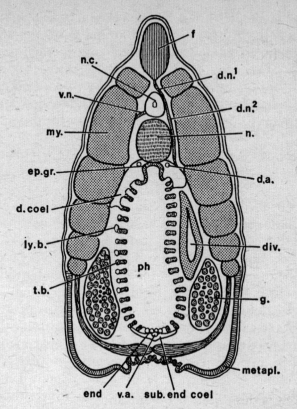

Fig. 209. Cross-section of *Amphioxus* in pharyngeal region. Dorsally, the muscle blocks (*my.*) surround the nerve cord (*n.c.*) with the notochord (*n.*) lying immediately in front. *ph.* is the cavity of the pharynx, which becomes continuous with the gut, and *g.* is the pair of gonads. (From Young's *Life of Vertebrates.*)

reach the surface and to browse on the ample food resources of the plankton, the first step from the sea bottom to dry land.

Vertebrates

The vertebrates are chordates, possessing at least the rudiments of a cranium and some evidence of a vertebral column. The forward end of the simple nervous system has become a complex brain, and in relation to this there have appeared sense organs—eye, nose and ear—permitting a much more elaborate pattern of response to environmental stimuli. Movements became more specialized to meet the needs of these stimuli, and the side-fins of fishes were made use of, when vertebrates eventually became land animals, as limbs for locomotion and ultimately as hands for manipulation. In this long development they probably passed through a freshwater stage and reduced the salt content of their blood. The earliest specimens of vertebrates, like our surviving lampreys, were without jaws; and the acquisition of jaws opened new possibilities in attack and food gathering, and lessened the need for heavy protective armament in the shape of ensheathing scales or bony plates. Typical also of vertebrate evolution was the acquisition of a chambered heart and of a respiratory pigment, *haemoglobin*, carried in little packets in the red corpuscles, not in simple solution in the blood plasma—two changes greatly improving the efficiency of the circulation. This, with the air-breathing lungs and warm blood of birds and mammals, permitted in these two vertebrate forms an extravagant expenditure of energy in movement which had been hitherto impossible—a change almost comparable, in biological progress, to the arrival of atomic energy on the industrial scene.

The head

As vertebrates evolved, more and more importance attached to central control from the enlarging brain. An elaborate skull developed to protect this and to house the sense organs. In the dogfish the skull is a simple collection of cartilaginous boxes serving this purpose, boxes developed as modifications of the vertebral segments at this end of the body. (It must be remembered that in more primitive forms there is no particular head or tail end.) The jaws developed as modifications of the gill arches just behind the skull, into which they were eventually incorporated.

In fact, the head of higher vertebrates, an apparently unique structure with a specialized neck to connect it to the trunk, is merely the end-result of continuous evolutionary change in the vertebral segments at one end of the body. Because brain, jaws and sense organs have developed here, the local segmentation of the skeleton has been correspondingly distorted to form the skull; and yet, despite their modern complexity, the bony skull and the brain with its pairs of cranial nerves correspond in their origin to the simple segmental elements of spinal column and spinal cord in the trunk proper.

Mammals

With the birds, the mammals represent the vertebrate form most adapted to life on land. Their peculiarities enable them to live in the most difficult conditions. As Professor J. Z. Young strikingly puts it, 'A camel and the man he is carrying may contain more water than is to be found in the air and sand wastes for miles around', an instance of the extreme *improbability* of mammalian life which is so typical. All life is improbable against the background of the non-living universe, in the sense that living creatures maintain themselves and temporarily put a halt, in their protoplasmic whirlpools, to the general running down of energy. Mammals do this to a greater degree than any other forms, save the birds, and have raised the energy flux through the living body to its height. They can maintain a constant temperature higher than that of their surroundings. They can live in cold polar regions, and can also live and grow through the winters of temperate zones when cold-blooded creatures must hibernate. This free life in the most diverse external conditions is due to their ability to regulate such factors of their internal state as body temperature and the composition of blood and tissue fluids at an amazingly constant level, no matter how the outside world changes physically or chemically.

At the same time, mammals have produced devices—limbs, hands, trunks—enabling them to alter their environment to suit themselves by digging or handling, by building dams—if they are beavers—or blasting pieces off the earth's surface if they are men. To quote Professor Young once more, 'Warmth, enterprise, ingenuity and care of the young have been the basis of mammalian success throughout their history. . . They might well be defined as highly percipient and mobile animals with large brains, warm blood, and waterproofed, usually hairy, skins, whose young are born alive.' They also have a diaphragm, a four-chambered heart and a single left-sided aortic arch. Primitive mammals, it is true, shared the characters of other, lower, vertebrates. The surviving platypus has the skeleton and egg-laying habits of a reptile and a bird's bill, an exception that may once have been the rule. But the typical mammal is *placental*, i.e. the young develop for a long period within the uterus and are fed during that time through a placenta attached to the inner lining of the uterus (see page 489). Things are quite different in the few non-placental, pouched mammals, mainly marsupials, where the young are born only a few days or weeks after conception and crawl precociously to the pouch, where they glue themselves to the mother's teats for many days until separately viable.

The true placental mammals, as they evolved, tended to become larger and their limbs grew longer; they walked on their digits, not on the soles of their feet; the teeth became specialized for predominantly herbivorous or carnivorous diet; and the brain expanded in certain regions—notably the frontal lobes—at the expense of the parts concerned with smell and so important to lower animals.

Primates

The original placentals were tree-living. The *primates*—and the monkeys, apes and men known as *anthropoids* within the

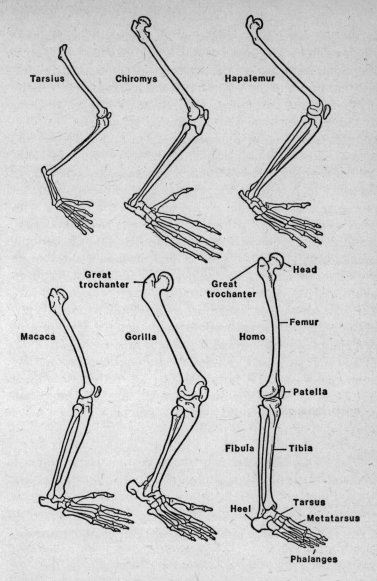

Fig. 210. The bones of the hind limbs of various primates arranged to suggest the evolutionary development of the characteristic features of the human leg—the vertical femur with its long neck and prominent great trochanter, the well-developed heel, and the non-opposable great toe. (From Young's *Life of Vertebrates*.)

larger primate group—remained so, and therefore retained
the five fingers and toes and the mobile clavicle which other
mammals have lost. In the active life of the treetops, quick-
ness of brain and movement, prehensile hands and feet and
grasping great toes, and the development of eyes and ears
at the expense of smell all progressed apace. The brain became enormously large, especially its occipital (visual) and
frontal (associational) lobes. Claws gave way to nails.
Mobility of the arm was greatly facilitated by retention of
the clavicle, by adding the range of a mobile scapula to that
of the true shoulder joint (see page 126), and by an increasing range of pronation and supination. In reproduction,
menstruation now appeared for the first time, a periodic
breakdown of the uterine lining after the failure of elaborate
preparations for the possible implantation of a fertilized
ovum and subsequent placentation. Parents began to care
for their young long after birth. The family arrived. Within the primates, the anthropoids continued the development
of a complex social life based on sight rather than on smell,
and accompanied by elaborate emotional expression. The
outer ears shrank, the large eyes faced directly forward, the
enormous cerebral hemispheres overhung the hindbrain
and the rounded head topped a mobile neck.

Man

Man's brain is larger than that of any ape and differs from
it more than the ape's, in its turn, differs from the brain of
other anthropoids. The swollen frontal and occipital lobes
point to the greater importance of caution and restraint,
reflection and vision in human life. We have developed
awareness of ourselves and of our world, conscious thought
and, above all, speech.

Man walks and balances on two legs routinely, not
occasionally, and his spinal column is correspondingly
modified into a sequence of alternating curves under the
vertical stresses (see Fig. 151). The intervertebral discs

become more important in size and function—and fail oftener. The architecture of pelvis, hip joints, thigh and leg bones also changes; and the internal structure and trabeculation of these bones reflect a purely mechanical solution to the engineering problems presented to them. *Balance*, the achievement of temporary stability in standing on one leg in running or walking, is a problem that produces great hypertrophy of the gluteal muscles in the buttock, whose function it is to fix the pelvis over the weight-bearing hip (see page 180). The long neck of the human femur increases the leverage available to these muscles; and, although this elongation keeps the upper ends of the femora well away from the body, the inward slope of the thighs brings the knees and feet back under the centre of gravity. The arched system of weight distribution in the feet is peculiar to man, and so is the abandonment of the short mobile first metatarsal segment and the opposable great toe for a long fixed segment strong enough to take the propulsive thrust of walking. This change is not always satisfactorily accomplished; in some individuals, the first metatarsal remains short and loose, and the great toe adducted, an atavism that produces bunions in young people and makes it difficult to wear ordinary shoes.

The human *arm* and its muscles have become shorter and less powerful than in the apes, as they do not have to support the body weight in swinging. And the chest, from which many of the shoulder girdle muscles arise, is correspondingly puny. In the hand, the small intrinsic muscles controlling the digits have greatly developed, and a truly opposable thumb is seen for the first time in the animal kingdom. All the differences between upper and lower limbs in apes and men result from the single original difference that the ape uses his arms and man his legs to get about.

The characteristics of the human *skull* are the result of changes in brain size, changes in dentition and of a new poise of head on neck. The growth of the frontal lobes produced the vertical forehead and face, the more apparent

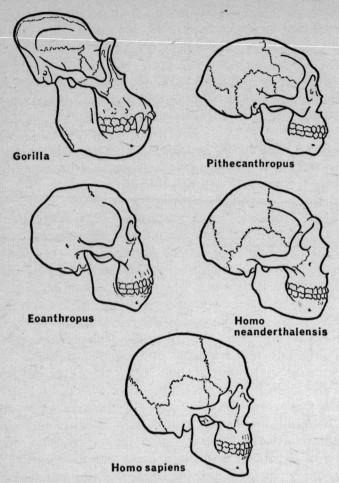

Gorilla

Pithecanthropus

Eoanthropus

Homo neanderthalensis

Homo sapiens

Fig. 211. The skulls of the gorilla and of progressive stages in the evolution of man. Note the general enlargements of the cranium, the rounding of contour, the development of the frontal region and the recession of the lower jaw. (From Young's *Life of Vertebrates*.)

since, at the same time, the jaws have receded (though the chin itself is a human phenomenon). The great brow and occipital ridges of the gorilla's skull have disappeared, and a general rounding of contour taken place. Vertical balance of the skull on the cervical spine is more perfect and the sternomastoid muscles, which turn the well-poised head from side to side, have pulled out the prominent mastoid processes from the base of the skull (Fig. 211).

The general rate of growth and development of the human body is considerably slower than that of the apes, whose long bones have their epiphyses finally closed by the age of twelve to fourteen instead of fourteen to eighteen as in man. But it is this long period of receptive immaturity, mental as well as physical, during which the young human being remains in constant need of care and protection, that also gives him an opportunity for learning and attainment unrivalled by any other creature.

Index

Index